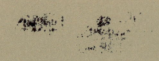

Communication and
Behavior of Whales

AAAS Selected Symposia Series

Published by Westview Press, Inc.
5500 Central Avenue, Boulder, Colorado

for the

American Association for the Advancement of Science
1776 Massachusetts Ave., N.W., Washington, D.C.

Communication and Behavior of Whales

Edited by Roger Payne

AAAS Selected Symposium **76**

To NIXON GRIFFIS and the late LANDON K. THORNE for making so much of this work possible.

To WILLIAM CONWAY for midwifery during a protracted period of labor, and in appreciation of the affection he also feels for the people and wildlife of Argentina.

QL
737
C4
C65

AAAS Selected Symposia Series

This book is based on a symposium that was held at the 1980 AAAS National Annual Meeting in San Francisco, California, January 3-8. The symposium was sponsored by AAAS Section G (Biological Sciences).

Copyright © 1983 by the American Association for the Advancement of Science.

Published in 1983 in the United States of America by
 Westview Press, Inc.
 5500 Central Avenue
 Boulder, Colorado 80301
 Fredrick A. Praeger, President and Publisher

Library of Congress Catalog Card Number LC 83-50721
ISBN 0-86531-722-4

Printed and bound in the United States of America

About the Book

Whales, perhaps more than any other animals, have in the last decade caught the attention of the world at large, both at a scientific and a popular level. This seems to have transpired as much through a series of unexpected and intriguing discoveries about their way of life and their communicative behavior as by other efforts. This volume documents some of the core studies that ushered in a dramatic change in focus and methodology. Research on dead whales gave way to studies of live populations in which identifications of natural markings of individual animals made possible long-term observations of communication and behavior. The authors present new findings on the changing songs of humpback whales, now considered evidence for the vocal transmission of a cultural trait in a non-human animal; the population dynamics of southern right whales, including the discovery of an unexpected function of callosities; the pod-specific vocalizations of killer whales; the behavior of gray whales and how it relates to tides; the behavior and migratory destinations of humpback whales; and several proven benign techniques for studying the biology of free-ranging marine mammals. An exhaustive annotated bibliography covers the literature on humpback and right whales from 1864 to the present. The volume demonstrates how broader knowledge of whales can come from these new research techniques, knowledge that is vital for preservation of the vast habitats required for the survival of these animals.

About the Series

The *AAAS Selected Symposia Series* was begun in 1977 to provide a means for more permanently recording and more widely disseminating some of the valuable material which is discussed at the AAAS National Meetings. The volumes in this Series are based on symposia held at the Meetings which address topics of current and continuing significance, both within and among the sciences, and in the areas in which science and technology impact on public policy. The Series format is designed to provide for rapid dissemination of information, so the papers are reproduced directly from the camera-copy submitted by the authors. The papers are organized and edited by the symposium arrangers who then become the editors of the various volumes. Most papers published in this Series are original contributions which have not been previously published, although in some cases additional papers from other sources have been added by an editor to provide a more comprehensive view of a particular topic. Symposia may be reports of new research or reviews of established work, particularly work of an interdisciplinary nature, since the AAAS Annual Meetings typically embrace the full range of the sciences and their societal implications.

WILLIAM D. CAREY
Executive Officer
American Association for
the Advancement of Science

Contents

PART 3: MIGRATORY DESTINATIONS
AND STOCK IDENTIFICATION

PART 4: RESEARCH TECHNIQUES

PART 5: BIBLIOGRAPHY

About the Editor and Authors

Roger Payne is a research scientist at the World Wildlife Fund U.S., the director of the Center for Long Term Research in Lincoln, Massachusetts, and an adjunct associate professor at The Rockefeller University. During progress of the work reported here, he was a research zoologist at the New York Zoological Society's Animal Research and Conservation Center. Since 1966, his research has focused on the vocalizations and other behaviors of baleen whales. He has also been active in conservation of whales and has received several awards and distinctions for his work. He is a knight in the order of the Golden Arc (Netherlands) and is a recipient of the Animal Welfare Institute's Albert Schweitzer Medal (shared with Katharine Payne).

James E. Bird, a librarian by training, has been working with Roger Payne as a research assistant since 1978 at the Center for Long Term Research in Lincoln, Massachusetts. He has been involved in studies of bowhead, gray, humpback, and right whales and has compiled annotated bibliographies on gray, humpback, and right whales and on bird-whale interactions.

Oliver Brazier did research on whales with Roger Payne from 1970 to 1977. He was subsequently in charge of the Woods Hole-based Platform of Opportunity Program for the Cetacean and Turtle Assessment Program (CETAP) and later did research at the Woods Hole Oceanographic Institution. Currently he is working in a marine engineering firm, of which he is cofounder.

Kevin Chu, a graduate student and teaching fellow in the Department of Biology at Boston University, has specialized in behavioral ecology. He has done research on the structure of humpback whale songs and on the distribution of humpback whales off the west coast of Greenland.

Christopher W. Clark is an assistant professor at The Rockefeller University Field Research Center in Millbrook, New York. He is studying the processes of acoustic production, recognition, and perception and their relationship to communication and the maintenance of social stability of animal groups.

James D. Darling, a doctoral candidate at the University of California, Santa Cruz, has done research on the abundance, behavior, and migration of humpback whales. In 1978, he received his master's degree from the University of Victoria, Canada, for his work on the abundance and behavior of gray whales off the coast of British Columbia.

Eleanor M. Dorsey is a research fellow in the Animal Research and Conservation Center at the New York Zoological Society. She has done research on the behavior of humpback, right, bowhead, and minke whales.

H. Dean Fisher is professor of zoology at the University of British Columbia, Vancouver. His major interest is in the functional anatomy and ecology of marine mammals, and he has been involved in studies of bio-acoustics and of dialects of gray whales, killer whales, and narwhals.

John K.B. Ford is a graduate student at the University of British Columbia, Vancouver, and a research associate of the Vancouver Public Aquarium. A specialist in marine mammal acoustics and behavior, he has published on underwater vocalizations of narwhals in the Canadian Arctic.

Peter Frumhoff, formerly a research assistant at the Center for Long Term Research in Lincoln, Massachusetts, is a graduate student in zoology at the University of California, Davis. His specialties include cognition and communication in mammals, song structure of humpback whales, language acquisition in humans, and behavioral development in vertebrates.

Kimberly M. Gibson, formerly assisted in research of the New York Zoological Society. Her work was on the abundance and behavior of humpback whales and the recording of their songs.

Deborah A. Glockner is director of humpback whale research at the California Marine Mammal Center in Fort Cronkhite, California. She has been studying humpback whales since 1975, focusing on the identification, sexing, and observation of behavior of individual whales. Currently she is analyzing the reproductive cycle of humpback whales. In

1981, she received the Society for Marine Mammalogy Award for outstanding research and best scientific presentation at the Fourth Biennial Conference on the Biology of Marine Mammals.

Linda N. Guinee is a research assistant at the Center for Long Term Research in Lincoln, Massachusetts. Since 1979, she has participated in field and laboratory studies of the songs of humpback whales and has carried out computer analyses of humpback whale song data and of data on the normal behavior of bowhead whales.

Charles M. Jurasz, director of Sea Search, Ltd., and an instructor in advanced biology and oceanography at Juneau-Douglas High School in Juneau, Alaska, has been doing research on humpback whales in southeast Alaska since 1966. He initiated the first Hawaiian census of humpback whales and documented their migration in Hawaii and Alaska. He has also studied feeding, territoriality, techniques for identification, inter- and intra-species interactions, and whale/vessel interactions.

Karen Miller has specialized in the behavior and ecology of marine mammals. She has done field research on the behavior patterns, distribution, and abundance of humpback and gray whales.

George Nichols, a physician by training, is president of the Ocean Research and Education Society, Inc., in Gloucester, Massachusetts. As master of the *Regina Maris*, a field research vessel, he has been involved in numerous studies of the ecology and population dynamics of humpback whales in the North Atlantic, particularly on the banks of the Dominican Republic, where the animals breed and calve, and off the coast of Greenland.

Kenneth S. Norris is professor of natural history, coordinator of the Environmental Field Program, and deputy director of the Center for Coastal Marine Studies at the University of California, Santa Cruz. He has been doing research on marine mammals for more than thirty years, and his work has earned him numerous awards and fellowships.

Katharine Payne is a research fellow at the Animal Research and Conservation Center of the New York Zoological Society. Her background is in music and biology. Since 1969, she has been studying the songs and behavior of humpback whales in the waters off Bermuda, the West Indies, Alaska, Hawaii, and Baja California; and the behavior of southern right whales off the coast of Argentina.

Judith S. Perkins is a graduate student in marine ecology specializing in natural resources management at the Yale School of Forestry and Environmental Studies. She conducted research on coral reefs in Belize and on seabirds and humpback whales in Greenland and Newfoundland.

Victoria J. Rowntree, a research assistant at the Center for Long Term Research in Lincoln, Massachusetts, has been involved in ecological and behavioral studies of right whales since the mid-1970's. She has also assisted in research on bowhead whales, gray whales, owls, animal energetics, wasps, and dinosaurs.

Gregory K. Silber is a graduate student at the Moss Landing Marine Laboratories in Moss Landing, California, where he is studying the behavior and acoustics of humpback and bowhead whales.

Alan Titus, a teacher of science and mathematics at Dr. Norman Bethune C.I. in Toronto, Ontario, was involved in whale research with Roger Payne at the Center for Long Term Research in Lincoln, Massachusetts.

Peter Tyack, a specialist in animal behavior and a postdoctoral scholar in biology at Woods Hole Oceanographic Institution, is studying cetacean communication and social behavior. He has done research on the role of humpback song and social sounds in mediating interactions between humpbacks on the Hawaiian wintering grounds.

Spearous C. Venus is vice president for research at the Mammals of the Sea Research Association in Lahaina, Maui.

Bernardo Villa-Ramirez, an educator and biologist, is professor of Zoology at Instituto de Biologia, UNAM, Mexico. The former head of Fauna y Silvestre, the agency responsible for wild resources in Mexico, he played a leading role in the establishment of that country's national park system.

Bernd Würsig is assistant professor of marine birds and mammals at Moss Landing Marine Laboratories in Moss Landing, California. His specialties include the movement and migration patterns of dolphins and whales, the behavior and ecology of cetaceans and pinnipeds, and their interactions with birds, fish, and marine invertebrates.

Introduction

As recently as ten years ago, very little was known about the lives of baleen whales. The information that did exist was mostly derived from corpses provided by the whaling industry. A few scientists such as W.E. Schevill and W.A. Watkins had started recording sounds from whales at sea, but people working with live whales were few and far between.

That picture has changed dramatically in the past decade. Many of the people responsible for the change are presenting their work in this volume.

One of the main features common to all of the studies in this volume is that they are all based on passive observation techniques. There is no result in this book that was derived from killing, capturing, confining, or even touching a whale. In some cases, observations of corpses were included, but while they may have provided a useful confirmation, they were in no case required for the conclusions that were drawn. Two of the papers in this volume, in fact, give evidence contradicting old assumptions based on whaling studies: the paper by Darling and Jurasz and that by Payne and Guinee demonstrate migratory paths of humpback whales (*Megaptera novaeangliae*) in the eastern North Pacific that are very different from what has been assumed from southern hemisphere whaling studies.

One of the rewards I have experienced during the editing of this book has been a growing confidence in the practicality of benign research techniques for studying whales. The work reported here demonstrates that basic science can be done at a useful level of rigor (and usually for much less money) without resorting to intrusive techniques or commercial whaling operations. I hope that one of the main values of this book will be to demonstrate the value of benign research techniques for studying the basic biology and population structure of whales.

I appreciate fully, of course, that there are many worthwhile techniques for studying whales which are intrusive and that in some cases they may be the only practical route to obtaining answers necessary for species management. However, I assert that such questions are few and that a great deal more can be learned from non-intrusive techniques than is generally realized. The few examples given here barely scratch the surface of what is possible.

According to the accepted doctrine of 20 years ago, most major advances in cetology were made by taking a broad stance on the flensing platform and dissecting ever deeper into the abundant corpses. The argument for more research on cadavers was self-perpetuating because it had a neat catch, sometimes put forward by whaling industry spokesmen as an argument in favor of whaling: since serious science cannot be done without dead whales, the industry should be retained if only for its valuable contribution to knowledge. On the theory that any science solely dependent on the remains of the deceased is, by definition, moribund, I decided to ignore the accepted doctrine. Others had reached the same conclusion. We joined forces and this book is the result.

Population estimates, based on population models, have played an important role in the recent efforts to bring whaling under rational control. One of the main difficulties in gaining agreement between members of the Scientific Committee of the International Whaling Commission has been the lack of information from which to derive values for some of the basic biological parameters upon which the population models depend. Perhaps even more important is that we do not have the biological information to put into the models about the dependence of vital rates (e.g., natural mortality, rate of reproduction, etc.) on population density and on the age of animals. The techniques which have been used to pursue answers to these questions mostly derive from techniques used by the first scientists who worked with the whaling industry -- techniques which rely upon measurements and samples from dead whales.

Blue (*Balaenoptera musculus*), fin (*Balaenoptera physalus*), and sei (*Balaenoptera borealis*) whales were three of the most hunted, and thus most studied, species during the modern whaling period. Yet, some of the gaps in our knowledge of their life histories were still with us in the 1960's and 70's when these species were declining precipitously and when better answers were urgently needed to stop that decline.

In other words, even when we had hundreds of thousands of corpses from which to derive data, the answers to some very simple questions (the consequences of which are nevertheless very important) were not forthcoming. I believe that the reason these gaps are still with us is because of a shortfall in imagination, coupled with dogged adherence to an approach long ago demonstrated to be inadequate to the task of providing the required information. The mephistophelean conclusion is that as long as a species is hunted commercially we will remain in relative ignorance about it. (This conclusion is supported by the recent unprecedented growth in the science of cetology at a time when the whaling industry is collapsing.) Once the hunt stops and we cannot rely on the old techniques any more, we are forced by necessity to use our imaginations to develop new approaches. These new approaches are bound to bring new understanding. Far from nurturing the growth of knowledge in whale biology, I feel that the availability of large numbers of corpses, and thus the possibility of more years of the old study methods, has actually held back the growth of this branch of science. Now that whaling is dying, we can look forward to solving some of the old problems as well as to gaining important new insights, not just into the remains of the whale but into the rest of the whale, the part that has escaped the whalers for so long.

Of the 14 papers in this book, 8 report on work performed entirely in my laboratory in Lincoln and/or the New York Zoological Society field station at Peninsula Valdés. Five cover research by members of a large student project studying humpback whales in Hawaii and working in connection with the laboratory in Lincoln. Only two papers were done entirely outside this sphere. The number of people in the main research group waxed and waned in response to the availability of funds and there were times when many of us were scavenging for funds and lived a cryptozoic existence (like raccoons) in the shadow of more affluent laboratories and scientific disciplines. Yet through these strains, all concerned managed to maintain a strong *comaraderie* within which it was always a pleasure work. As a result, there is a cross-fertilization or connectedness to many of these papers which is somewhat unusual for a symposium volume.

The papers on vocalizations of humpback whales start with an analysis (by Payne, Tyack, and Payne) of how the rapid and synchronous progressive changes occur in the songs of humpback whales. They show that the songs change in complex and unanimous ways when the whales are singing and are stored without much change during the relatively quiet feeding season. Guinee, Chu, and Dorsey then demonstrate that individual humpback whales modify their songs in accordance with the changes going on around them. Frumhoff follows with a paper showing that the aberrant songs of this species provide important clues as to which song elements are most unalterable and thus what the underlying structure of the song is like.

Ford and Fisher, in the only paper included here that is not on a mysticete, demonstrate that individual pods of killer whales (*Orcinus orca*) have their own dialects. The part of the sound repertoire that is unique to each pod remains highly stable over periods of at least ten years and probably longer. In addition, the extent of the vocal differences between pods appears to correlate with the degree of associations between the pods.

Clark then tackles the most complicated question regarding whale sounds, looking for indications of meaning in specific sounds. His subject is the southern right whale, *Eubalaena australis*. He finds that the acoustically simplest sounds are associated with contact over long distances while the most complex are associated with groups of socially active whales. The complexity of the sounds correlates with the complexity of the social context: this suggests that right whales may be using sounds to communicate a variety of messages.

The work dealing with behavior of mysticetes on their breeding grounds starts with a paper on humpback whales by Darling, Gibson, and Silber who demonstrate that, although individual humpbacks visiting the Hawaiian Islands each year may stay for 11 weeks, much of the population is not a static group but a passing parade. This has an important consequence -- the population of humpbacks visiting the Hawaiian assembly grounds is larger than had been thought previously. Another interesting finding from this work is the evidence for quite violent physical encounters between animals competing to escort females with calves -- encounters which may even result in bleeding wounds.

These results interact with the studies of humpback songs. Competition between individuals during the singing season provides evidence that song in whales may be part of a behavioral complex similar to that of other breeding, singing animals. The rapid replacement of individuals in a population makes it even more remarkable that whole groups of singers are able to concur at any time on the direction and extent of the changes that they are incorporating into their song.

The work of Glockner and Venus with the same species in the same area offers information on the behavior of female humpbacks and their calves, their preference for shallow water, along with new data on growth rate of the calf, external body features, and natural markings not formerly used to distinguish between individuals. Glockner and Venus also show that more than two thirds of calving females are accompanied by an adult-sized escort who exhibits protective behavior toward the female and calf.

Norris, Villa-Ramirez, Nichols, Würsig, and Miller offer interesting data on another species in its breeding lagoons, the gray whale (*Eschrichtius robustus*). They demonstrate that the behavior in lagoon entrances includes courtship and considerable feeding. They also suggest that the lagoon entrances serve as a concentration mechanism for the whales' prey.

Payne and Dorsey present the results of a close examination of the callosities -- thickened patches of very rough skin -- on the heads of southern right whales and conclude that males have more and larger callosities than females. On their winter assembly areas in Argentina, the males are probably using their callosities rather like underwater horns to fight other males for access to females.

On the question of migration, Payne and Guinee offer evidence that the song of humpback whales can serve as a means of identifying stocks. The fact that humpbacks in two areas 4700 km apart are singing similar, if not virtually identical, songs at the same time suggests that (unlike South Atlantic right whales) the eastern and central Pacific humpback whales are a single population which must be mixing considerably every year. This speculation is confirmed by Darling and Jurasz, who compare photographs of naturally marked whales in the North Pacific. They find the same individual wintering, in two different years, in waters near Hawaii and near the Revillagigedo Islands 4700 km away. They also found seven individuals that were seen both in southeast Alaska and Hawaii. Since whales from both wintering areas feed in Alaskan waters, it may be there that the mixing postulated by Payne and Guinee is taking place.

The next section of the volume is given over to some of the new research techniques which have been useful in studies of the behavior of baleen whales. The paper by Payne, Brazier, Dorsey, Perkins, Rowntree, and Titus reports on the characteristics of callosities as a means of identifying right whales. An old spectre has always haunted studies based on natural markings -- what if there are look-alikes in the study group? These authors offer a new technique by which one can determine what the probability is that there are two or more animals with the same natural markings in a given population. The technique is broadly applicable and should be useful with other species. Payne et al. also argue that, contrary to the usual perceptions, natural markings are probably more useful than artificial tags for studies of large populations.

Glockner offers a new technique for determining sex of humpback whales from lateral views of their bodies. Using this technique, she presents evidence that both singers and the whales that escort females with calves are males.

The final section of the book is an annotated bibliography by Bird on right and humpback whale literature from 1864 to 1980.

I have had the pleasure of working directly with 20 of the 25 other authors who have contributed to this book. Of these, 12 were colleagues and/or students in the field and 17 worked in my laboratory in Lincoln, Massachusetts. The laboratory received its major support for all the years that this work was in progress from the New York Zoological Society. The Society also funded a field station at Peninsula Valdes, Argentina from which most of the data on right whales came.

I wish to acknowledge my debt to the members of my laboratory who have helped with the production of this volume. Jim Bird and Linda Guinee bore the major brunt of the effort. In addition to endless proofreading, correspondence, and cross-checking, they also produced the index. Linda typeset the entire book, a job that would have daunted anyone with less energy and good will. Victoria Rowntree and Eleanor Dorsey both assisted in ways too numerous to enumerate.

William Schevill reviewed the entire manuscript and made many helpful suggestions for which I and the other authors are grateful. I am also grateful to Eleanor Dorsey, Steven Green, Donald Kroodsma, Peter Major, James Mead, Katharine Payne, Colin Pennycuick, John Smith, Peter Tyack, and Hal Whitehead for giving critical readings to one or more of the manuscripts. Misstatements which remain are my responsibility, not theirs.

I am also obliged to Anne Monk for proofreading the whole manuscript.

My final acknowledgement of gratitude goes to my family and to the New York Zoological Society who separately and collectively had the patience to see this project through to this conclusion.

## Part 1

Vocalizations

Katharine Payne, Peter Tyack,
Roger Payne

1. Progressive Changes in the Songs of Humpback Whales (*Megaptera novaeangliae*): A Detailed Analysis of Two Seasons in Hawaii

Abstract

Previous studies have shown that humpback whales sing long, complex songs which are shared by all animals on the same calving ground but which change with time. The present study analyzes in detail the process by which the rapid and synchronous changes throughout two singing seasons in Hawaii occurred. For purposes of statistical analysis, each season is divided into six equal periods. 151 songs from 1976–77 and 159 songs from 1977–78 are dissected and compared in terms of the duration, number, frequency, spacing, and configuration of units, phrases, themes, and songs. The mean results from all measurements in each period are compared to those from other periods, revealing dramatic monthly evolution following set rules of change. Substitution, omission, and addition occur at different rates in different themes, but at any one time all songs are similar.

The mechanism by which the whales achieve this, and the adaptive significance of such an elaborate feature of a display remains unclear. Little change occurs between the form of songs at the end of one singing season and the start of the next (six months later), discrediting the theory that forgetfulness during the months on the feeding grounds (where singing rarely occurs) might be the main source of song change. Major trends of song change in one year do not repeat in the next, thus the changes may be thought of as primarily cultural rather than being linked to natural rhythms or periodic phenomena. A classification of all animal songs based on how variety is introduced is proposed, in which humpback whales fill a missing link between some birds that increase their repertoires all their lives by mimicry, and human beings, who create songs *de novo*, without confining themselves to the reworking of existing material, as humpbacks do.

Introduction

While on their breeding grounds, humpback whales (*Megaptera novaeangliae*) show an unusual vocal behavior: they sing long and complex songs (Payne 1968, 1970; Payne and McVay 1971; and Winn, Perkins, and Poulter 1971). Humpback whales in different oceans sing different songs (Payne 1978) even though at any one time and in any one area, all individuals sing similar songs (Payne 1978; Winn and Winn 1978; Payne and Payne in press). However, whales in each broad area of ocean (perhaps even over entire oceans) change their songs together over time (Winn and Winn 1978; Payne and Guinee 1983).

Large scale progressive changes were observed in humpback songs from Bermuda in a sample containing 13 of the 18 years between 1957 and 1975 (Payne and Payne in press). In all but one year, the songs were collected only in April and May. At any given moment, all singing whales sang the same version of their song, but after a year or more, that version had disappeared and was replaced by a new version. As time went on, the song became increasingly different from the earliest version recorded. The extent of the changes from one year to the next varied, but in times of rapid change, it only took a few years for every part of the song to be altered beyond recognition.

It is inconceivable that such rapid and complete turnover of the song material could reflect genetic changes. Guinee, Chu, and Dorsey (1983) show that the explanation is not to be found in a turnover of individuals, for the songs of individuals change just as the songs of the group do. New variations in the song must be transmitted by learning. The progressive changes in whale song can thus be seen as a form of cultural evolution, in the sense that the song is a learned trait which evolves.

When we started the research reported here, we were testing a simple hypothesis for what caused the observed progressive changes. Our own experience and that of others indicates that during the summer months when the whales are on their feeding grounds, singing is extremely rare (pers. comm. with P. Beamish, E. Guthrie, D. McSweeney, J. Perkins, W. Schevill, W. Watkins and H. Whitehead). We know of only three recordings of more than a few isolated sounds from feeding grounds (McSweeney and Payne in prep). It therefore seemed reasonable that the whales might forget some details of their complex song during their almost silent summers. Upon returning to their breeding grounds in the following winter, they might piece together a new song out of whatever they remembered. The result would be a modified version of the old song, which would hold throughout the singing season. Through this mechanism, humpback song would evolve from year to year.

The Hawaiian Islands provide ideal conditions to test this hypothesis while examining the process of song change. Because there are humpbacks in Hawaiian waters for six months during the singing season, one can record throughout the time. Further, the waters on the lee side of the Hawaiian Islands allow recording from small boats during most days of the season.

Our first Hawaiian recordings were made in the 1975-76 singing season. In 1976-77, we recorded throughout the full season, and we have continued making season-long samples each year through the 1981-82 season. This paper reports the results of detailed analysis of recordings from the first two complete singing seasons (1976-77 and 1977-78).

It provides evidence that the changes in song are not simply due to forgetfulness between singing seasons. Most changes did not occur between seasons. Instead, they occurred during the time when the whales were singing, developing their songs methodically in measurable steps. Furthermore, the types of change varied from season to season, and so could not be attributed to repeating seasonal factors.

We know of no other animal where whole populations introduce such complex, rapid and non-reversing changes into their vocal displays, abandoning old forms and replacing them with new. It is not clear what selective advantage would be obtained by changing songs continuously.

Materials and Methods

All but two of the humpback song sessions discussed in this paper were recorded off the leeward coast of Maui in the Hawaiian Islands (the two exceptions were recorded approximately 150 km away, near the big island of Hawaii). Humpbacks are sighted in the Hawaiian Islands from mid-November to mid-May. Singing is heard sporadically, becomes almost continuous, and dies out again during those six months, in the shallow waters bounded by the four islands of Maui, Kahoolawe, Lanai, and Molokai. We made recordings in daylight from a 16' motorboat using several tape recorder and hydrophone combinations, in all cases with a frequency response uniform within 5 db for frequencies between 50 and 10,000 Hertz. In order to avoid recording in the same area on consecutive days, we left our base in Lahaina, Maui each day on a different bearing. We would periodically listen for song until we got a good enough signal-to-noise ratio from one singer to record its song

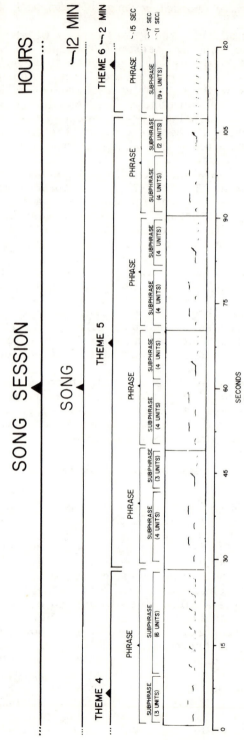

Figure 1. Diagram of hierarchical structure of all humpback whale songs, using a tracing of a spectrogram to illustrate. Times given are rough indicators.

without interference from other singers or from ambient ocean noise.

During the 1976-77 season (19 December 1976 to 18 May 1977), 111 hours of recordings were made. During the 1977-78 season (25 November 1977 to 27 April 1978), 121 hours of recordings were made. In order to obtain these recordings, observers spent over 1000 hours in boats actively searching for, observing, photographing (for identification of individuals) and recording humpbacks.

Upon return from the field, spectrograms of all good recordings were made using a Spectral Dynamics SD-301C real time analyzer, a Tektronix 604 monitor oscilloscope, and a 35mm Nihon Kohden oscilloscope camera. Frequency was displayed on the y-axis of the oscilloscope. The sweep was held stationary and the film (actually a continuous strip of 35 mm photographic paper) was moved parallel to the x-axis at a rate of 0.5 cm per second. The sampling rate was once per 10 msec and sound intensity at each point of the frequency spectrum was roughly indicated by modulating the intensity of the electron beam. This spectrum intensity was logarithmic. Frequency display was linear; the analysis range was from 40 - 2500 Hz; and the effective bandwidth was 10 Hz.

The resulting reels of photographic paper were developed, cut into strips 60 cm long and mounted in sequence on large sheets of paper, each sheet displaying from 2 to 6 songs. Songs that were incompletely recorded (due to temporary electronic or mechanical disturbances or the need to change tape) were not included in the analyzed songs. However, complete themes and their component phrases from these song fragments were counted when tallies of total themes and phrases were being considered. (Terminology is explained in Figure 1). A total of 151 complete songs from 1976-77 and 159 complete songs from 1977-78 made up the final sample of analyzed songs. Twelve sessions contained aberrant songs. The study of these songs was carried out by Frumhoff (1983).

The duration of songs, themes, and phrases was measured directly from the strips of photographic paper (spectrograms) with a ruler or with dial calipers when unit durations were being considered. A set of fixed definitions of different categories of units, phrases, and themes was made. This allowed us to make an objective analysis of any song; to make counts of specific elements of which it was composed; and to subject the results to statistical tests.

All themes and phrases and many individual units in each song were categorized both by inspection and by measurements of the spectrograms. Only a few examples of each type of change which we discovered will be presented in this paper. Over 5000 hours were spent making and analyzing the spectrograms.

Table 1. Distribution of sample of analyzed songs. "Periods" are consecutive portions of the singing season 31 days long. Periods having the same number occur at the same time each year.

	1976-77			1977-78		
	# Sessions	# Songs	# Songs/Session	# Sessions	# Songs	# Songs/Session
PERIOD I (Nov 14 - Dec 14)	-	-	No Data	7	20	1,8,2,1,4,2,2
PERIOD II (Dec 15 - Jan 14)	4	23	5,4,11,3	9	28	1,1,7,1,2,12,1, 1,2
PERIOD III (Jan 15 - Feb 14)	9	36	3,2,5,6,4,6,2,4,4	10	45	5,4,3,6,10,5,1,5, 1,5
PERIOD IV (Feb 15 - Mar 17)	12	45	1,4,3,2,6,4,1,6,1, 11,1,5	5	19	1,4,2,8,4
PERIOD V (Mar 18 - Apr 17)	4	19	7,7,2,3	15	47	5,1,1,5,2,4,5,1,2, 2,5,3,1,7,3
PERIOD VI (Apr 18 - May 18)	11	28	4,3,3,2,4,2,2,1, 1,5,1	-	-	No Data
TOTAL	40	151		46	159	

TOTAL (BOTH YEARS) 310 Songs
86 Song Sessions

In order to analyze statistically the changes that we measured, we divided each singing season into six, 31-day time periods. The placement of the six successive periods within the 1976-77 singing season was adjusted so as to distribute recording days (see Table 1 for the number of songs recorded each day) as evenly as possible within each period. This same distribution of periods gave a good fit to the days on which we recorded during the 1977-78 singing season. As a result, periods indicated by the same number in different years cover exactly the same days of the year. Table 1 gives the starting and ending dates of the six periods. It should be noted that we failed to get any recordings in all of the first period (Period I) of the 1976-77 season and during most of the last period (Period VI) of the 1977-78 season. In other years, a few whale songs have been heard and recorded in Hawaii both earlier and later than the time encompassed by our six periods, but the great majority of the singing occurs within them.

Mean values of several parameters of themes and songs were calculated for each of the six time periods. Unless otherwise indicated, these "period" means are means of means (since they are the means of the mean values for each song session rather than for each song). We chose to base our analysis on song sessions, rather than on songs, so as to give equal weight to the contributions of different whales. During the years this research has been carried out, we have had the benefit of knowing the identifications of many humpback whales thanks to the concurrent work with this same population by Darling, Gibson, and Silber (1983). They identified many singing humpbacks by photographing distinctive pigmentation patterns on their flukes (see Katona, Baxter, Brazier, Kraus, Perkins, and Whitehead 1979 for a description of this technique). Out of a large number of singers identified, Darling has resighted very few (see also Guinee et al. 1983). We believe, therefore, that most of the song sessions recorded come from different whales.

Results

Song Structure

The structure of humpback songs recorded from Bermuda in 1961-64 was described by Payne and McVay (1971) and is summarized diagrammatically in Figure 1. All songs we have studied from all times and places adhere to this format. Songs are composed of a series of discrete notes, or units. We define units as the shortest sounds in the song which seem continuous to the human ear.[1] Small repeated groups of units are called phrases. Phrases are usually uniform in duration. Many phrases consist of two

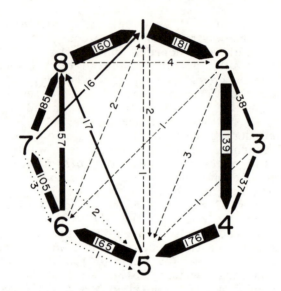

Figure 2. Transition probabilities from one theme to another
in 1976-77. (All transitions from one theme to any
other theme are included). The relative thickness
of the arrows pointing from one theme to the next
reflect the number of transitions from one theme
to the next. The three types of reversal of
theme order (7-6, 7-5, 6-5) are shown as dotted
lines. These lines, as well as the dashed lines,
are indicating transitions that were so rare that
the arrows had to be drawn disproportionately
thicker for the sake of visibility. The strictness
with which whales maintain theme order, even
when omitting themes, is obvious. Numbers
indicate how many transitions of each type
occurred in our 1976-77 sample.

subphrases, which may in themselves contain repeated sounds. All phrases of one kind make up a theme. Themes may contain any number of phrases and so their length is extremely variable.

We say a whale is singing when we hear groups of units repeated. A song is a series of different themes given in a predictable order. Successive songs are sung without pauses between them, and we refer to all songs in an unbroken sequence as a song session. The longest continuous excerpt from a song session which we recorded (in this case from an identified individual) lasted 10.5 hours; however, the complete session was longer, as the whale was already singing when we found it and still singing when we were forced to leave it.

Themes occur in an invariant order. This is a powerful constraint and makes the song highly ordered. Figure 2 shows the transition probabilities from one theme to another during the 1976-77 season. While it is obvious that deletions of one or more themes were common, there were only 5 reversals in the 1196 transitions. All but one of these reversals came from a single, highly aberrant song session which was highly aberrant in other aspects as well (Frumhoff 1983).

Evolution of the Song: Gradual Modification of Themes

General changes observed in two years. In this section we will give a preview of the kinds of gross changes we have observed and measured. In the following section we will present a more detailed analysis of our whole data set in terms of certain easily measured parameters.

Figure 3 shows a spectrogram of representative Hawaiian humpback songs from the middle of the 1976-77 and 1977-78 seasons, identifying the themes from each year and illustrating the differences between March songs of one year and the next. The tracing omits all extraneous sounds (e.g., ocean, ships, other whales, underwater echoes). Vertical lines running through the spectrograms from top to bottom of each one designate divisions between phrases.

Figure 4 presents a sample phrase from each theme of seven representative songs that were recorded at the beginning, middle, and end of the 1976-77 and 1977-78 singing seasons. It demonstrates the ways each theme changed during the two years.

---

[1] Some units can be subdivided into subunits: sounds which, although they sound continuous to the human ear, can be seen on a spectrogram to be discrete. Grating or rasping sounds, composed of pulsive subunits, are examples.

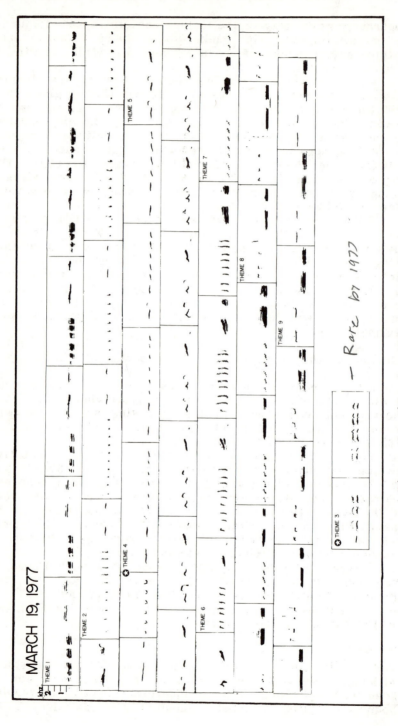

Figure 3A. (See caption on page 24.)

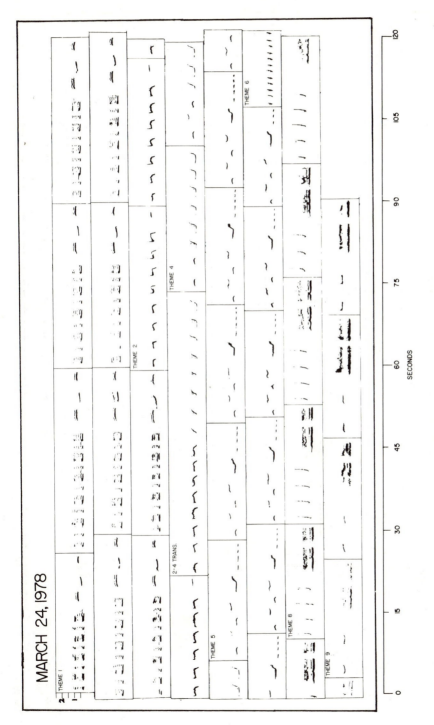

Figure 3B. (See caption on page 24.)

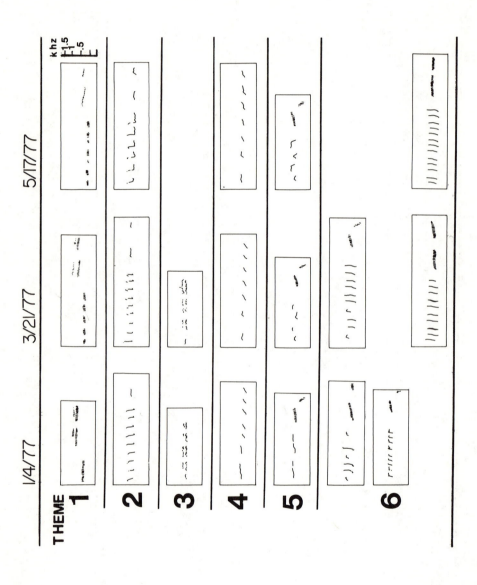

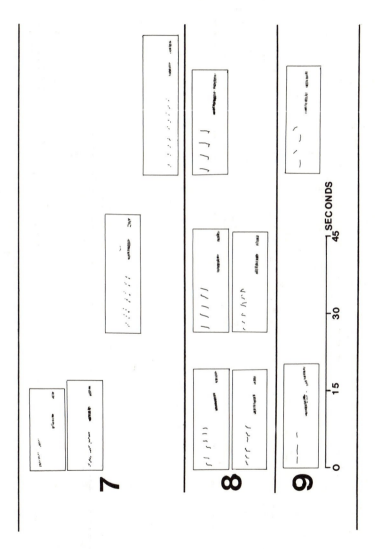

Figure 4A. (See caption on page 24.)

12/10/77   3/31/78   4/25/78

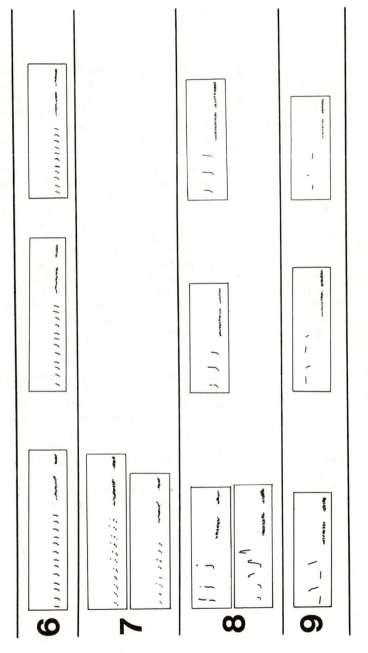

Figure 4B. (See caption on page 24.)

Figures 3A and B. Tracings of spectrograms of representative songs from March 1977 and March 1978. We selected songs which contained all possible themes. In 1977, theme 3 was rare. Although the sample song shown here omitted it, it was included in the next song sung by that same whale. The ✪ in the tracing indicated where theme 3 is placed when it is sung. Two phrases of theme 3 are then shown under the song. They are phrases by the same whale and came from the song it sang following the one fully traced here. The tracings omit all extraneous sounds (e.g. ocean noise - like ships, other whales, underwater echoes, etc.) as well as harmonics. Pulsive sounds are represented diagrammatically by closely spaced vertical lines which occupy the space on the spectrogram that showed the dense harmonics. The spacing of these traced lines is arbitrary and does not necessarily represent the repetition rate of the pulses.

Figures 4A and B. Sample phrases traced from spectrograms of representative songs throughout the 1976-77 and 1977-78 seasons showing the evolution of the themes, and of the songs. When the sample song omitted a theme characteristic of the period, we traced a phrase from another song. Such phrases are labeled with a ✪. (For explanation of tracings, see Figure 3). Graphs in later sections will demonstrate the degree to which observed changes are typical of all whales recorded.

It is apparent at once from Figure 4 that the last recordings in the 1976-77 singing season and the earliest in the 1977-78 winter were remarkably similar. The time when songs changed most was not during the relatively silent summer, but during the singing season itself. Thus our original hypothesis that the annual differences in song might be due to errors in memory was not confirmed. In fact, many of the changes progressed in such a predictable fashion that far from looking like accidents of forgetfulness, they appeared to follow set rules of progressive change.

Although we found that all themes changed, the characteristics which were changing varied from one theme to the next. The changes occurred at different rates and the peaks of change occurred in different time periods for different themes. It is remarkable that although the changes were so complex and asynchronous from theme to theme, all songs sung at any one time are very similar.

Figure 5 illustrates the evolution of theme 5 between 1976 and 1980. Even though the basic phrase structure remained recognizable through the 5 years, phrases changed conspicuously in the frequency, duration, spacing, configuration, and number of units, as well as in the duration of the average phrase. These processes did not occur all at once but in stages. The splitting of two units in the first subphrase to become four started in 1976 and was completed in 1977. But it wasn't until 1978 that the frequency range and unit duration in this same subphrase began expanding rapidly. The number of short units at the end of the second subphrase also increased between 1978 and 1980, while the duration of the whole phrase was increasing gradually throughout the five years from 1976 through 1980.

Analysis of specific changes in several themes over two years. Up to this point, we have selected examples from our data to illustrate changes in humpback songs. In the remainder of this paper, we will demonstrate changes by mathematical analyses of the complete data set, following the process described in the "Materials and Methods" section. We will start with Theme 6.

Theme 6 exhibited a change which was different from those described above. In this case, one type of unit slowly replaced another until the first was lost entirely. The rate at which the replacement occurred was surprisingly regular. To see how this worked, let us look at the structure of theme 6.

Figure 6 shows sample phrases of theme 6 from songs recorded about one month apart starting in February and ending in May of 1977. Theme 6 phrases consist of rapidly rising frequency sweeps followed by a long and then a

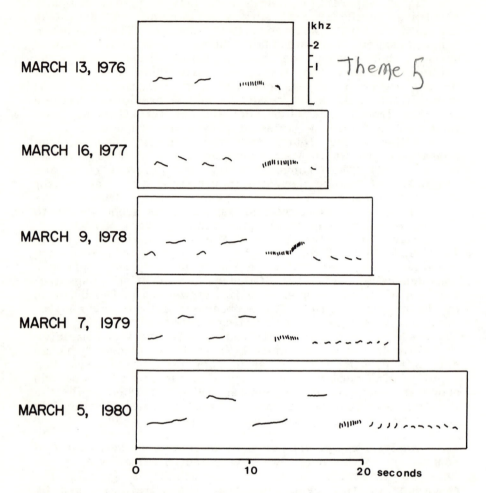

Figure 5. Sample phrases showing the evolution of theme 5 over five years. Note that phrases changed duration gradually. The units changed in frequency, duration, spacing, configuration and numbers. (For explanation of tracings see Figure 3). Figure 15 demonstrates the degree to which such changes are typical to our whole sample of songs.

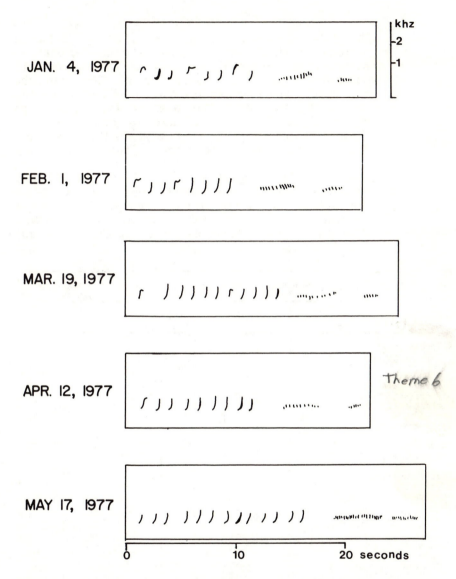

Figure 6. Tracings of single phrases of theme 6 showing that the first part consists of two different units, **rs** and **js** (see text) given in various mixtures. Note that the **rs** are replaced by **js**.

Table 2. This table compares the degree of change between periods to find when the largest change, marked with an X, occurred in each of the parameters described in this paper. While few of the parameters are completely independent of others in the table, the list clearly demonstrates that there is no trend for more change to occur between seasons than during any one period within a season. This result is even more dramatic than is indicated by the table, because the whales kept on singing after the last period we studied in 1976-77 and resumed before the first one in 1977-78. Moreover, the total time between seasons is greater than six times the duration of a period. (The duration of theme 3 is too variable to permit comparison.)

	Data Periods in 1976-77						Data Periods in 1977-78				
	2	3	4	5	6	June-Nov.	1	2	3	4	5
# Themes/Song		X									
Theme 5											
# Units/Subphr 1									X		
Phrase Duration	X										
# Phrases/Theme									X		
Theme Duration									X		
Song Duration			X								
Theme 6											
% r vs. % j				X							
Phrase Duration											
Theme 1						X					
Theme 2		X									
Theme 3		?									
Theme 4		X									
Theme 5		X									
Theme 6				X							
Theme 7								X			
Theme 8		X									
Theme 9									X		
% Songs with both Themes 8 & 9									X		
% Theme 1 in which subphrase 1 alternates long and short						X					

shorter pulse train. The frequency sweeps are of two types and differ principally by whether or not the highest frequency in the sweep is sustained. When it is sustained, the sweep has the form similar to the letter **r** and when it is not it looks more like a **j**. It is the ratio of **r**'s to **j**'s which interests us here. **r**'s are often scattered as if at random among **j**'s and sometimes a whale rendered each successive phrase of theme 6 as a different mixture of **r**'s and **j**'s.

In an effort to make some kind of sense out of so much variation, we found that a very simple form of analysis gave a striking result. We ignored all information on relative placement of **r**'s and **j**'s in a phrase and simply counted the total of each in the theme. Each was then averaged for every song session and then expressed as a percentage of the total. The result is shown in Figure 7. It is apparent that however else they may have been manipulating **r**'s and **j**'s, the singers were rapidly replacing **r**'s with **j**'s in theme 6. It is interesting to note that although **r**'s were almost gone by the end of the first singing season, they were still present in Period I of the following year. There is a 6 month period between Period VI of one singing season and Period I of the next, during which the whales are principally feeding and are singing very little. During this hiatus, the evolution of theme 6 slowed down, so that when the next singing season started, the whales were still mixing a few **r**'s with their **j**'s during the first period. This, along with other examples (see Table 2), suggests that songs evolve principally when they are being sung and very little during periods of silence. The regularity of the rate of replacement of **r**'s and **j**'s, in spite of there being no obvious ordering of **r**'s and **j**'s in successive phrases of theme 6 during the whole replacement, suggests that the whales were somehow attending to a simple law of ratios during the replacement process.

The history of theme 7 exhibits a type of change based on substitution of phrases rather than of units. There were four common alternate types of phrases in theme 7 which were used in 1976-77. We classified these alternate phrase types by applying three criteria to the first subphrase (Figure 8). The first criterion discriminated on the basis of the frequency of the units. There were two possibilities, those units with a fundamental frequency <900 Hz and those with a fundamental frequency >900 Hz. The second criterion discriminated on the basis of the number of units in each "cluster" of units. Subphrase 1 of theme 7 consisted of groups of two or more units given in rapid sequence with pauses between the groups. We call these little bursts of short units "clusters". A complete phrase of theme 7 consists of between two and six clusters followed by a long

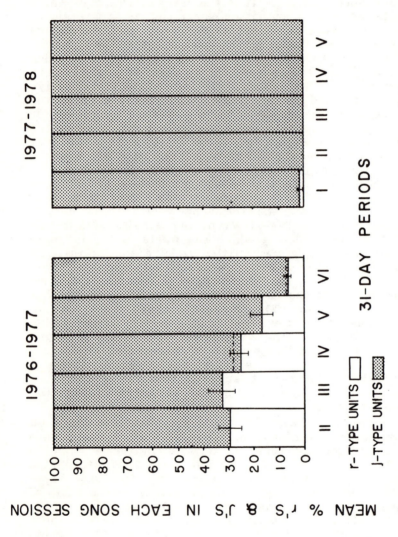

Figure 7. % of **r** and **j** units per song session versus time. The time periods (labeled 2-6 and 1-5 in the two different seasons, 1976-77 and 1977-78) are equal 31-day time periods. Periods having the same number start and stop on the same dates of their respective years. Notice that **j** units replace **r** units. Standard error is indicated by vertical lines. Dashed lines indicate means when aberrant songs are included in the calculations.

ALTERNATE FORMS OF THEME 7 SUBPHRASE I

	MINIMUM FREQUENCY	NUMBER OF UNITS IN EACH CLUSTER	RISE IN FREQUENCY WITHIN CLUSTER ?	
7A	>900 Hz	N>4	NO	
7B	<900 Hz	N>4	NO	
7C	<900 Hz	2<N≤4	YES	
7D	<900Hz	N=2	YES	
RARE INTERMEDIATE FORMS				
7BC	<900 Hz	2<N≤4	NO	
7BC	<900 Hz	N>4	YES	

Figure 8. Alternate forms of the first subphrase of theme 7. Alternate forms are labeled 7A through 7D. They were all common at some point of the 1976–77 season. There were also two rare alternate forms of theme 7 which were intermediate between forms 7B and 7C. The tracings of spectrograms to the right of the alternate phrases 7A–7D depict examples of each type.

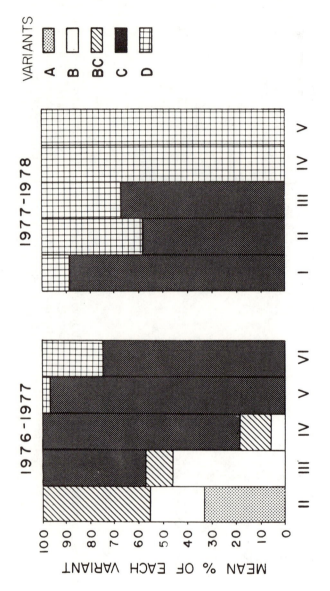

Figure 9. Several changes in the alternate phrases of theme 7. This figure shows the percentage of occurrence of each alternate form of theme 7 throughout each period of the 1976-77 and 1977-78 seasons. Only one song session in each of the last 2 1/2 periods of 1977-78 contain theme 7 because this theme was dying out. The percentages in each case thus reflect just this one song session.

and a short pulse train (Figure 8). Of the number of clusters possible, we chose three arbitrary categories: N=2, N=3 or 4, N>4. (Thus in the second criterion there were three possibilities.) Our third criterion discriminated on the basis of whether the units within a cluster were constant or rising in frequency. There were two possibilities in this case -- rising or constant. Taking our three criteria together, there were thus 2 x 3 x 2 = 12 possible phrase forms of theme 7. Yet, we saw only 6 out of these 12 phrase types. One half of the possible combinations simply did not occur, demonstrating that the alternate forms of phrases did not occur randomly but were selected by the whales.

The four common alternate phrases were not randomly dispersed through the season. Figure 9 shows the frequency of occurrence of the 4 alternate phrase forms of theme 7 which we labelled A, B, C, and D. Alternate phrase 7A, which was the most common form in the singing season, only occurred in the beginning of the 1977-78 season. Even in the first period, alternates 7B and 7C as well as the intermediate BC were also present but there was a steady progression of alternates, with different alternates predominating as the year progressed.

By closely examining the ways in which themes change, we found that new forms of a theme could be introduced at any point in the singing season. For example, the 7C alternate first appeared in February, 7D first appeared in April. We have found only two possible exceptions to the rule of gradual adoption of alternate phrases. In the very end of the 1976-77 singing season, two striking new phrase forms were adopted: in theme 2, phrases with elongated units began to appear, and in theme 1 phrases with alternating short and long units in the first subphrase took the song by storm. (Example phrases are shown in Figure 4).

In this last change, a theme which had been relatively chaotic became more organized. Phrases of theme 1 always contained in their first subphrase both long and short units (short units, < 0.5 length of the long ones). Until Period IV of 1977 the placement of long and short units appeared random, but in the last few recordings of the year, the two kinds of units alternated regularly (long-short-long-short, etc.). This new phrase form was precipitously adopted. By the start of the next period (Period I in 1977-78, following the migration to the summer feeding grounds and back), this form of theme 1 was almost universal, and the change was maintained (see Figure 10). This is one of the few changes we observed which occurred more between, than within, singing seasons: but even in this case, the onset of the change occurred within the singing seasons.

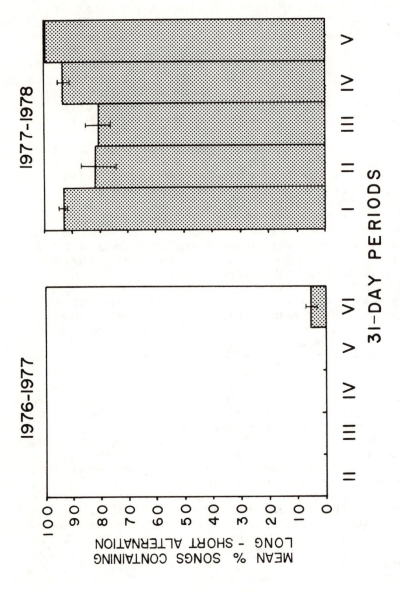

Figure 10. Adoption of a new phrase form which was introduced at the end of a season.
Rhythmically organized phrases were observed only in the last few songs of 1976-77,
but by the start of the next season, had become the prevailing form (see text). ▨ =
songs in which long and short units alternated at least 50% of the time.
Standard error is indicated by vertical lines.

Table 3. Types of phrases in themes 8 through 9 in 1976-77.

Phrase Type	Subphrase 1	Example
8	a series of high, short units, each lasting <1 sec. and in which most rise in frequency, the frequency change spanning at least 500 Hz.	
9	a series of high units in which at least one lasts ≥1 second. The units may rise or fall in frequency but the frequency change spans less than 500 Hz.	
8-9	a series of high units of which at least one lasts ≥1 second and at least one rises, changing in pitch by ≥500 Hz.	
misc.	a series of high units which falls into none of the above categories.	

Evolution of the Song: Birth and Death of Themes

Not only is there a turnover of units or alternate phrases within each theme, but entire themes gradually die out and new ones appear. Two themes -- themes 3 and 7 -- were sung less and less frequently; both were finally eliminated from the song (Figure 13). As of this writing (1981), we have never heard the sounds of either theme since in the Hawaiian song.

We have also witnessed the birth of new themes. This occurred in two different ways: 1) In 1979, a completely novel sound entered the song, became a pattern of phrases and established its own place in the orderly progression of themes (Payne and Guinee 1983). Whether this sound was spontaneously invented and then imitated, whether it developed from material in the songs that we did not happen to record, or whether it reflects interchange with whales that had learned phrases from another dialect, we do not know. To date, this is the only example of apparent *de novo* innovation that we have discovered. 2) More commonly, a theme which had been stable became variable and differentiated into two themes. The sequence of the variable phrase forms was at first unpredictable. Later, the phrases took on a definite sequence and eventually became two clearly defined themes sung in consistent order. An example of this was a section of the song following theme 7 which we called theme 8 in 1976-77, but which split to become themes 8 and 9 in 1977-78. In the first periods of 1976-77, this section of the song contained many kinds of loosely organized phrases that occurred in no particular sequence. All phrases could, however, be divided into four categories on the basis of units of the first subphrase (Table 3).

Figure 11A shows how these phrase types were distributed in the two seasons. In the start of the 1976-77 season, the two phrase forms which were to become themes 8 and 9, together constituted only 63% of the theme, while the remaining 37% was devoted to hybrid and miscellaneous phrases. As time went on, the miscellaneous and then the hybrid phrases dropped out, and eventually in the last period of 1978, only the 8 and 9 forms existed.

At the same time, the sequence in which the whales sang these phrases was also stabilizing, so that all phrases of type 8 were more likely to be completed before the whale began singing phrases of type 9 (Figure 11B). The number of songs containing both themes 8 and 9 was somewhat coordinated with this development. In Period I of 1976-77, 8 and 9 were sung in only 35% of all songs but by early 1978, they appeared in 100% of all songs, after which the percentage fell off somewhat (Figure 11C).

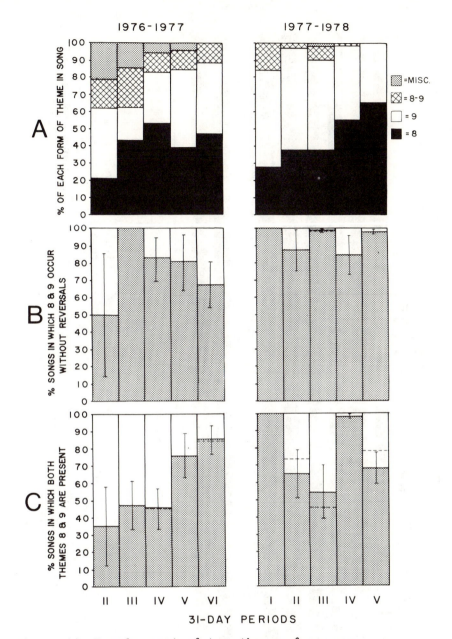

Figure 11. Development of two themes from one.
    A % occurrence of four alternate phrase forms from themes 8 & 9 (two seasons).
    B % songs containing no reversal of order of themes 8 & 9.
    C % songs containing phrases of both themes 8 & 9.
    In B and C, standard error is indicated by a vertical line. Dashed lines indicate means when aberrant songs are included in the calculations.

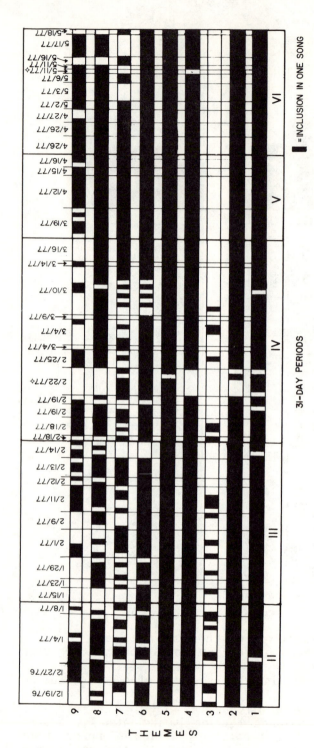

Figure 12A. The inclusion or exclusion of all themes in our total sample of 310 songs. Song sessions are identified by date. Each column within a given song session represents one song, with the nine themes occupying rows and reading from bottom to top. The themes are always given in the order sung. With any reversal of order by the whale, the record is advanced to the next song column. When a whale sang at least one phrase of a theme, the corresponding box was blackened. The sporadic nature of theme inclusion is obvious. Note also the loss over time of theme 3 and the blooming and fading of theme 7. A ⟡ is used to signify aberrant song sessions.

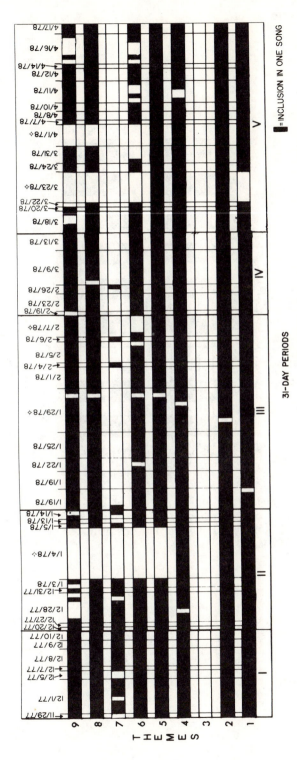

Figure 12A (continued).

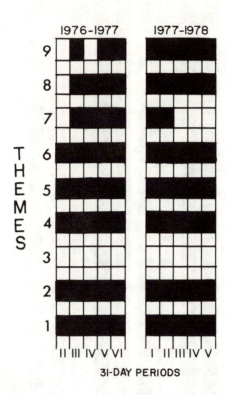

Figure 12B. A summary of 12A. Now the columns represent periods in which the majority of the songs in the majority of the song sessions contained a given theme.

The Changing Constitution of the Song as a Whole

We have presented the separate history of several themes through the two years. We will now consider the changing constitution of the song as a whole. We will examine this in terms of two features: 1) the amount of time devoted by the whales to each theme in the song; and 2) the changes in average duration of phrases, themes, and songs.

Figure 12A shows which themes were included in each song of every song session in our sample. For a theme to be counted as being present, at least one complete phrase had to be included in a song. It is immediately apparent that most songs are incomplete. For example, out of our total sample of 310 songs, only 3 included all 9 themes (these songs were recorded on 4 January, 11 February and 18 February 1977). This poor showing is of course largely an artifact related to the constant changing of the song. For example theme 3, which was never very strong in our sample, was dropping out in the 1976-77 singing season and was last heard 10 March 1977. Theme 7 also dropped out about a year later than theme 3: it was last heard 26 February 1978.

By examination of Figure 12A, one sees an apparent instability early in the 1976-77 season. The reason for this, as will be demonstrated below, is that themes 7, 8, and 9 were all undergoing development during these early months. It was a time when they were not sung by all whales and only sporadically by those that did sing them. Theme 7 dropped out the following year, but themes 8 and 9 became a part of almost every singer's repertoire. Three of the four singers that left these themes out entirely sang songs that were otherwise aberrant (Frumhoff, 1983).

Figure 12B, using broader criteria, gives a simplified overview of theme replacement.

The question of the relative popularity of the different themes -- i.e., the degree to which whales adopt and sing themes -- is worth pursuing. Figure 13 shows the percentage of songs in each song session which included each theme. It demonstrates an interesting point. The only stable themes are ones that are sung in close to 100% of the songs. Strangely, no theme seems to survive at some intermediate value, say in 50% of the songs. Rather, if it is not a part of virtually all songs it is either increasing or decreasing (e.g., themes 3,6,7,8,9).

Adoption of a theme by all of the singers for all of their songs during some period does not guarantee its tenure (e.g., themes 6 and 7), but so far, a theme which was observed to increase in popularity over two or more periods never failed to gain adoption by all singers during at least one period. Thus, one can predict that if the popularity of a new motif increases significantly, it will probably soon enjoy acceptance by the entire singing population. Its

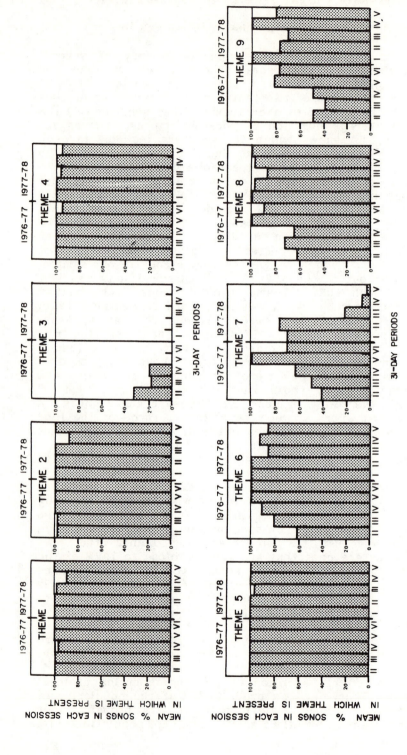

Figure 13. The percentage of song sessions which included each theme. It demonstrates several interesting points. The only stable themes are the ones that are sung in close to 100% of the song sessions. Strangely, no theme seems to survive at some intermediate value, say in 50% of the sessions. Rather, if it is not a part of virtually all sessions, it is either increasing or decreasing.

popularity will not remain at some fixed, intermediate value. Apparently its status in the song can only be stable for long periods when it has reached full acceptance by the entire population.

Progressive Changes in Duration

There was a great range in the length of humpback songs during the 1976-77 and 1977-78 Hawaiian seasons, from a minimum of 4.4 minutes to a maximum of 26.4 mins. (not including a few sequences of themes for which it was impossible to define what constituted a song). This variation followed clear trends throughout the song seasons. Figure 14A shows that in 1976-77, the songs more than doubled in average length between Periods II (7.5 mins.) and V (16 min.). The change is so marked that there is no overlap in song durations within the standard error for Periods II and V. A small decrease in duration occurred between Periods V and VI.

The dramatic increase in song length seen in 1976-77 was not repeated in the 1977-78 season. On the contrary, while song length increased in the first part of 1977 and decreased in the last part, exactly the opposite occurred in the 1977-78 season.

As we have seen in Figure 12, most songs do not include all possible themes. It seems likely that the variable number of themes per song might be largely responsible for the radical changes in song length. In Figure 14B, we have tabulated the average number of themes included in songs of each period, and indeed, the result is a curve remarkably parallel to that of song length: in early 1977 when many songs were incomplete, the average song was short. In Period V, when most songs included all possible themes, the song was more than twice as long. In Period VI, however, one theme (theme 7) was omitted, and the average song length again decreased. In 1977-78 as well, the number of themes included allows one to predict song duration (except in Period V, when some other factor must have come into play).

Song length is affected not only by the number of themes but also by the average duration of each theme. This in turn is the product of the number of phrases in the theme and the average duration of each phrase. In Figure 15, we see five different measures of theme 5. Theme 5 is an example of how changes at each level of the song affect the durations of other levels. At the most detailed level of the song, we can see that the increase in phrase length parallels the increase in number of units per phrase. In 1976-77, the average duration of phrases was increasing and the average number of phrases in the theme was also increasing. Theme length, being a product of these two

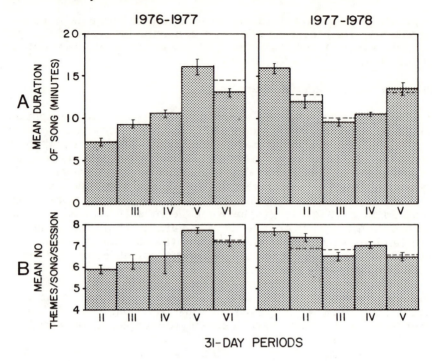

Figure 14. Parameters of song that changed together.
A: Average duration of songs through two singing seasons.
B: Average number of themes per song in each song session over two singing seasons.
Standard errors indicated. Dotted lines represent means when aberrant song sessions were included in the calculations.

Figure 15. Some changing parameters of theme 5 over two singing seasons.

A: mean number of units per phrase in each song session.
B: mean phrase duration in each song session.
C: mean number of phrases per theme in each song session.
D: mean theme duration in each song session.
E: % song occupied by theme 5.

Dashed lines indicate means when aberrant songs from these periods were included in the calculations. Vertical lines indicate standard errors.   ⟶

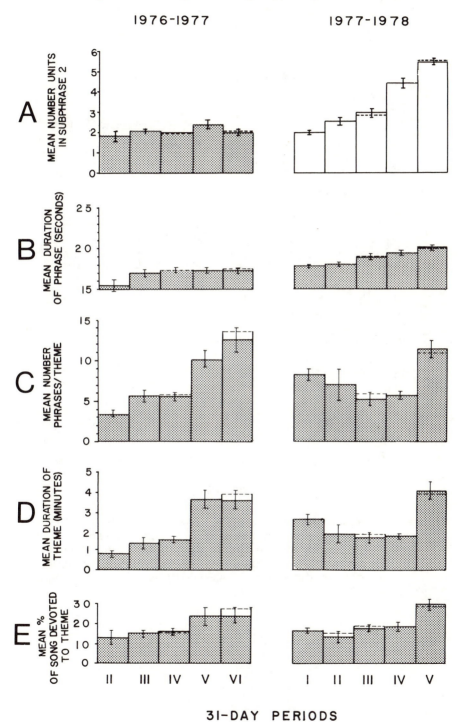

1976-1977   1977-1978

31-DAY PERIODS

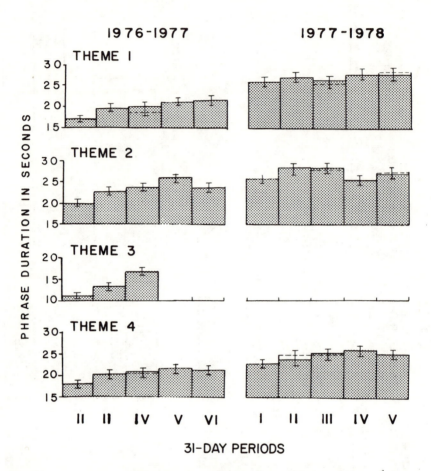

Figure 16. Mean phrase duration for every theme in two singing seasons showing: 1) most phrases increase in length as the theme gets older, and 2) there is no apparent annual cycle underlying changes in phrase duration. Dashed lines are as in Figure 15.

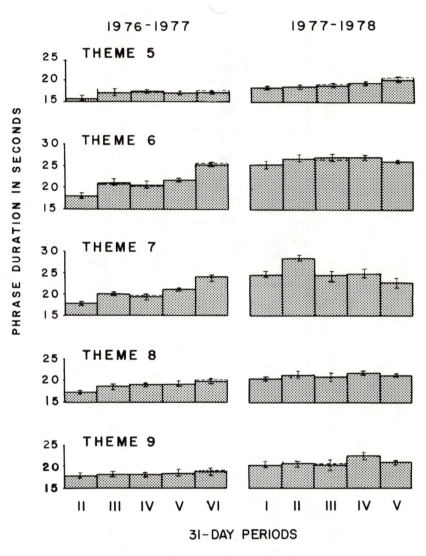

Figure 16 (continued).

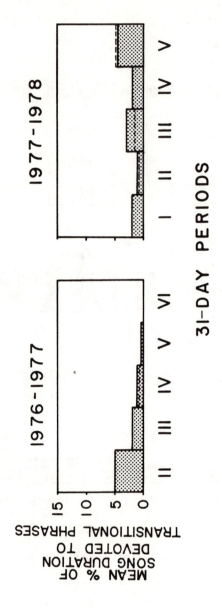

Figure 17. The presence of transitional phrases in the song. Their omission reflects an increased stability in the song as the 1976-77 season progressed. The trend reversed in the 1977-78 season. Dotted lines are as in Figure 15.

factors, shows, of course, an even more prominent increase. The only exception occurred in Period V, when (as we saw in Figure 11) the entire song was slightly declining in duration. However, even in that period, the growth rate of theme 5 was proportionately greater than the growth rate of other themes, with the result that theme 5 occupied an increasing percentage of the song in each time period in 1977. Although in 1976-77 the duration and number of phrases per theme worked together to lengthen the theme, the story was more complex in 1977-78, when the two factors affecting theme duration had opposite effects.

Figure 16 shows the average duration of all phrases in all themes for each period in the 1976-77 and 1977-78 seasons. A continuing tendency for phrases to lengthen with time is conspicuous in every theme (with a partial reversal in the last part of 1978). Similar increases in phrase duration were repeatedly noted in the Bermuda song from our 22 year sample (Payne and Payne, in press).

The tendency for phrases to expand not only within each season but continuously over almost two seasons provides clear evidence that these changes are not under control of environmental parameters such as water temperature or day length, which fluctuate on an annual or seasonal basis. Presumably, changes involving duration must have some sort of periodicity, but apparently a two-year sample is not long enough to discover a complete cycle.

Synchronization of Trends that Increase or Decrease:
Variability in the Whole Song

It is interesting to observe that the progressive stabilizing of themes 8 and 9 in the 1976-77 singing season coincided with a general increase in stability of several other aspects of the song as well. As noted above, the whales were the most consistent in terms of which themes they included between late 1977 and early 1978 (Figures 12A and 12B): a time when the number of themes in each song was highest (Figure 14B). It was also the time when the fewest transitional phrases were present (Figure 17).

A transitional phrase is a phrase occurring between two themes which combines features of both (Payne and McVay, 1971). The 24 March 1978 song presented in Figure 3 shows a very simple one made by joining of incomplete phrases from themes 2 and 4. Some transitional phrases mix units from two adjoining themes in a more complex way. Transitional phrases may occur between any two themes, or occasionally replace one or more themes that are omitted from the song.

During the times when the song was least stable in terms of which component themes were present, it contained transitional phrases between all themes: but at the end of the 1976-77 season, when all themes were firmly established, there were no transitional phrases left. In Period IV of 1977, the song structure was rigid and unambiguous, nearly every possible theme was included in each song by each whale, and every phrase could be clearly assigned to one theme, with no miscellaneous, hybrid, or transitional forms. In other words, the song was maximally compartmentalized, organized, and predictable. Thus stabilization in many factors of the song occurred synchronously.

To return to the thesis with which we opened this study, we had expected that in the end of the season, when whales in Hawaii had been listening to each other for several months, the song would be maximally stereotyped. As a corollary, we had expected the least stereotypy in the beginning of a season (at the end of the relatively silent feeding period when the whales might be expected to have difficulty recalling the song). The 1977 data fulfilled these expectations completely, but in the 1978 data, all the trends reversed! Thus although tendencies for stabilizing the song do seem to rise, peak, and decline together, the influence which drives them does not operate on a seasonal cycle and remains to be discovered.

Further evidence that the stabilizing tendencies are not on a seasonal cycle is obtained by looking for periods of greatest change in the song and paying particular attention to whether these changes occur more between singing seasons or within them. In Table 2, we have noted between which two periods the largest changes occurred in every parameter presented in this paper. In only 2 out of the 27 parameters measured did the greatest change occur during the six months of non-singing. Frumhoff (1983) notes that the only period in which aberrant songs were not recorded was in Period I -- the time when one would expect the most aberrancy in songs, were memory failure a major factor.

These facts suggest that song change is not merely a consequence of forgetfulness, or of other accidents during the summer or during the migrations to and from the singing grounds. On the contrary, the overwhelming majority of major changes occurred while singing was in full swing. Thus, the song is probably not mixed up or changed significantly during the relatively silent summer months. Rather, it is retained intact in the singer's memory. It appears that for a humpback whale, to remain silent is to keep the song in a fixed state, whereas to sing is to change.

Discussion

We have seen that the songs of humpback whales are at once highly organized and also labile. During the time of year when they are heard, they are constantly changing in a wide variety of ways. The most pronounced changes are adopted by all singing whales, while old forms of phrases and

themes become obsolete and disappear in a coordinated
fashion shared by all singers.

One might postulate that humpback songs convey
information about the environment, but because they are so
repetitious, it seems unlikely that much new information is
conveyed with each repeat. However, it is possible that
changes in the song might in some way reflect changes in the
environment. Most environmental changes of a kind likely to
be of importance to the lives of whales (e.g., changes in the
food supply, the time to migrate, etc.) are cyclical in nature
and likely to repeat on some regular basis. If the changes
in the whales' songs are related to these phenomena, they
would probably repeat in synchrony with them. But they do
not repeat at all.

Three basic cycles affect living systems the most.
They are based on the day, the lunar month, and the year.
Since several days must pass before even rapid song changes
can be detected, a change based on daily cycles is ruled out.
But lunar and annual cycles are also ruled out because the
changes made by humpback whales do not correlate with
either of these cycles (indeed, they do not seem to repeat at
all). Therefore, it seems unlikely that songs are detailed
comments on changing features in the natural environment,
and more likely that they are displays and thus internally
controlled.

Ever since Darwin, it has been recognized that one of
the processes which can drive evolution of displays is sexual
selection. Large repertoires in songbirds seem to have
evolved by sexual selection. In test situations, female
canaries tend to select males which are singing a more
varied repertoire. Kroodsma (1976) has shown that not only
are female songbirds reproductively primed by the song of
conspecific males, but exposure to more varied song
repertoires stimulates female canaries (*Serinus canarius*) to
engage in more nest building and egg laying behavior (he
was, however, unable to rule out the possible special effect
of particular songs). Studies in the field by Yasukawa,
Blank, and Patterson (1981) indicate that male Red-winged
Blackbirds (*Agelaius phoeniceus*) with large repertoires tend
to have higher reproductive success than males with small
repertoires, although this finding is complicated by the fact
that males with large repertoires tended to have other
attributes like greater age that might make them more likely
to succeed in reproduction regardless of song repertoire.
McGregor, Krebs, and Perrins (1981) have shown the same in
Great Tits (*Parus major*) in a study which was controlled for
age.

We have noted that the changes in the songs of
humpback whales which we have described cannot be caused
only by accumulation of copying errors between seasons,
since most of the greatest changes in the parameters of song

Table 4. Five hypothetical stages in the evolution of singing, using as a criterion the mechanisms by which variability is introduced into the songs. Humpback songs can be seen as an advanced stage in the increase of variability in a hypothetical continuum between songs of crickets and those of humans.

SINGING TYPE	REPRESENTATIVE SPECIES	MECHANISM OF INTRODUCING VARIABILITY	RESULT
1	Cricket, Doves	No vocal learning variability introduced genetically, by mutation or by hybridization	Rigidly fixed songs with little variability
2	Chaffinch, White-crowned sparrow	Learning from conspecifics during sensitive period in early life.	Modest song repertoires.
3	Mockingbird, Canary, Red-winged blackbird	Learning throughout lifetime from conspecifics or from other species	Larger song repertoires present at a given time.
4	Humpback whale possibly Yellow-rumped Cacique	Continuous learning from conspecifics of new versions of song, as it changes constantly within set laws of form.	Rapid, continuous song evolution: small song repertoire at any given time but enormous repertoire over many years.
5	Human being	Learning *de novo* compositions governed only by laws of form.	Rapid, discontinuous, largest and most varied song repertoires.

that we measured occurred during the singing seasons. Recent studies by Tyack (1981) indicate that the songs of humpback whales play a role in their reproductive behavior similar to that in songbirds. Female choice may drive the changes we have observed in humpback song the way female choice drives some songbird species to large vocal repertoires.

The peculiar characteristics of humpback whale songs and their unique attributes when compared with the songs of other animals suggest that they may demonstrate an important step in the evolution of singing and provide an interesting point along a continuum of song development from invertebrates to humans. In order to explain this point further, we will broaden the definition of singing to mean a display employing repeated sounds. We will then look for common aspects in the many kinds of singing found throughout the animal kingdom.

If we focus on the mechanisms by which variability is introduced into the singing performances of different species, we find that singing falls rather naturally into five distinct categories or singing types (Table 4). We suspect that the higher the number of the singing type, the fewer the number of species in it.

Type 1 Singing. Songs that are characteristic of this type are genetically determined and often relatively simple. Learning seems to play no role but variations can be introduced genetically through mutation or hybridization. Doves (Lade and Thorpe 1964) and crickets (*Teleogryllus* sp.) are examples of this type. Bentley and Hoy (1972) showed that in F_1 cricket hybrids, intervals between calls are intermediate between corresponding parental intervals.

Type 2 Singing. In singers of this type, song variation is introduced through learning, but only during a critical period, and in some cases the song which is learned must be a song for which the animal has a predisposition. The White-crowned Sparrow (*Zonotrichia leucophrys*) is an example (Marler 1970). In some species, for example the Chaffinch (*Fringella coelebs*), several songs may be learned by the same male (Marler 1956; Hinde 1958) during a critical period (Thorpe 1958), but after the critical period is over, the song repertoire remains fixed.

Type 3 Singing. In this type, song variability is introduced by learning throughout much or all of the animal's life. The mimicry of Mockingbirds (*Mimus polyglottos*) is an example (Wildenthal 1965), as are Red-winged Blackbirds (Yasukawa et al. 1981). Other probable examples are canaries (Nottebohm and Nottebohm 1977). Singers of this type have various different songs in their repertoire at any time. Old songs are not necessarily lost when new songs are added, but may be retained.

Type 4 Singing. Song variability in this type also reflects continuous learning; however, a population abandons songs it learned first and only sings one song at a time. This song undergoes constant changes which rapidly spread through the population. Trends in the changes are subject to certain set rules of form and nearly always involve modifications of pre-existing material. Past versions of the song do not recur - thus if, as seems likely in humpback whales, an individual male sings for many years (see Guinee et al. 1983), his repertoire over his whole life is very large, but at any given time consists only of the single current version of the song.

Humpback whales are, to date, the best documented and certainly the most extreme example of type 4 singing, but there is a bird, the Yellow-rumped Cacique (*Cacicus cela*) which may constitute a second example (Feekes 1977).

Type 5 Singing. Singers of this type combine some of the features of types 3 and 4 and add new complexities. The only limits on song variability in type 5 are learned conventions (e.g. scales, rhythmic patterns, rules of harmony, forms, etc.). Like type 3 singers, individuals have a large repertoire of songs at any time. Like type 4 singers, they have the capacity to create new songs adhering to certain fixed rules of form. However, in type 5 singing, new songs do not have to be derived from existing ones but can be created *de novo*. Humans seem to be the sole practitioners of this kind of singing.

In summary, the study of whale songs has unexpectedly demonstrated a kind of missing link in the continuum of sound display leading from simple stereotyped singing to the full complexity of human song. Instead of viewing human song as an isolated phenomenon, as has often been done in the past, we see now that it may well have developed through a simple step-by-step evolution, the stages of which can still be studied by listening to several disjunct species, each singing to us from their respective branches of the phylogenetic tree.

Acknowledgements

We have been helped by generous friends in all phases of this work; in the logistics of living while in the field, in preparing and maintaining equipment, in gathering data, and in the long process of analyzing it. The New York

Zoological Society gave encouragement and support throughout the project. In addition, we owe a debt of gratitude to many others. Valuable help was provided by the Eppley Foundation, the Lahaina Restoration Foundation, the National Geographic Society, the World Wildlife Fund, Shirley M. Herpick, James Luckey, John and Walter McIlhenny, and Drake and Maureen Thomas. We are grateful to Donald Kroodsma for his critical reading and especially for his help in the comparison of bird and whale song. For help with recording songs, we thank also Jim Darling, Kim Gibson, Ted Goodspeed, Jan Heyman-Levine, Fred Levine, Beth Matthews, and Greg Silber. Many friends helped in analyzing data, including James Bird, Eleanor Dorsey, Peter Frumhoff, Victoria Rowntree, and Daphne Stuart. But our major thanks for help of this sort are warmly given to Linda Guinee and Jan Heyman-Levine, who spent thousands of hours poring over whale songs.

Literature Cited

Bentley, D.R. and R.R. Hoy.
  1972. Genetic control of the neuronal network generating cricket (*Teleogryllus gryllus*) song patterns. Anim. Behav. 20: 478-492.

Darling, J.D., K.M. Gibson and G.K. Silber.
  1983. Observations on the abundance and behavior of humpback whales (*Megaptera novaeangliae*) off West Maui, Hawaii 1977-79. *In* R. Payne (ed.), Communication and behavior of whales, AAAS Selected Symposia Series, p. 201-222. Westview Press, Boulder, Colo.

Feekes, F.
  1977. Colony-specific song in *Cacicus cela* (Icteridae, Aves): The pass-word hypothesis. Ardea 65: 197-202.

Frumhoff, P.
  1983. Aberrant songs of humpback whales (*Megaptera novaeangliae*): Clues to the structure of humpback song. *In* R. Payne (ed.), Communication and behavior of whales, AAAS Selected Symposia Series, p. 81-127. Westview Press, Boulder, Colo.

Guinee, L.N., K. Chu, and E.M. Dorsey.
  1983. Changes over time in the songs of known individual humpback whales (*Megaptera novaeangliae*). *In* R. Payne (ed.), Communication and behavior of whales, AAAS Selected Symposia Series, p. 59-80. Westview Press, Boulder, Colo.

Hinde, R.A.
1958. Alternate motor patterns in chaffinch song. Anim. Behav. 6: 211-218.

Katona, S., B. Baxter, O. Brazier, S. Kraus, J. Perkins, and H. Whitehead.
1979. Identification of humpback whales by fluke photographs. *In* H.E. Winn and B.L. Olla (eds.), Behavior of marine animals - Current perspectives in research, vol. 3: Cetaceans, p. 33-44. Plenum Press, N.Y.

Kroodsma, D.E.
1976. Reproductive development in a female songbird: Differential stimulation by quality of male song. Science (Wash. D.C.) 192: 574-575.

Lade, B. and W. Thorpe.
1964. Dove songs as innately coded patterns of specific behavior. Nature (Lond.) 202: 366-368.

Marler, P.
1956. Behavior of the chaffinch (*Fringilla coelebs*). Behaviour Suppl. 5, 184p.

1970. A comparative approach to vocal learning: Song learning in White-crowned sparrows. J. Comp. Physiol. Psychol. Monogr. 71, 25p.

McGregor, P., J. Krebs, and C. Perrins.
1981. Song repertoires and lifetime reproductive success in the Great Tit (*Parus major*). Am. Nat. 118 (2): 149-159.

McSweeney, D. and R. Payne.
In prep. Songs of a humpback whale recorded in Alaska, 1979.

Nottebohm, F. and M. Nottebohm.
1977. Relationship between song repertoire and age of the canary, *Serinus canarius*. Z. Tierpsychol. 46(3): 298-305.

Payne, K. and R. Payne.
In press. Large-scale changes over 17 years in songs of humpback whales in Bermuda. Z. Tierpsychol.

Payne, R.
1968. Among wild whales. The New York Zoological Society Newsletter, 6p.

1970. Songs of the humpback whale. Phonograph record with accompanying 36p. book. Del Mar, CA: CRM Books; New York: Capitol Records, SWR-11.

1978. Behavior and vocalizations of humpback whales (*Megaptera* sp.). *In* K.S. Norris and R.R. Reeves (eds.), Report on a workshop on problems related to humpback whales (*Megaptera novaeangliae*) in Hawaii. U.S. Dep. Commer. NTIS PB-280 794, p. 56-78.

Payne, R. and S. McVay.
    1971. Songs of humpback whales. Science (Wash. D.C.) 173: 585-597.

Payne, R. and L.N. Guinee.
    1983. Humpback whale (*Megaptera novaeangliae*) songs as an indicator of "stocks". *In* R. Payne (ed.), Communication and behavior of whales. AAAS Selected Symposia Series, p. 333-358. Westview Press, Boulder, Colo.

Thorpe, W.H.
    1958. The learning of song patterns by birds, with especial reference to the song of the chaffinch *Fringilla coelebs*. Ibis 100: 535-570.

Tyack, P.
    1981. Interactions between singing Hawaiian humpback whales and conspecifics nearby. Behav. Ecol. Sociobiol. 8: 105-116.

Wildenthal, J.L.
    1965. Structure in primary song of the mockingbird (*Mimus polyglottos*). Auk 82: 161-189.

Winn, H.E., P.J. Perkins, and T.C. Poulter.
    1971. Sounds of the humpback whale. Proceedings of the 7th annual conference on biological sonar and diving mammals, Menlo Park, CA, p. 39-52.

Winn, H.E. and L.K. Winn.
    1978. The song of the humpback whale (*Megaptera novaeangliae*) in the West Indies. Mar. Biol. (Berl.) 47: 97-114.

Yasukawa, K., J.L. Blank, and C.B. Patterson.
    1981. Song repertoires and sexual selection in the Red-winged Blackbird (*Aselaius phoenicus*). Behav. Ecol. Sociobiol. 7: 233-238.

Linda N. Guinee, Kevin Chu,
Eleanor M. Dorsey

2. Changes Over Time in the Songs of Known Individual Humpback Whales (*Megaptera novaeangliae*)

Abstract

Humpback whale songs change over time. Is this due to changes in the membership of a singing population or do individual humpbacks change their songs? We studied this in Hawaii during the 1977-78, 1978-79, and 1979-80 singing seasons. During this time, three known individuals were recorded on two occasions each. The first of these three whales was re-recorded after 54 days, the second after 39 days, and the third after 329 days. We analyzed spectrograms of the songs sung by each whale on each of the occasions it was recorded and compared them to spectrograms of the songs of other whales recorded during the same time periods. We found that individuals changed their songs in the same ways the rest of the group did. It is therefore unlikely that overall changes in humpback song simply reflect changing membership in a singing population. Rather, individual humpbacks keep updating their songs.

Introduction

Since the discovery that humpback whales (*Megaptera novaeangliae*) change their songs (Payne 1978; Winn and Winn 1978; Payne, Tyack, and Payne 1983), it has been a matter of speculation as to what causes these changes. Work describing the nature of the changes shows that there is probably no correlation between song change and either time of year or environmental factors, indicating that changes in humpback song can be attributed to cultural evolution (Payne et al. 1983).

Without knowing over what span of time individuals sing or how the songs of an individual change in relation to the songs of the rest of the population, we cannot be sure that the overall changes in humpback song do not simply reflect changing membership in a group of singing whales. There might be a high turnover rate of members (Darling, Gibson, and Silber 1983), and song change might result from "incomplete" learning by new singers. Given the length and complexity of songs, it would not be surprising to find that new singers would make some "mistakes" when learning the song. If still later arrivals copied these errors while adding new ones of their own, the songs might change without individual singers keeping pace with the changes. It is therefore important to learn whether individual humpbacks continue singing over an extended period of time and whether they mimic the most current version of the song sung by the rest of the population.

Through photo-identification of individuals (Katona, Baxter, Brazier, Kraus, Perkins, and Whitehead 1979; Katona and Kraus 1979; Katona and Whitehead 1981; Darling and Jurasz 1983; Darling et al. 1983), we know that we have recorded the songs of at least three whales on two occasions each. Their songs provide an opportunity to address the above questions.

Materials and Methods

All of the humpback songs discussed in this paper were recorded around the four island area in the Hawaiian Islands during the breeding season (lasting approximately from mid-November to mid-May) of 1977-78, 1978-79, and 1979-80. Recordings were made from a 16' powerboat using a number of tape recorder and hydrophone combinations which in all cases had a frequency response uniform within 5 db for frequencies between 50 and 10,000 Hz. Songs were analyzed and terms describing song structure were used in the manner described by Payne et al. (1983). A song session, for example, consists of all songs recorded of a single whale which were sung in an unbroken sequence.

The three twice-recorded whales were recorded on the following dates:

WHALE	FIRST RECORDED	RERECORDED
A	5 February 1978	31 March 1978
B	10 March 1979	18 April 1979
C	11 April 1979	5 March 1980

In addition to these six song sessions, we analyzed 41 song sessions from other singers in the three singing seasons. The cataloguing of individual identifications of the other singers was not complete at the time of this analysis, making it impossible to know at all times which individual was singing. However, we considered all singers recorded on different dates to be different individuals unless there was information to the contrary. We feel this assumption is justified because the number of known repeat singers is approximately as many as would be expected in our sample given the low observed resighting rate for singers. In the 1976-77, 1977-78, and 1978-79 seasons, Jim Darling and his co-workers took 56 identification photographs of singers in Hawaii. Four of these were re-identifications of singers who had already been photographed (J. Darling, pers. comm.). If we find a rate from this data which would give us an estimate of the number of resightings we can expect from all observations (4/56), the observed rate is 7.14%. At the 7.14% rate, we would expect 3.36 of our 47 song sessions to be re-recordings of individuals already in our sample. This is about as many as we have detected, so it is likely that, if this rate is accurate, the other sessions were all sung by different individuals.

Payne et al. (1983) divided the singing season of humpbacks in Hawaii into six 31-day periods for the purpose of analysis. For this paper, we compared the songs of our repeat singers with those of other singers in the corresponding 31-day periods. For the two 31-day periods used from the 1977-78 season, we made and analyzed spectrograms of all recordings of good quality, a total of 28 song sessions, with a mean of 14 sessions per period. (These were some of the same songs analyzed for the Payne et al. paper, 1983). For the following two seasons, we devised a method of subsampling the recorded song sessions in order to reduce the lengthy analysis time (at least four hours for every hour of tape). Songs for the subsampling were selected from the 8th and 23rd days of each period or as close as possible. In this way, 12 evenly spaced samples resulted from each singing season and both early and late portions of each 31-day period were represented. We wanted a minimum of three songs from each song session. In practice, we supplemented the sample with songs from other whales that were recorded on days as close to the target date as our sample permitted, until we had at least three songs from either one whale or a combination of whales. We tested the reliability of our subsampling method against the fully analyzed 1977-78 season and found it to be adequate.

Spectrograms were made using a Spectral Dynamics SD-301C real time analyzer, a Tektronix 604 monitor oscilloscope and a 35mm Nihon Kohden oscilloscope camera. This produced reels of photographic paper which we developed, cut into strips approximately 60 cm long, and mounted in order on sheets of paper. We measured durations of song sessions, songs, themes, and phrases directly from the spectrograms using a ruler. We then counted the number of songs, themes,

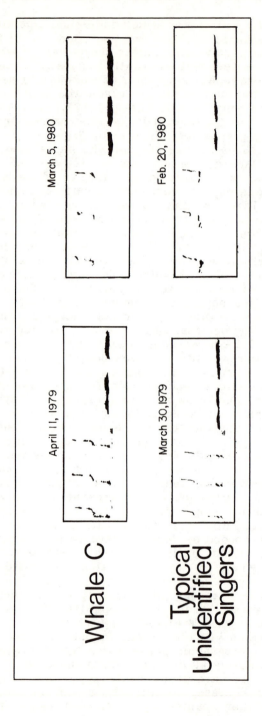

Figure 1. Sample phrases of theme 8 sung by whale C and by other whales recorded in the same two periods during which whale C was recorded. Note that the phrases of theme 8 sung by whale C are more similar to those of its contemporaries at a given time than they are to its own renditions from an earlier or later date.

phrases, and units and calculated a percentage of songs containing each theme.

Once the variables were measured for each song, we calculated song session means for each of the twice-recorded individuals. We then did the same for the other singers in the corresponding 31-day periods, using the mean of their session means to represent the period. Then we compared the session means of each twice-recorded whale with the corresponding period means. In each case, the twice-recorded whale appeared to change its song in the same ways as the other whales. Figure 1 shows sample phrases from Theme 8 sung by Whale C and by other whales recorded in the same two time periods during which Whale C was recorded, an example showing that the songs of an individual are, at any given time, qualitatively closer to those of its contemporaries than to its own songs from an earlier or later date. To test this quantitatively, we applied two different methods of analysis. First, we calculated the probability of the observed similarities occurring by chance if the changes over time in the individual songs were random with respect to the changes made by the other singers (Method 1). Then we did discriminant analyses on the data (Method 2). We will describe the two methods in turn.

Method 1. Almost all of the variables that we measure in humpback song can be shown either to increase or to decrease in value over time if they are measured precisely enough. There are consequently four possible outcomes when comparing the songs of an identified singer, recorded twice, with those of the other whales singing during the same time periods. The possible outcomes are listed below, using "+" to indicate an increase in the measured variable and "-" to indicate a decrease:

OUTCOME	IDENTIFIED SINGER	OTHER SINGERS
1	+	+
2	+	-
3	-	+
4	-	-

We define two of these outcomes, 1 and 4, as "successes" in that the identified singer changed its song over time in a way similar to the way the songs of other whales changed. If the direction of change of the individual is random with respect to the direction of change of the other whales, then the probability of success p(++ or --) will be equal to the probability of failure p(+- or -+) and both will be equal to 1/2 for any one variable. If we find, however, that the probability of success is not equal to the probability of failure, then we can conclude that the direction of change of

Table 1. Variables considered in both analysis methods. The mean %'s of song devoted to each theme were not considered because they do not continuously change.

VARIABLE

Mean song duration (sec.)
Mean number of themes per song
Theme 1: mean duration of phrases (sec.)
Theme 2: mean duration of phrases (sec.)
Theme 4: mean duration of phrases (sec.)
Theme 5: mean duration of phrases (sec.)
Theme 6: mean duration of phrases (sec.)
Theme 8: mean duration of phrases (sec.)
Theme 1: mean number of phrases per theme
Theme 2: mean number of phrases per theme
Theme 5: mean number of phrases per theme
Theme 8: mean number of phrases per theme
Theme 1: mean number of units per phrase
Theme 5: mean number of units per phrase
Theme 5: mean duration of theme (sec.)

---

the individual is not random. In order to test this, we state our hypothesis as follows:

Null hypothesis: $p(++ \text{ or } --) = p(+- \text{ or } -+)$

Alternate hypothesis: $p(++ \text{ or } --) \neq p(+- \text{ or } -+)$

If we are able to measure more than one independent variable, the probability of success in all of them, given the null hypothesis, is the product of the probability of success in each of them, or 1/2 raised to the nth power, where n equals the number of independent variables. More generally, the probability of r successes in n independent variables is equal to $_nC_r \div 2^n$, $_nC_r$ being the combination of n things taken r at a time.

We therefore set out to identify as many independent variables from the song data as possible, then looked to see how each variable changed over time in the songs recorded. We then calculated the probability of the observed outcome occurring by chance, given our null hypothesis. A probability of less than 0.05 for the observed outcome would cause us to reject our null hypothesis that the resighted singers do not change as the other whales do.

We considered only the 15 variables which displayed a continuously varying quantity (Table 1). To identify which of those were independent, we generated a correlation matrix for each pair of periods during which the twice-recorded whales were singing. We calculated the correlation coefficient r for each pair of variables and for each r calculated the two-tailed significance level α_2, the probability of r occurring by chance, assuming no correlation. We accepted as uncorrelated only those pairs for which α_2 was greater than 0.65. In most cases it was greater than 0.80.

Method 2. As another approach to the same problem, we performed a series of discriminant analyses on our data. Discriminant analysis is designed to assign an observation to one of two or more distinct groups on the basis of multivariable observations. Thus, we treated different periods as distinct groups and classified each song session sung by an individual into the period it was most like in order to discover whether the individuals sing more like their contemporaries than like other whales recorded at different times.

For each song session in our sample, we had at least 15 variables (Table 1). Because there are fewer than 15 analyzed song sessions in each period, and because discriminant analysis cannot use a greater number of variables than observations, we needed to select a subset of the variables. We did the discriminant analyses on three different combinations of variables. Our first set of variables was chosen by scanning a correlation matrix for our sample and picking the variables with the consistently highest α_2, those which appeared to be the most independent measures of humpback song. The second group of variables consisted of phrase durations for each theme and the number of themes in each song. According to Frumhoff (1983), there is little variability in these measurements within periods but considerable variability between periods. These variables, we felt, would differentiate well between periods. The third group consisted of the number of phrases in each theme and the number of units in each phrase, both of which are extremely variable within period measurements.

With each set of variables, we performed an overall discriminant analysis on all the song sessions in the six periods of our sample, omitting the twice-recorded whales. We selected for further analysis only the set of variables that best distinguished between periods for these singers. Using the discriminant function generated with this set of variables, we did discriminant analyses restricted to the two periods in which each twice-recorded whale was recorded. We then classified each song session of the twice-recorded whales into one of the two periods. Because results from this analysis were so clear cut, we thought it would be interesting to see into which period the song sessions of the twice-recorded whales would be classified if we considered all six periods in our sample. This method of classification

Table 2. Uncorrelated pairs of variables used in Method 1.

WHALE	TIME 1	TIME 2	VARIABLE 1	VARIABLE 2	r	df	α_2
A	1978 Per 3	1978 Per 5	Th. 8 # Phr.	Th. 4 # Phr.	.0339	22	.8750
B	1979 Per 4	1979 Per 6	Th. 1 # Phr.	Th. 1 # Units	.1436	9	.6736
	"	"	Th. 1 # Phr.	Th. 5 # Units	.0598	9	.8614
	"	"	Th. 1 # Units	Th. 5 # Units	.0544	9	.8738
C	1979 Per 5	1980 Per 4	Th. 1 # Units	Th. 1 # Phr.	.0414	6	.9225
	"	"	Th. 1 # Units	Th. 5 # Phr.	.1203	4	.8204
	"	"	Th. 1 # Phr.	Th. 5 # Phr.	.2080	4	.6925

Table 3. Change over time in means for the twice-recorded individuals and their comparison group of singers for the uncorrelated variables used in Method 1.

WHALE	VARIABLE	TIME 1 MEAN INDIVIDUAL	GROUP	TIME 2 MEAN INDIVIDUAL	GROUP	PARALLEL CHANGE?
A	Th. 8 # Phr.	1.93	2.04	5.40	3.95	Yes
	Th. 4 # Phr.	2.00	2.49	2.25	2.95	Yes
B	Th. 1 # Phr.	7.00	7.96	8.00	11.50	Yes
	Th. 1 # Units	7.33	7.63	7.61	7.13	No
	Th. 5 # Units	8.91	8.97	12.50	9.97	Yes
C	Th. 1 # Units	5.59	7.54	8.93	9.73	Yes
	Th. 1 # Phr.	6.67	11.83	5.10	11.39	Yes
	Th. 5 # Phr.	3.00	7.43	1.00	1.00	Yes

really asks a more difficult question of our data than we need to ask. Instead of asking merely, "Is whale A's song session from Period 5 of 1978 more like 1978 Period 3 song sessions or more like the other song sessions in Period 5, 1978", this test asks "Is whale A's 1978 Period 5 song session more like the other sessions in that period or is it more like the sessions from any other period examined in our sample?"

Results

Method 1. We found four pairs of statistically uncorrelated variables for whale A and the whales singing in the same periods, eight for whale B, and 15 for whale C. In order to use more than two variables, all had to be mutually uncorrelated for each test. This narrowed the number of usable variables to two for whale A and the surrounding periods and three each for whales B and C and their surrounding periods (see Table 2). Although this may seem a small number, it is probably explained by the progressive nature of change for many of the variables tested (Payne et al. 1983).

It is also interesting to note that, for whale A, 16 out of 28 pairs of variables were statistically correlated ($\alpha_2 < 0.05$). For whale B, 2 out of 21 pairs were and for whale C, only 3 out of 64 were.

Looking at the eight variables that were uncorrelated, we compared mean values to see if the twice-recorded individual was changing in the same direction as the other contemporaneous singers. In all cases but one, the individual and the group changed in the same direction over time (Table 3). The probability of getting this result, only one failure out of eight trials, is $_8C_7 \div 2^8$ or 0.0313, which is less than the usually accepted level of statistical significance. (The probability may actually be somewhat higher than 0.0313 since not all the variables show complete independence. However, the trend is clear).

We should note that two of the variables used for whale B are also used for whale C. This should not present any problems since for each whale we are considering a different set of song sessions, which should therefore be independent of each other. This expectation is supported by the results in Table 3. One of the repeated variables, the number of units in subphrase one of Theme 1, changed differently from the rest of the singers with one whale and similarly to it with the other. This independent behavior of the same variable indicates that it is proper to multiply probabilities, as we have done above.

Method 2. Tables 4, 5, and 6 show how well each of the three discriminant functions was able to classify the song sessions in our sample. Analysis using the first set of

Table 4. Results of the discriminant analysis on all whales recorded only once in all six periods, using the variables: mean number of units per phrase of Theme 1, mean phrase duration of Theme 1, mean number of phrases of Theme 1, mean phrase duration of Theme 8, mean song duration, mean phrase duration of Theme 4, mean number of themes per song, mean phrase duration of Theme 5, and mean phrase duration of Theme 6. Nine of the 38 song sessions are misclassified.

OBS	FROM PERIOD	CLASSIFIED INTO PERIOD	Posterior Probability of Membership in Period:					
			1978-3	1978-5	1979-4	1979-6	1979-5	1980-4
1	1978-3	1978-3	0.6935	0.3065	0.0000	0.0000	0.0000	0.0000
2	1978-3	1978-3	0.8150	0.1850	0.0000	0.0000	0.0000	0.0000
3	1978-3	1978-3	0.6598	0.3402	0.0000	0.0000	0.0000	0.0000
4	1978-3	1978-5*	0.3665	0.6335	0.0000	0.0000	0.0000	0.0000
5	1978-3	1978-3	0.7676	0.2324	0.0000	0.0000	0.0000	0.0000
6	1978-3	1978-3	0.7065	0.2934	0.0001	0.0000	0.0000	0.0000
7	1978-3	1978-3	0.8239	0.1761	0.0000	0.0000	0.0000	0.0000
8	1978-3	1978-5*	0.4779	0.5212	0.0009	0.0000	0.0000	0.0000
9	1978-3	1978-3	0.6781	0.3219	0.0000	0.0000	0.0000	0.0000
10	1978-5	1978-3*	0.5424	0.4576	0.0000	0.0000	0.0000	0.0000
11	1978-5	1978-3*	0.7753	0.2247	0.0000	0.0000	0.0000	0.0000
12	1978-5	1978-5	0.1640	0.8360	0.0000	0.0000	0.0000	0.0000
13	1978-5	1978-5	0.3612	0.6388	0.0000	0.0000	0.0000	0.0000
14	1978-5	1978-5	0.0409	0.9591	0.0000	0.0000	0.0000	0.0000
15	1978-5	1978-5	0.1955	0.8045	0.0000	0.0000	0.0000	0.0000
16	1978-5	1978-5	0.3021	0.6979	0.0000	0.0000	0.0000	0.0000
17	1978-5	1978-5	0.0720	0.9280	0.0000	0.0000	0.0000	0.0000
18	1978-5	1978-3*	0.5658	0.4340	0.0002	0.0000	0.0000	0.0000
19	1978-5	1978-3*	0.6967	0.3033	0.0000	0.0000	0.0000	0.0000
20	1978-5	1978-5	0.0612	0.9387	0.0000	0.0000	0.0000	0.0000
21	1978-5	1979-4*	0.3686	0.2335	0.3979	0.0000	0.0000	0.0000
22	1978-5	1978-3*	0.5668	0.4332	0.0000	0.0000	0.0000	0.0000
23	1979-4	1979-4	0.0000	0.0000	0.7475	0.0002	0.2523	0.0000
24	1979-4	1979-4	0.0000	0.0000	0.6996	0.0001	0.3003	0.0000
25	1979-4	1979-4	0.0000	0.0000	0.9999	0.0000	0.0001	0.0000
26	1979-4	1979-5*	0.0000	0.0000	0.4828	0.0001	0.5171	0.0000
27	1979-4	1979-4	0.0000	0.0000	0.9608	0.0000	0.0392	0.0000
28	1979-4	1979-4	0.0003	0.0004	0.9990	0.0000	0.0002	0.0000
29	1979-4	1979-4	0.0200	0.0258	0.9541	0.0000	0.0000	0.0000
30	1979-4	1979-4	0.0001	0.0000	0.9188	0.0002	0.0809	0.0000
31	1979-6	1979-6	0.0000	0.0000	0.0001	0.9996	0.0004	0.0000
32	1979-6	1979-6	0.0000	0.0000	0.0000	0.9980	0.0020	0.0000
33	1979-5	1979-5	0.0000	0.0000	0.0299	0.0048	0.9653	0.0000
34	1979-5	1979-5	0.0000	0.0000	0.2222	0.0000	0.7777	0.0000
35	1979-5	1979-5	0.0000	0.0000	0.0001	0.0036	0.9963	0.0000
36	1980-4	1980-4	0.0000	0.0000	0.0000	0.0000	0.0000	1.0000
37	1980-4	1980-4	0.0000	0.0000	0.0000	0.0000	0.0000	1.0000
38	1980-4	1980-4	0.0000	0.0000	0.0000	0.0000	0.0000	1.0000

* = misclassified observation

Table 5. Results of the discriminant analysis on all whales recorded only once, in all six periods, using the variables: mean phrase duration of Theme 1, mean phrase duration of Theme 2, mean phrase duration of Theme 5, mean phrase duration of Theme 6, mean phrase duration of Theme 8, and mean number of themes per song. Eight of the 38 song sessions are misclassified, and many are classified correctly with a probability only a little better than 0.5.

OBS	FROM PERIOD	CLASSIFIED INTO PERIOD	Posterior Probability of Membership in Period:					
			1978-3	1978-5	1979-4	1979-6	1979-5	1980-4
1	1978-3	1978-3	0.6389	0.3603	0.0007	0.0000	0.0000	0.0000
2	1978-3	1978-3	0.6242	0.3758	0.0000	0.0000	0.0000	0.0000
3	1978-3	1978-3	0.6864	0.3075	0.0061	0.0000	0.0000	0.0000
4	1978-3	1978-3	0.5086	0.4914	0.0000	0.0000	0.0000	0.0000
5	1978-3	1978-3	0.5293	0.4701	0.0006	0.0000	0.0000	0.0000
6	1978-3	1978-3	0.7458	0.2541	0.0001	0.0000	0.0000	0.0000
7	1978-3	1978-3	0.5711	0.4289	0.0000	0.0000	0.0000	0.0000
8	1978-3	1978-5*	0.3519	0.6480	0.0001	0.0000	0.0000	0.0000
9	1978-3	1978-3	0.4763	0.4190	0.1046	0.0000	0.0001	0.0000
10	1978-5	1978-3*	0.5093	0.4885	0.0022	0.0000	0.0000	0.0000
11	1978-5	1978-5	0.4891	0.5109	0.0000	0.0000	0.0000	0.0000
12	1978-5	1978-5	0.4835	0.5159	0.0006	0.0000	0.0000	0.0000
13	1978-5	1978-5	0.4205	0.5794	0.0001	0.0000	0.0000	0.0000
14	1978-5	1978-5	0.0586	0.9412	0.0002	0.0000	0.0000	0.0000
15	1978-5	1978-5	0.4422	0.5578	0.0001	0.0000	0.0000	0.0000
16	1978-5	1978-5	0.4135	0.5864	0.0000	0.0000	0.0000	0.0000
17	1978-5	1978-5	0.2677	0.7323	0.0000	0.0000	0.0000	0.0000
18	1978-5	1978-3*	0.7503	0.2436	0.0061	0.0000	0.0000	0.0000
19	1978-5	1978-5	0.4809	0.5173	0.0018	0.0000	0.0000	0.0000
20	1978-5	1978-5	0.4781	0.5219	0.0001	0.0000	0.0000	0.0000
21	1978-5	1978-5	0.2968	0.5356	0.1674	0.0000	0.0002	0.0000
22	1978-5	1978-3*	0.5404	0.4590	0.0006	0.0000	0.0000	0.0000
23	1979-4	1979-5*	0.0000	0.0000	0.0996	0.0483	0.8522	0.0000
24	1979-4	1979-5*	0.0000	0.0000	0.1620	0.2187	0.6193	0.0000
25	1979-4	1978-5*	0.2545	0.3876	0.3578	0.0000	0.0001	0.0000
26	1979-4	1979-5*	0.0000	0.0000	0.0906	0.0607	0.8487	0.0000
27	1979-4	1979-4	0.0000	0.0000	0.9079	0.0027	0.0894	0.0000
28	1979-4	1979-4	0.0056	0.0053	0.9884	0.0001	0.0005	0.0000
29	1979-4	1979-4	0.0353	0.0256	0.9388	0.0000	0.0003	0.0000
30	1979-4	1979-4	0.0660	0.0437	0.7899	0.0088	0.0917	0.0000
31	1979-6	1979-6	0.0000	0.0000	0.0007	0.9919	0.0074	0.0000
32	1979-6	1979-6	0.0000	0.0000	0.0231	0.8952	0.0817	0.0000
33	1979-5	1979-5	0.0000	0.0000	0.0830	0.0171	0.8999	0.0000
34	1979-5	1979-5	0.0000	0.0001	0.0381	0.0013	0.9605	0.0000
35	1979-5	1979-5	0.0000	0.0000	0.0375	0.3839	0.5786	0.0000
36	1980-4	1980-4	0.0000	0.0000	0.0000	0.0000	0.0000	1.0000
37	1980-4	1980-4	0.0000	0.0000	0.0000	0.0000	0.0000	1.0000
38	1980-4	1980-4	0.0000	0.0000	0.0000	0.0000	0.0000	1.0000

* = misclassified observation

Table 6. Results of the discriminant analysis on all whales recorded only once, in all six periods, using the variables: mean number of phrases per theme in Theme 1, mean number of phrases per theme in Theme 2, mean number of phrases per theme in Theme 5, mean number of phrases per theme in Theme 8, mean number of units per phrase in Theme 1, and mean number of units per phrase in Theme 5. Six song sessions are misclassified. Half of these are misclassifications between the three adjacent periods in 1979.

			Posterior Probability of Membership in Period:					
OBS	FROM PERIOD	CLASSIFIED INTO PERIOD	1978-3	1978-5	1979-4	1979-6	1979-5	1980-4
1	1978-3	1978-3	0.9902	0.0063	0.0000	0.0000	0.0000	0.0035
2	1978-3	1978-3	0.9475	0.0491	0.0000	0.0000	0.0000	0.0033
3	1978-3	1978-3	0.9502	0.0472	0.0000	0.0000	0.0000	0.0026
4	1978-3	1978-3	0.8592	0.1408	0.0000	0.0000	0.0000	0.0000
5	1978-3	1978-3	0.9891	0.0032	0.0000	0.0000	0.0000	0.0077
6	1978-3	1978-5*	0.4438	0.4962	0.0088	0.0000	0.0000	0.0510
7	1978-3	1978-3	0.9673	0.0322	0.0000	0.0000	0.0000	0.0005
8	1978-3	1978-3	0.9612	0.0388	0.0000	0.0000	0.0000	0.0001
9	1978-3	1978-3	0.9758	0.0040	0.0000	0.0000	0.0000	0.0202
10	1978-5	1978-5	0.3736	0.6260	0.0000	0.0000	0.0000	0.0004
11	1978-5	1978-5	0.1833	0.8165	0.0000	0.0000	0.0000	0.0000
12	1978-5	1978-5	0.0066	0.9934	0.0000	0.0000	0.0000	0.0000
13	1978-5	1978-5	0.0847	0.9148	0.0000	0.0000	0.0000	0.0005
14	1978-5	1978-5	0.0018	0.9981	0.0000	0.0000	0.0000	0.0001
15	1978-5	1978-5	0.0562	0.9438	0.0000	0.0000	0.0000	0.0000
16	1978-5	1978-5	0.0699	0.9300	0.0000	0.0000	0.0000	0.0001
17	1978-5	1978-5	0.0011	0.9978	0.0000	0.0000	0.0000	0.0011
18	1978-5	1978-5	0.0158	0.9528	0.0305	0.0000	0.0008	0.0001
19	1978-5	1978-3*	0.7938	0.1098	0.0002	0.0000	0.0000	0.0963
20	1978-5	1978-5	0.0005	0.9995	0.0000	0.0000	0.0000	0.0000
21	1978-5	1978-5	0.0852	0.9099	0.0036	0.0000	0.0000	0.0014
22	1978-5	1978-5	0.0177	0.9807	0.0015	0.0000	0.0000	0.0001
23	1979-4	1979-4	0.0000	0.0000	0.9605	0.0326	0.0068	0.0000
24	1979-4	1979-5*	0.0000	0.0033	0.4862	0.0009	0.5096	0.0000
25	1979-4	1979-4	0.0000	0.0000	0.6239	0.0831	0.2930	0.0000
26	1979-4	1979-4	0.0000	0.0000	0.9490	0.0168	0.0342	0.0000
27	1979-4	1979-4	0.0000	0.0000	0.7237	0.1697	0.1066	0.0000
28	1979-4	1979-4	0.0000	0.0000	0.6986	0.0212	0.2802	0.0000
29	1979-4	1979-4	0.0000	0.0000	0.9093	0.0814	0.0092	0.0000
30	1979-4	1979-4	0.0000	0.0000	0.6312	0.1452	0.2237	0.0000
31	1979-6	1979-4*	0.0000	0.0000	0.7820	0.2058	0.0121	0.0000
32	1979-6	1979-6	0.0000	0.0000	0.0005	0.9677	0.0317	0.0000
33	1979-5	1979-5	0.0000	0.0000	0.1076	0.0043	0.8881	0.0000
34	1979-5	1979-5	0.0000	0.0000	0.0642	0.0008	0.9350	0.0000
35	1979-5	1979-6*	0.0000	0.0000	0.0079	0.9467	0.0454	0.0000
36	1980-4	1980-4	0.0003	0.0000	0.0000	0.0000	0.0000	0.9997
37	1980-4	1979-4*	0.0002	0.0067	0.6387	0.0006	0.0019	0.3518
38	1980-4	1980-4	0.0266	0.0000	0.0000	0.0000	0.0000	0.9733

* = misclassified observation

Table 7. Results of a discriminant analysis using the variables in Table 6 on the song sessions in 1978 Period 3 and in 1978 Period 5, not including those of Whale A. Only one whale out of 22 was misclassified.

			Posterior Probability of Membership in Period:	
OBS	FROM PERIOD	CLASSIFIED INTO PERIOD	1978-3	1978-5
1	1978-3	1978-3	0.9999	0.0001
2	1978-3	1978-3	1.0000	0.0000
3	1978-3	1978-3	1.0000	0.0000
4	1978-3	1978-3	0.9994	0.0006
5	1978-3	1978-3	1.0000	0.0000
6	1978-3	1978-3	1.0000	0.0000
7	1978-3	1978-3	1.0000	0.0000
8	1978-3	1978-3	1.0000	0.0000
9	1978-3	1978-3	1.0000	0.0000
10	1978-5	1978-5	0.0000	1.0000
11	1978-5	1978-5	0.0000	1.0000
12	1978-5	1978-5	0.0000	1.0000
13	1978-5	1978-5	0.0000	1.0000
14	1978-5	1978-5	0.0000	1.0000
15	1978-5	1978-5	0.0000	1.0000
16	1978-5	1978-5	0.0000	1.0000
17	1978-5	1978-5	0.0000	1.0000
18	1978-5	1978-5	0.0000	1.0000
19	1978-5	1978-3*	0.7270	0.2730
20	1978-5	1978-5	0.0000	1.0000
21	1978-5	1978-5	0.0000	1.0000
22	1978-5	1978-5	0.0000	1.0000

* = misclassified observation

Table 8. Results of a discriminant analysis using the variables presented in Table 6 to compare the song sessions of Whale A to the other sessions in the period during which Whale A was recorded. Both song sessions are classified correctly.

Posterior Probability of Membership in Period:

OBS	FROM PERIOD	CLASSIFIED INTO PERIOD	1978-3	1978-5
1	1978-3	1978-3	1.0000	0.0000
2	1978-5	1978-5	0.0000	1.0000

Table 9. Results of discriminant analysis using the variables in Table 6 on the song sessions in 1979 Period 4 and in 1979 Period 6, not including those of Whale B. Every whale is properly classified.

Posterior Probability of Membership in Period:

OBS	FROM PERIOD	CLASSIFIED INTO PERIOD	1979-4	1979-6
1	1979-4	1979-4	1.0000	0.0000
2	1979-4	1979-4	1.0000	0.0000
3	1979-4	1979-4	1.0000	0.0000
4	1979-4	1979-4	0.9999	0.0001
5	1979-4	1979-4	0.9628	0.0372
6	1979-4	1979-4	0.9843	0.0157
7	1979-4	1979-4	0.9858	0.0142
8	1979-4	1979-4	1.0000	0.0000
9	1979-6	1979-6	0.0094	0.9906
10	1979-6	1979-6	0.0000	1.0000

Table 10. Results of a discriminant analysis using the variables in Table 6 to compare the song sessions of Whale B to the other song sessions in the periods during which Whale B was recorded. Both song sessions are properly classified.

Posterior Probability of Membership in Period:

OBS	FROM PERIOD	CLASSIFIED INTO PERIOD	1979-4	1979-6
1	1979-4	1979-4	1.0000	0.0000
2	1979-6	1979-6	0.0000	1.0000

variables (Table 4) misclassifies 9 individuals out of 38.
Analysis using the second set (Table 5) misclassifies eight
individuals. In both of these discriminant analyses, many of
the song sessions which were classified correctly have a
probability of belonging to the group assigned which was
barely greater than 0.5. Neither of these sets of variables
was deemed satisfactory for distinguishing between periods in
our sample. The third set of variables (Table 6)
misclassifies six individuals. The probabilities of "belonging" to the proper class are generally much higher using this
set of variables than using either of the other sets. Since
we are looking for a function which will separate each
period as clearly as possible, we will only use the function
generated by this set of variables for further analysis.

Table 7 shows the results of discriminant analysis
between the two periods in which whale A was recorded,
Period 3, 1978 and Period 5, 1978, omitting the song sessions
by whale A. Only 1 out of 22 observations is misclassified,
and all those classified correctly have extremely high
probabilities of belonging to the period to which they are
assigned. Likewise, when we use the discriminant function
thus generated to classify song sessions from whale A, each
session is classified into its proper period with a probability
of 1.0000 (Table 8). Thus the song of whale A has changed
with the general population between the two periods.

Table 9 shows the results of discriminant analysis
between the two periods in which whale B was recorded,
1979 Period 4 and 1979 Period 6, omitting whale B's song
sessions. The separation between the periods is excellent.
No sessions are misclassified. When we apply this
discriminant function to the song sessions of whale B, each
one is perfectly categorized (Table 10).

Table 11 shows the results of discriminant analysis
between Period 5, 1979 and Period 4, 1980, the two periods
during which whale C was recorded, omitting the sessions sung
by whale C. Every observation is correctly classified.
Obviously, this discriminant function gives excellent
separation between the periods. Table 12 presents the
results of classifying the two song sessions of whale C using
this discriminant function. Both song sessions are classified
correctly.

Thus, each time we use only two periods in the
discriminant analysis, the results are unambiguous. Each
song session of each twice-recorded whale is much more like
the song sessions of the period during which it was recorded
than it is like the period during which it was recorded at an
earlier or later date.

However, when we consider all six periods in our sample
to categorize the twice-recorded individuals into the period
they were most like, our results are not as consistent.

Table 11. Results of a discriminant analysis using the
variables listed in Table 6 on the song sessions in
1979 Period 5 and in 1980 Period 4, not including
the song sessions of Whale C. Every whale is
properly classified into the period in which it was
recorded.

			Posterior Probability of Membership in Period:	
OBS	FROM PERIOD	CLASSIFIED INTO PERIOD	1979-5	1980-4
1	1979-5	1979-5	1.0000	0.0000
2	1979-5	1979-5	1.0000	0.0000
3	1979-5	1979-5	1.0000	0.0000
4	1980-4	1980-4	0.0000	1.0000
5	1980-4	1980-4	0.0000	1.0000
6	1980-4	1980-4	0.0000	1.0000

Table 12. Results of a discriminant analysis using the
variables in Table 6 to compare the song sessions
of Whale C to other sessions in the periods during
which Whale C was recorded. Both sessions are
properly classified.

			Posterior Probability of Membership in Period:	
OBS	FROM PERIOD	CLASSIFIED INTO PERIOD	1979-5	1980-4
1	1979-5	1979-5	1.0000	0.0000
2	1980-4	1980-4	0.0000	1.0000

Both observations of whale A are very well classified with probabilities over 0.9 (Table 13). Likewise both song sessions of whale B are well classified, with probabilities over 0.8 (Table 14). Thus each song session by both whale A and whale B is much more like the other song sessions of the period in which it was recorded than it is like song sessions of any other period in our sample.

In contrast, both of the song sessions of whale C are misclassified (Table 15). The first song session, recorded in Period 5, 1979, is classified into Period 6, 1979 and the second recording, made in Period 4, 1980, is classified into Period 4, 1979. These unexpected results neither confirm nor deny our hypothesis. The fact that both song sessions of the same individual were misclassified suggests that some whales may have idiosyncracies in their songs or that the variables we used are accurate discriminators for the first two seasons, but are not as accurate for the third season.

The misclassification of whale C's 1979 song session does not concern us as much as the misclassification of the 1980 session. The 1979 session was recorded at the end of Period 5 and was classified into Period 6. The 1980 session, however, was misclassified by an entire year. That these misclassifications of the song sessions of whale C by the overall discriminant function do not necessarily have a direct bearing on our investigation is clear given the results of the discriminant analysis on just the two periods in which a known whale was recorded (Table 12). When faced with a dichotomous choice, the song sessions of whale C were both categorized into the correct period, confirming our results from Method 1.

Discussion

Until this point, there has not been any evidence that individual humpback whales continue singing over an extended period of time. The fact that whale C was recorded singing approximately one year after we first recorded it establishes that some humpbacks sing in at least two seasons. At this time, information about whether individuals continue singing over a longer period of time than this is unavailable. However, because singing whales appear to be mature whales (P. Tyack, pers. comm.) and because the songs appear to play some function in reproduction (Tyack 1981), we see no reason why singers should not continue singing for the duration of their reproductive lives.

Within one season, we do not know whether we are recording one stable population or a continuous stream of migrating singers. Of singers recorded twice within one season, the span between the first and second recordings was approximately eight weeks for whale A and six weeks for whale B. These were unusually long spans compared to the

Table 13. Results of the comparison of the song sessions by Whale A with all the song sessions of all other whales in our sample. Both song sessions are correctly assigned to the period in which they were sung despite the relatively large number of possible categories into which they could have fallen.

Posterior Probability of Membership in Period:

OBS	FROM PERIOD	CLASSIFIED INTO PERIOD	1978-3	1978-5	1979-4	1979-6	1979-5	1980-4
1	1978-3	1978-3	0.9970	0.0017	0.0000	0.0000	0.0000	0.0013
2	1978-5	1978-5	0.0886	0.9111	0.0001	0.0000	0.0000	0.0002

Table 14. Results of the comparison of the song sessions by Whale B with all the song sessions of all other whales in our sample. Both song sessions are correctly assigned to the period in which they were recorded.

Posterior Probability of Membership in Period:

OBS	FROM PERIOD	CLASSIFIED INTO PERIOD	1978-3	1978-5	1979-4	1979-6	1979-5	1980-4
1	1979-4	1979-4	0.0000	0.0000	0.8778	0.0881	0.0341	0.0000
2	1979-6	1979-6	0.0000	0.0000	0.0718	0.8415	0.0866	0.0000

Table 15. Results of the comparison of the song sessions by Whale C with all the song sessions of all other whales in our sample. Neither song session is correctly classified, and each is misclassified into a different period. See text for a discussion of these results.

Posterior Probability of Membership in Period:

OBS	FROM PERIOD	CLASSIFIED INTO PERIOD	1978-3	1978-5	1979-4	1979-6	1979-5	1980-4
1	1979-5	1979-6*	0.0000	0.0000	0.1522	0.8468	0.0010	0.0000
2	1980-4	1979-4*	0.0000	0.0000	0.9014	0.0300	0.0687	0.0000

* = misclassified observation

spans over which other humpbacks have been resighted in Hawaii. Of all whales resighted during one year in Hawaii, 26 out of 28 were resighted less than six weeks after the initial sighting (see Darling et al. 1983). But there is, at the time of this writing, no conclusive information showing either change or lack of change in the singing population in one season.

The misclassification of whale C in the overall discriminant analysis raises some questions we were unable to answer. It is possible that, given the dynamic nature of humpback song, it is inappropriate to try to find a single set of variables which will discriminate in all years. For two of the three sets of variables used, those presented in Tables 4 and 6, there appears to be an uneven distribution of the misclassified songs over time. It is also possible that we have not yet discovered variables which will continue to be good discriminators over time and that analysis of individual units in the songs will provide a better understanding of this. Another possibility is that whale C's songs had idiosyncracies which threw the analysis off. This seems unlikely, however, since, according to Frumhoff's (1983) guidelines for aberrancy in Hawaiian humpback song during this time, whale C sang normal songs.

The results of this study show clearly that individual humpback whales change their songs over time and that the changes occurring in the songs of individuals tend to parallel the changes made in the population as a whole. Whether or not there is a stable population of singers, individuals who are singing are continuously updating their songs. The probability that whales modify their songs in the same way at the same time by chance is staggeringly small, and we must conclude once again that they listen to each other and learn the changes in the song. This, most certainly, is another clue into the mental capabilities of the species. In order to learn and adopt the current version of the song at all times, humpbacks must continuously be attending to a very large collection of details.

Acknowledgements

This paper is the product of a group effort, requiring the work of many people over time. We are grateful to Roger and Katharine Payne for advice, support, and much encouragement. Hal Whitehead, James Bird, Patricia Harcourt, Roger Payne, Victoria Rowntree, and Peter Tyack read and made comments on the paper. Hal Whitehead, Ronald Christensen, and Ralph D'Agostino helped considerably with suggestions on the statistical analysis. James Darling, Deborah Glockner, Jan Heyman-Levine, Fred Levine, Gregory Silber, and Peter Tyack made the tapes and

took the photographs which we analyzed for this paper.
Greg Silber helped make the spectrograms. Jim Darling made
available information about the resighting rate of singers and
provided us with identification photographs of singers from
all seasons. James Bird helped with the bibliography and
with final manuscript preparation. Valuable help of many
kinds was provided by the Eppley Foundation, the Lahaina
Restoration Foundation, the National Geographic Society, the
New York Zoological Society, the World Wildlife Fund,
Shirley M. Herpick, James Luckey, John and Walter
McIlhenny, and Drake and Maureen Thomas.

Literature Cited

Darling, J.D., K.M. Gibson, and G.K. Silber.
   1983. Observations on the abundance and behavior of
   humpback whales (*Megaptera novaeangliae*) off West
   Maui, Hawaii, 1977-79. *In* R. Payne (editor),
   Communication and behavior of whales. AAAS Selected
   Symposia Series, p. 201-222. Westview Press, Boulder,
   Colo.

Darling, J.D. and C.M. Jurasz.
   1983. Migratory destinations of North Pacific humpback
   whales (*Megaptera novaeangliae*). *In* R. Payne (editor),
   Communication and behavior of whales. AAAS Selected
   Symposia Series, p. 359-368. Westview Press, Boulder,
   Colo.

Frumhoff, P.
   1983. Aberrant songs of humpback whales (*Megaptera
   novaeangliae*): Clues to the structure of humpback song.
   In R. Payne (editor), Communication and behavior of
   whales. AAAS Selected Symposia Series, p. 81-127.
   Westview Press, Boulder, Colo.

Katona, S., B. Baxter, O. Brazier, S. Kraus, J. Perkins, and
   H. Whitehead.
   1979. Identification of humpback whales by fluke
   photographs. *In* H.E. Winn and B.L. Olla (editors),
   Behavior of marine animals - Current perspectives in
   research, vol. 3: Cetaceans, p. 33-44. Plenum Press,
   N.Y.

Katona, S. and S. Kraus.
   1979. Photographic identification of individual humpback
   whales (*Megaptera novaeangliae*): Evaluation and
   analysis of the technique. U.S. Dept. Commer., NTIS
   PB-298 740, 29p.

Katona, S. and H. Whitehead.
    1981. Identifying humpback whales using their natural
    markings. Polar Record 20: 439-444.

Payne, K., P. Tyack, and R. Payne.
    1983. Progressive changes in the songs of humpback
    whales (*Megaptera novaeangliae*). *In* R. Payne (editor),
    Communication and behavior of whales. AAAS Selected
    Symposia Series, p. 9-57. Westview Press, Boulder,
    Colo.

Payne, R.
    1978. Behavior and vocalizations of humpback whales
    (*Megaptera* sp.). *In* K.S. Norris and R.R. Reeves
    (editors), Report on a workshop on problems related to
    humpback whales (*Megaptera novaeangliae*) in Hawaii.
    p. 56-78. U.S. Dept. Commer., NTIS PB-280 794.

Tyack, P.
    1981. Interactions between singing Hawaiian humpback
    whales and conspecifics nearby. Behav. Ecol. Sociobiol.
    8: 105-116.

Winn, H.E. and L.K. Winn.
    1978. The song of the humpback whale, *Megaptera
    novaeangliae*, in the West Indies. Mar. Biol. (Berl.) 47:
    97-114.

3. Aberrant Songs of Humpback Whales *(Megaptera novaeangliae):* Clues to the Structure of Humpback Songs

Abstract

Variation in the songs of humpback whales was analyzed in 123 song sessions, containing 215 songs, recorded in Hawaii during the 1976-77, 1977-78, and 1978-79 singing seasons. While humpback song undergoes constant and progressive change, two basic song characteristics were unaltered in most song sessions in this sample: 1) Themes were consistently sung in the same order, and 2) Certain themes, termed <u>fundamental</u> themes, were consistently included in each song. Fourteen, or 11.4%, of the song sessions included deviations from this typical pattern and were termed <u>aberrant</u>. Each of the 12 aberrant song sessions of sufficient length for analysis was described by 1 of 3 mutually exclusive categories of aberrant song. The singers of 5 of the 14 aberrant song sessions have been photo-identified as different individuals. The identities of the others are unknown, as are the ages, sexes, and reproductive success of all singers of the fourteen aberrant song sessions. Variation in evolving features of humpback song was analyzed for typical and aberrant song sessions recorded during the 1976-77 and 1977-78 singing season. The standard error of the mean was used to compare variability in phrase duration, number of units within one subphrase of the phrase, theme duration, number of phrases in the theme, and song duration. Of these measures, phrase duration was the most consistent feature of both typical and aberrant humpback song. Song duration tended to be less variable than the duration of individual themes. The analysis of aberrant song structure yields insight into the "rules" of humpback song structure and the limits of song variability.

Introduction

One of the most striking findings of the long-term study of humpback song is that, at any given point in time and place, most humpback singers sing essentially the same song, although the song itself is undergoing constant and progressive change (Payne, Tyack, and Payne 1983). There are, however, occasional whales whose songs are strikingly different from those of the great majority of singers. We call such songs "aberrant", using the term to indicate only that they are atypical, not incorrect or inferior. The aberrant songs deviate from typical humpback song structure in very clear and predictable ways. Aberrant songs thus allow us to gain insight into the "rules" of humpback song structure and the limits of song variability.

Definitions

Because aberrancy can be defined only in terms of deviation from the norm, a set of guidelines are here proposed to set up a formal distinction between typical and aberrant song structure.

A humpback song characteristically consists of a set of discrete themes which are organized in a predictable sequence. The theme sequence (song) is repeated several times without a pause in a song session. Each theme consists of several phrases of a characteristic type. The transition from one theme to another is marked by an abrupt change in the type of phrase which is sung. Each phrase may be further divided into two or more subphrases, each consisting of a characteristic sequence of discrete sounds called units. A diagram illustrating the levels of humpback song structure and a detailed description of typical song structure and change is presented in Payne et al. (1983).

The songs of humpback whales have been recorded in the Hawaiian Islands for several consecutive years. Each year in Hawaii, humpbacks can be heard singing from approximately mid-November to mid-May. The present analysis of aberrant song focuses on recordings from three consecutive singing seasons: 1976-77, 1977-78, and 1978-79. At each point in time within these three seasons, most humpback songs recorded from different singers were remarkably similar to one another. The structure and sequence of individual units within phrases was highly consistent from one song session to the next; only rarely were new or unusual sounds incorporated into any song. As humpback song progressively changed over the three study seasons, two basic song characteristics remained remarkably unaltered: 1) Themes were consistently sung in the same order, and 2) Certain themes were consistently included in each song. These two stable facets of song structure provide the basis by which typical and aberrant songs are here distinguished.

The preservation of an invariant theme order over time has been described both in the Hawaiian humback song (Payne et al. 1983) and over several seasons of humpback song recorded in the western North Atlantic (Winn and Winn 1978; Payne and Payne in prep.). Of the complete "repertoire" of themes that humpbacks sing in a given season, several themes are frequently omitted from individual songs. Consecutive songs in a song session may include different selections of these themes, but the order in which themes are sung is consistently unaltered. Some themes, however, are rarely deleted from any songs for consecutive seasons. These are labeled __fundamental__ themes and are distinguished from frequently deleted themes by a simple criterion: a theme is considered fundamental in a given season if it is present in all songs of at least 90% of the song sessions recorded in both that season and at least one contiguous season.[1]

Themes, therefore, are fundamental only if they are consistently sung for at least two successive seasons. Since the percentage of songs in which each theme is included fluctuates over time (Payne et al. 1983), a given theme is fundamental only for the seasons in which it meets the above criteria.

Sound spectrograms of humpback songs representative of the 1976-77 and 1977-78 Hawaiian singing seasons are presented in Payne et al. (1983). A sound spectrogram of a humpback song representative of the 1978-79 season is presented in Payne and Guinee (1983). A typical sequence of themes (constituting a typical song) sung in each season is presented below:

Season	Typical Theme Sequence
1976-77	1-2-3-4-5-6-7-8-9
1977-78	1-2-4-5-6-8-9
1978-79	1-2-4-5-6-8-9

[1] Song sessions are treated as though each was sung by a different individual. This is based on the low resighting rate of photo-identified singers. Of 50 individual singers photographed from 1977 to 1979, only 4, or 8.0%, were identified singing on more than one occasion (J. Darling pers. comm.). A detailed description of humpback photo-identification is presented in Katona, Baxter, Brazier, Kraus, Perkins, and Whitehead (1979).

Table 1. Percentage of song sessions containing each theme
in all songs of the session.   The theme frequencies
of fundamental themes are underlined.   Both
aberrant and typical song sessions are included.

Theme	1976-77 %	1977-78 %	1978-79 %
1	88	96	100
2	98	96	100
3	3	---	---
4	95	94	97
5	98	96	97
6	78	79	42
7	43	21	---
8	73	90	90
9	38	81	26
# of sessions	40	52	31

For each of the study seasons, the following themes
were fundamental (see Table 1):

Season	Fundamental Themes
1976-77	Themes 2, 4, 5
1977-78	Themes 1, 2, 4, 5, 8
1978-79	Themes 1, 2, 4, 5, 8

The consistency with which the great majority of song
sessions contain fundamental themes sung in typical sequence
highlights the rare song sessions which deviate from the
norm.   An aberrant song session is one in which one or more
aberrant theme transitions are sung.   An aberrant theme
transition occurs when two themes that are typically
separated by one or more fundamental themes are sung in
succession.   The sequences of themes sung in aberrant song
sessions thus sharply differ from the typical theme sequence
and in order to study aberrant songs it is necessary to set the
minimum conditions for what constitutes a humpback song.
As a working definition, a theme sequence constitutes a song
when that sequence consists of at least three themes which
are repeated in the same order two or more times during a
recorded song session.   Given these definitions, a song
session in which the recorded theme sequence was 1-2-4-5-1-
2-4-5 would contain two songs (each consisting of Themes 1,
2, 4, and 5).   It would be considered aberrant if it were sung

in the 1977-78 or 1978-79 seasons.   This is because Theme 8, a fundamental theme in these two seasons, is omitted in the Theme 5 - Theme 1 aberrant theme transition.   The same song session would not be considered aberrant in the 1976-77 season because Theme 8 was not fundamental in that season. A 1-5-8-9-1-5-8-9 sequence would contain two songs (each consisting of Themes 1, 5, 8, and 9) and would be aberrant if sung in any of the three study seasons.   This is because both Theme 2 and Theme 4, fundamental in all three seasons, are omitted in the Theme 1 - Theme 5 aberrant theme transition.

Methods and Materials

The database for this study was a collection of 215 humpback whale songs, contained within 123 song sessions, recorded off Hawaii in the 1976-77, 1977-78, and 1978-79 singing seasons.   Songs were recorded from a 16' motorboat through several hydrophone/preamplifier/tape recorder combinations which in all cases gave a flat response (to within at least 5 db) to all frequencies between 50 and 10,000 Hz.   Spectrograms were made using a Spectral Dynamics SD 301C real time analyzer, a Tektronix 604 monitor oscilloscope and a Nihon Kohden oscilloscope camera.   In the first two seasons, every usable song was analyzed (Payne et al. 1983), whereas in the last season only a subsample was used (Guinee, Chu, and Dorsey 1983).   These analyses were used as a basis of comparison for this study of aberrant songs.

The duration of songs, themes, and phrases was measured directly from the spectrograms with a ruler and the number of units in each analyzed subphrase was counted.   As an aid in the analysis of within-season changes in humpback song, each season has been divided into the following six consecutive 31-day time periods, as was done by Payne et al. (1983).

Period	Dates
I	11/14-12/14
II	12/15-1/14
III	1/15-2/14
IV	2/15-3/17
V	3/18-4/17
VI	4/18-5/18

Humpback songs were recorded over the following dates:

1976-77: 19 December 1976 - 3 May 1977 (Periods II - VI)
1977-78: 29 November 1977 - 27 April 1978 (Periods I - VI)
1978-79: 7 December 1978 - 23 April 1979 (Periods I - VI)

Results

Fourteen (11.4%) of the songs recorded during the three study seasons were aberrant. Aberrant song sessions were recorded in all three seasons and in all but the earliest 31-day period (Period I) into which each season has been divided (see Table 2).[2]

The singers of five of the fourteen aberrant song sessions have been photo-identified as different individuals (J. Darling pers. comm.). Their ages and sexes are unknown, and none have been resighted on any other occasion. While the identities of the singers of nine other aberrant sessions are unknown, the low resighting rate of identified singers makes it unlikely that any two sessions were recorded from the same whale.

The sequence of themes sung in the aberrant song sessions are listed in Appendix Table 1. In the fourteen aberrant sessions, there were thirteen different aberrant theme transitions. Two of these, a Theme 5 – Theme 1 transition and a Theme 4 – Theme 1 transition, were sung in four different sessions. Every theme sung during the three study seasons was deleted in at least one of the fourteen aberrant sessions. Thus, no single theme appears to be an indispensable component of humpback song. No unusual behavior is correlated with any of the aberrant song sessions (P. Tyack pers. comm.). Present knowledge of what constitutes "typical behavior" of singers, however, is based primarily on surface observations and is not yet complete (Tyack 1981).

Categories of Aberrant Song: Preliminary Findings

Comparison of the order in which themes were sung during the aberrant song sessions yields three mutually exclusive categories of aberrant song (see Table 3).[3] In all

---

[2] This may reflect the low number of song sessions recorded each year in the first period. In all three seasons, a total of only seven song sessions were recorded during all three Period I's, making Period I by far the most poorly represented period.

[3] To avoid using short and potentially unrepresentative samples of aberrant song, only sessions containing a minimum of three songs are categorized. Two of the fourteen aberrant sessions (recorded on 14 March 1977 and 22 February 1979) are too short to be categorized and are not included in this analysis.

Table 2. Frequency of aberrant song sessions per period.

Period	I	II	III	IV	V	VI	Total
			1976-77				
# Aberrant	---	0	0	2	0	1	3
Total #	---	4	9	12	4	11	40
% Aberrant	---	0	0	16.7	0	9.1	7.5
			1977-78				
# Aberrant	0	1	2	0	2	4	9
Total #	7	9	10	5	15	6	52
% Aberrant	0	11.1	20.0	0	13.3	66.7	17.3
			1978-79				
# Aberrant	---	0	0	1	0	1	2
Total #	---	2	10	10	6	3	31
% Aberrant	---	0	0	10.0	0	33.3	6.5
Totals: All Years (1976-77, 1977-78, 1978-79)							
# Aberrant	0	1	2	3	2	6	14
Total #	7	15	29	27	25	20	123
% Aberrant	0	6.7	6.9	11.1	8.0	30.0	11.4

examples given below, we have put complete songs or repeated groups of themes on separate lines. This punctuation is provided only to highlight the repeated sequences and is not meant to indicate any break in singing between themes.

Type A. One or more aberrant theme transitions are sung in the song session, but at some point in the session a typical humpback song (i.e., all fundamental themes in typical order) is sung. The following example of a Type A aberrancy was included in a song session recorded on 7 February 1978:

1-2-4-5-6-8-9-   (Typical theme sequence)

1-2-4-   (Aberrant theme transition: Theme 4-
                  1 transition)

1-2-4-5-8-9   (Typical theme sequence)

Type B. Two or more different aberrant theme transitions are sung in the song session and the session does not include a typical humpback song. The following sample

Table 3. Characteristics of aberrant sessions: 1976-77, 1977-78, 1978-79.

Date Recorded	# of Aberrant Theme Transitions	# of Songs	Typical Song Present	Aberrant Category	ID'ed[o] Indiv.
2/22/77	4	6	no	B	no
3/14/77	1	1	yes	X*	no
5/11/77	2	3	no	B	no
1/4/78	2	11	yes	A	no
1/29/78	1	11	yes	A	no
2/7/78	1	5	yes	A	yes
3/23/78	5	7	no	B	yes
4/1/78	1	5	no	C	yes
4/19/78	1	3	no	C	yes
4/22/78	1	4	no	C	yes
4/25/78	1	3	no	C	no
4/27/78	1	3	no	C	no
2/22/79	1	0	no	X*	no
4/19/79	1	5	yes	A	no

Aberrant Song Categories

	A	B	C
Typical song present?	Yes	No	No
Number of Aberrant Theme Transitions[o]	$\geq$1	>1	1

X* = Song session is too short to categorize

[o] = An aberrant theme transition occurs when two themes that are typically separated by one or more fundamental themes are sung in succession (See text.).

[o] = whether or not an individual singer was identified.

of a Type B aberrancy was included in a song session recorded on 23 March 1978:

```
1-5-
1-5-
2-4-5-
4-
2-4-5-
2-4-5-
1-5-
```

Type C. A single aberrant theme transition is repeatedly sung throughout the song session and the session does not include a typical humpback song. The following example of a Type C aberrancy was included in a song session recorded on 22 April 1978:

```
...4-9-
1-2-4-9-
1-2-4-9-
1...          (Ellipses, "...", represent a break of a
1-2-4-9-      few seconds in the recording of the song
1-2-4-9-      session.)
1-2...
```

Appendix Figure 1 presents continuous sound spectrograms from the above sessions as examples of each of the three categories of aberrant song.

As noted above, singers of Type A aberrant song demonstrate an ability to sing the complete theme sequence of typical humpback song. In our sample of four Type A sessions, only three different aberrant theme transitions were sung (a Theme 2 - Theme 1 transition, a Theme 4 - Theme 1 transition, and a Theme 5 - Theme 1 transition). Thus, at some point in their song sessions, each singer abridged the song and returned prematurely to Theme 1. The theme that we call "Theme 1" was so labeled arbitrarily since humpback singers stop and start singing at various points in their songs. It is therefore interesting to note that these Type A singers seem to treat Theme 1 as an integral component of their theme sequences.

Type B aberrant song is characterized by a highly variable theme order. Each of the three Type B sessions in our sample include typical sequences of themes, which are frequently repeated (e.g., the 23 March 1978 singer often sang a 2-4-5 theme sequence). In addition, at least one fundamental theme is always omitted in the two recorded Type B sessions.

Singers of Type C aberrant song also omit at least one fundamental theme, but their sessions are marked by a highly

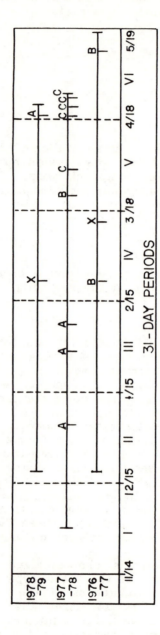

Figure 1. The distribution of aberrant song sessions recorded in 1976-77, 1977-78, and 1978-79. Aberrant sessions analyzed here are indicated by their corresponding song type (A, B, or C -- see text for details). Aberrant sessions too short for analysis are marked by an 'X'.

consistent theme order. Each of the five Type C sessions in the sample consists of a unique aberrant theme sequence sung continuously throughout the session. Like the singers of typical humpback song, the singers of Type C sessions maintain an internally consistent theme sequence.

The distribution of each type of aberrant song within each season is presented in Figure 1. Type A and Type B aberrant sessions have both been recorded in two of the three analyzed seasons. Sessions of these two aberrant song types are distributed throughout the singing season; both Type A and Type B sessions have been recorded in three of the six 31-day periods into which we have divided each season.

In contrast, all five Type C sessions were heard during the final 28 days of the 1977-78 singing season. Four of the five sessions were sung in the final nine days of recording (19 April to 27 April). During that 9-day period, these four Type C sessions comprised 67% (4 out of 6) of all song sessions recorded. The singers of three of the five Type C sessions have been identified as separate individuals and, as noted above, each Type C session was comprised of a unique theme sequence. Such a concentration of Type C aberrant sessions cannot therefore be understood as the re-recording of a single individual.

Variation in Evolving Features of Humpback Song

The description of three types of aberrant song sessions is based on their variation from the typical sequence of fundamental themes, a characteristic that is consistently unaltered in typical humpback song. As Payne et al. (1983) point out, many of the song's elements (e.g., units, phrases, and themes) are undergoing constant and progressive change. Even though they are singing themes out of the typical sequence, do the singers of aberrant song sessions attend to the evolving elements of humpback song? Do they sing units, phrases, and themes which are the same as those sung in typical (i.e., non-aberrant) song sessions recorded during the same 31-day period?

Variation in typical humpback song. Answering the above question requires first examining how similar to one another typical song sessions are at any given point in time. At the time of this writing, the 1978-79 season of Hawaiian humpback song has not been fully analyzed. The following analysis of variability in both typical and aberrant song sessions is therefore limited to the 1976-77 and 1977-78 singing seasons. Variability between typical song sessions in these two seasons is calculated for the following five measures:

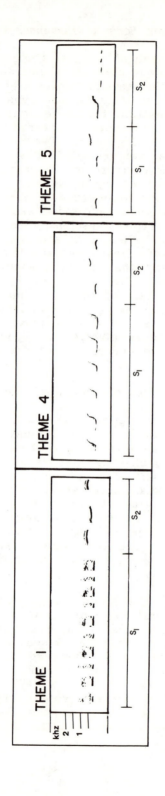

Figure 2. Typical phrases of Themes 1, 4, and 5 from a song session recorded on 28 March 1978. The number of units per subphrase was measured for subphrases 1 (S_1) of Themes 1 and 4 and for subphrase 2 (S_2) of Theme 5.

1. Song duration
2. Theme duration
3. Number of phrases in the theme
4. Phrase duration
5. Number of units within one subphrase of each phrase

These five measures provide a comparison of variability between four levels of song structure: 1) songs, 2) themes, 3) phrases, and 4) subphrases.

Themes 1, 4, and 5 were selected for this analysis. The same data for the same two singing seasons examined here were first examined by Payne et al. (1983), who provide a detailed discussion of the change in Theme 5 for the four measures involving theme structure (i.e., all of the measures listed except for song duration). For each measure, they plot the mean value of all sessions in each 31-day period. Theme 4 and Theme 1 are analyzed in the same way here. In counting the number of units per subphrase, we have measured for each theme only the subphrase which contains the most rapidly changing number of units (see Figure 2).

Table 4 shows the change in the mean value of each measure of song structure over the two study seasons. During this time, songs showed increases in each characteristic, but the extent of the increase varied widely.

The <u>Standard Error of the Mean</u> provides an index of the variability between song sessions for each 31-day period.[4] Table 5 presents the relative variability of each measure of song structure over the 10 periods of song recorded during the 1976-77 and 1977-78 singing seasons along with the results of testing for significant differences between these measures in their variability.[5]

The five analyzed measures of song structure are not independent (e.g., themes containing several phrases last

---

[4] This measure is analogous to the coefficient of variation but takes into account the variable number of song sessions recorded in each period:

$$\text{Average Standard Error of the Mean} = \frac{\sum_{i=1}^{10}\left(\frac{s_i/n_i}{\bar{x}_i} \times 100\right)}{10}$$

$\bar{X}$ is the mean of the 31-day period, s is the standard deviation, and n is the number of sessions included in the period.

[5] Period VI, 1978, is not included in this discussion of typical song variability because only two typical song sessions were recorded during only the first eight days.

Table 4. Measures of typical songs and their components (mean ± standard error): Period II, 1976-77 to Period V, 1977-78.*

		Period II 1976-77	Period V 1977-78	Percent Change
Song Duration (min)		7.2±1.0	13.4±1.3	+86.1
	Theme			
Theme	1	1.3±0.2	2.9±0.4	+123.1
Duration	4	0.8±0.1	1.2±0.2	+50.0
(min)	5	0.9±0.1	3.9±0.2	+333.3
Number of	1	4.9±0.7	6.1±0.6	+24.5
Phrases	4	2.7±0.4	2.9±0.5	+7.4
per Theme	5	3.5±0.5	11.1±1.3	+212.1
Phrase	1	17.3±0.3	28.1±0.5	+62.4
Duration	4	18.1±0.9	25.4±0.7	+40.3
(sec)	5	15.6±0.8	20.6±0.1	+31.8
Number of	1	3.0±0.5	8.7±0.3	+190.0
Units per	4	5.1±0.3	5.6±0.2	+9.8
Subphrase	5	1.9±0.2	4.6±0.2	+194.7

*In the 1976-77 season, typical song sessions were first recorded in Period II. In the 1977-78 season, only two typical song sessions were recorded after Period V.

longer than themes with only a few phrases). Given that these measures reflect a hierarchical view of song structure, one might expect to find either a steady increase in variability from the most specific level of the song (number of units per subphrase) to the most general (song duration), or the reverse, with variability increasing with increasingly detailed levels of song structure.

There is instead a more surprising result (see Table 5 and Figure 3). Phrase duration, an intermediate level of structure, is the most consistent feature of each of the three themes that we have analyzed (Themes 1, 4, and 5). At each point in time, phrases of each theme are of a remarkably uniform length. Features both higher and lower in the song hierarchy show greater variability than phrase duration. A more detailed level of song structure, the number of units within subphrases of each theme, tends to vary more than the phrase duration (significantly more in Themes 1 and 5). At a more general level of song structure, both the number of phrases per theme and the theme duration are significantly more variable than the duration of phrases of the same theme. Further, song duration tends to be <u>less</u> variable than theme duration (a significant difference in the comparison of song duration and Theme 1 duration). This is surprising because the number of themes within a song is variable at each point in time (i.e., non-fundamental themes are often omitted from songs). This suggests that singers may be modifying the theme duration to achieve a song length that is characteristic of songs of the period.

The variability in the duration of phrases of each theme in the complete song "repertoire" was calculated and found to be consistently low (see Table 6). This indicates that, at any given time in each season, humpbacks sang phrases of a highly uniform length. This consistent within-period phrase duration exists even though the mean duration of phrases of all themes increases rapidly over the two study seasons (see Table 7). Phrases of Theme 1, for example, increase in length from 17.3 seconds to 28.1 seconds (an increase of 62.4%) over the two study seasons. Yet for each of the 10 periods within these two seasons, variability in phrase duration of Theme 1 is limited to within 3.1 seconds of the mean Theme 1 phrase duration for all typical song sessions of the same period.

Variation in aberrant song sessions. As has been noted earlier, aberrant song sessions are few and are scattered throughout the singing season. In order to measure the extent to which singers of aberrant song sessions attend to the evolving characteristics of humpback song, the five parameters outlined above were calculated for each aberrant

Table 5. Variability in five measures of song structure in typical song sessions for Themes 1, 4, and 5, 1976-77 to 1977-78. Variability is given as a dimensionless number equal to the average standard error of the mean (see text). Phr. = phrase, Th. = theme, and Dur. = duration.

Song Dur.		Th. Dur.	Phr. per Th.	Phr. Dur.	Units per Phr.
	Theme				
	1	14.96	13.98	2.33	6.74
9.03	4	13.81	12.81	3.50	5.12
	5	13.90	12.58	1.94	8.35

F-test for Significant Differences in Variability*
(df=80,80; two-tailed test)

Th. Dur./Phr. Dur.
Th. 1: F=6.42, p<0.0001
Th. 4: F=3.95, p<0.0001
Th. 5: F=7.16, p<0.0001

Phr. per Th./Phr. Dur.
Th. 1: F=6.00, p<0.0001
Th. 4: F=3.66, p<0.0001
Th. 5: F=6.48, p<0.0001

Units per Phr./Phr. Dur.
Th. 1: F=2.89, p<0.0001
Th. 4: F=1.46, p=0.091●
Th. 5: F=4.30, p<0.0001

Th. Dur./Phr. per Th.
Th. 1: F=1.07, p=0.76●
Th. 4: F=1.08, p=0.73●
Th. 5: F=1.10, p=0.67●

Th. Dur./Song Dur.
Th. 1: F=1.66, p=0.025
Th. 4: F=1.53, p=0.059●
Th. 5: F=1.54, p=0.055●

Phr. per Th./Song Dur.
Th. 1: F=1.55, p=0.053●
Th. 4: F=1.42, p=0.12●
Th. 5: F=1.39, p=0.14●

*The F-test is used to compare relative variability because the distribution of standard errors of the mean for each measure approximates a chi-square distribution.

●Not significant

	MEASURE OF SONG STRUCTURE	LEVEL OF SONG STRUCTURE
Least Variable	Phrase Duration	Phrase
	Units per Subphrase	Subphrase
	Song Duration	Song
	Phrases per Theme	Theme
Most Variable	Theme Duration	Theme

Figure 3. Relative variability of five measures of typical song structure. Significant differences are presented in Table 5.

Table 6. Variability in phrase duration for each theme in typical song sessions, 1976-77 to 1977-78. Variability is given as a dimensionless number equal to the average standard error of the mean (See Table 5).

Th.	1	2	3[1]	4	5	6	7[2]	8	9
	2.33	3.18	7.23	3.50	1.94	2.42	2.95	1.96	2.13

[1] Theme 3 was recorded only in the first three periods of 1976-77.

[2] Measures of variability in Theme 7 were obtainable only through the first three periods of 1977-78 (only one session per period included Theme 7 in Periods IV and V, 1977-78).

Table 7. Average duration in seconds of phrases of each theme in typical song sessions, Period II, 1976-77[1] to Period V, 1977-78[2].

Theme	Ave. Phrase Dur. Per. I, 1976-77	Ave. Phrase Dur. Per. V, 1977-78	Percent Change
1	19.3	28.1	62.4
2	19.9	27.2	36.7
3	19.9	----	----
4	18.1	25.4	40.3
5	15.6	20.6	32.1
6	18.1	26.3	45.3
7	17.8	23.0	29.2
8	17.3	21.3	23.1
9	17.9	21.2	18.4

[1] In the 1976-77 season, typical song sessions were first recorded in Period II.

[2] In the 1977-78 season, typical song sessions were last recorded in Period V.

session and compared with the same parameters averaged over all typical song sessions in the same period.[6]

Table 8 presents the results of this comparison. Like singers of typical humpback song, singers of aberrant song sessions adhere closely to the duration of the phrase, and somewhat less closely to the number of units within a subphrase. For the three themes analyzed, phrase duration in aberrant song sessions deviated by an average of 7.0% from the mean phrase duration of typical sessions of the same period. (For each parameter, the average deviation changes only slightly if the measures from the typical songs sung in Type A aberrant sessions are excluded from the average of the aberrant sessions). In contrast to the consistent phrase duration, aberrant song sessions deviate, on the average, 50.4% from the mean theme duration and 49.1% from the mean number of phrases per theme sung in typical sessions of the same period. Finally, like singers of typical song sessions, singers of aberrant sessions tend to adhere more closely to the mean duration of typical song than to the duration of individual themes; aberrant sessions deviate, on the average, 30.1% from the mean song duration of typical sessions of the same period. Of the five analyzed measures of song structure, phrase duration emerges as the most consistent feature of both typical and aberrant humpback song.

More can be learned about the structure of aberrant songs by directly comparing aberrant and typical song sessions. Singers of Type C aberrant sessions, for example, characteristically delete one or more fundamental themes, yet sing a single consistent theme sequence. Do they therefore sing relatively short songs, or do they maintain typical humpback song length by increasing the duration of individual themes?

Appendix Figure 2 presents a series of histograms showing the distribution of all song sessions around the mean of typical songs for each of the five measures of song structure. (For each measure, this distribution is obtained by subtracting the mean of typical song sessions from the mean of each session within the period.) Because the variation of typical song sessions around each mean approximates a normal distribution, we can find any aberrant

---

[6] Two of the twelve aberrant sessions recorded in the 1976-77 and 1977-78 singing sessions are not included in this analysis; the short uncategorized session of 14 March 1977, and the Type B session of 11 May 1977 (high background noise made the detailed analysis of this session impossible). The typical theme sequences sung in Type A sessions (A_{typ}) are analyzed separately from the aberrant sequences (A_{ab}).

Table 8. Five measures of humpback song structure in aberrant song sessions: the deviation (in percent) of aberrant song sessions from average measures of typical song sessions* (1976-77 and 1977-78). A_{ab} refers to average measures from aberrant songs sung in Type A aberrant sessions. A_{typ} refers to average measures from typical songs sung in Type A aberrant sessions.

Aberrant Category	Song Dur.	Theme Dur.	Phrases /Theme	Phrase Dur.	Units/ Subphrase	# of Sessions[o]	Parameter Analyzed
A_{ab}	56.7%	----	----	----	----	3	
A_{typ}	24.0%	----	----	----	----	2	Song
B	32.5%	----	----	----	----	2	
C	15.5%	----	----	----	----	5	
A_{ab}	----	48.8%	42.1%	4.3%	15.7%	3	
A_{typ}	----	37.8%	28.6%	9.3%	21.7%	2	Theme
B	----	16.0%	4.2%	15.8%	11.8%	2	1
C	----	131.1%	127.8%	6.9%	10.9%	4	
A_{ab}	----	19.2%	16.7%	9.2%	14.5%	2	
A_{typ}	----	15.5%	17.1%	5.2%	5.2%	2	Theme
B	----	60.7%	45.7%	8.3%	7.3%	1	4
C	----	33.6%	36.1%	11.5%	9.2%	3	
A_{ab}	----	----	----	----	----	0	
A_{typ}	----	44.1%	46.1%	2.5%	22.1%	2	Theme
B	----	2.9%	0.7%	3.4%	13.7%	2	5
C	----	69.1%	86.2%	2.9%	18.4%	3	
A_{ab}	----	37.0%	31.9%	6.3%	15.2%	5	Averages of
A_{typ}	----	32.5%	30.6%	5.7%	16.3%	6	Themes 1,
B	----	19.7%	11.1%	9.3%	11.7%	5	4, & 5 by
C	----	83.3%	87.8%	7.1%	12.6%	10	Category

Average: 31.3% 55.8% 54.7% 7.5% 13.0%   10/20[⊕] (not incl. A_{typ})

Average: 30.1% 50.4% 49.1% 7.0% 13.8%   12/26 (incl. A_{typ})

* Aberrant sessions are compared to the mean values of typical sessions recorded in the same period. As no typical sessions were recorded in Period VI, 1977-78, aberrant sessions recorded in that period are compared to typical sessions recorded in Period V, 1977-78.

o The number of sessions counted for each meaure of song structure varies, as individual themes are never sung in some sessions.

⊕ The first number indicates the total number of aberrant sessions analyzed for song duration; the second number indicates the combined number of aberrant sessions analyzed for the structure of Themes 1, 4, and 5.

song sessions which are significantly different from the typical humpback song sessions in each measure. (Any aberrant session which deviates from the mean of typical song sessions beyond the 95% confidence interval, 1.96 standard deviations, is considered to be significantly beyond the range of typical humpback song variability.)

Table 9 presents the features of song which the singers of aberrant sessions render differently from singers of typical humpback song sessions. There are nine aberrant sessions that include Theme 1, six that include Theme 4, and six that include Theme 5. Of these 21 cases, only 4 include phrases that are significantly different from phrases sung in typical song sessions. Further, there are four cases in which the mean number of units in the measured subphrase deviates significantly from the mean number of units in typical subphrases. These deviations in phrase structure occur in sessions of each aberrant song type, but no aberrant session contains deviant phrases in more than one of the three themes. Aberrant sessions include phrases that are significantly longer and shorter than phrases sung in typical song sessions and subphrases with significantly larger and smaller numbers of units than are found in the same subphrases in typical songs. Thus, there is no consistent direction in these deviations. In contrast, the deviations in theme duration, number of phrases per theme, and song duration are markedly skewed both for direction and for aberrant song type.

Type A. Singers of Type A aberrant sessions never significantly altered the duration of any of the three themes that we analyzed. As a result, Type A aberrant songs are significantly shorter than typical songs from the same period. The inclusion of typical songs in Type A sessions further indicates that the aberrant sequences are simply incomplete versions of typical humpback song.

Type B. The singers of Type B aberrant sessions also sang themes of the same length as those sung in typical humpback songs. Here, however, the variable theme order precludes the possibility that what they sing are simply abridged versions of typical song.

Type C. Singers of Type C sessions tended to sing extremely long versions of both Theme 1 and Theme 5 (the duration of Theme 4 is unaltered). The marked increase in the length of these themes results in songs that have the same length as typical songs from the same period. Singers of Type C sessions thus appear to adhere closely to typical song length while singing unique versions of humpback song. As a further indication of this, there was one Type C session

Table 9. Five measures of humpback song structure: The number of aberrant song sessions that differ significantly from typical song sessions.* A_{ab} refers to the average measures from aberrant songs sung in Type A aberrant sessions. A_{typ} refers to the average measures from typical songs sung in Type A aberrant sessions. + = significantly greater (length or number), - = significantly less (length or number).

Aberrant Category	Song Dur.	Theme Dur.	Phrases /Theme	Phrase Dur.	Units/ Subphrase	# of Sessions[o]	Parameter Analyzed
A_{ab}	2-	---	---	---	---	3	
A_{typ}	0	---	---	---	---	2	Song
B	1-	---	---	---	---	2	
C	0	---	---	---	---	5	
A_{ab}	---	0	0	1+	1+	3	
A_{typ}	---	0	0	0	1+	2	Theme
B	---	0	0	1+/1-	0	2	1
C	---	2+	3+	0	1+	4	
A_{ab}	---	0	0	0	0	2	
A_{typ}	---	0	0	0	0	2	Theme
B	---	0	0	0	0	1	4
C	---	0	0	1-	0	3	
A_{ab}	---	---	---	---	---	0	
A_{typ}	---	0	0	0	0	2	Theme
B	---	0	0	0	1-	2	5
C	---	2+	2+	0	1+	3	
A_{ab}	---	0	0	1+	1+	5	Totals of
A_{typ}	---	0	0	0	1+	6	Themes 1, 4,
B	---	0	0	1+/1-	1-	5	and 5 by
C	---	4+	5+	1-	2+	10	Category
Total:	2-/1+	4+	5+	2+/2-	3+/1-	10/20[⊛]	(not incl. A_{typ})
Total:	2-/1+	4+	5+	2+/2-	4+/1-	12/26	(incl. A_{typ})

* Significant difference is defined as beyond the 95% confidence interval for typical song sessions.

[o] The number of sessions counted for each measure of song structure varies, as individual themes are never sung in some sessions.

[⊛] The first number indicates the total number of aberrant sessions analyzed for song duration; the second number indicates the combined number of aberrant sessions analyzed for the structure of Themes 1, 4, and 5.

(recorded on 25 April 1978) that did not have extra
renditions of Themes 1 and 5. But the songs in that session
contained almost all of their themes, with only Theme 4
being deleted. Presumably, it was not necessary for the
singer to compensate with extremely long renditions of
Themes 1 and 5 as did the Type C singers who sang fewer
themes.

Variation in Humpback Song: Further Notes

The distinction of three types of aberrant song has thus
far been based upon very large-scale, easily quantifiable
characteristics of humpback song. A detailed analysis of the
fine structure of song, such as the variability in the
characteristic frequency of each unit, is beyond the scope of
this study. Within each theme, however, the individual units
are highly consistent in structure from one song to the next.
The introduction of new or unusual sounds into any song is
extremely rare and is found primarily in transitional phrases.
A transitional phrase is occasionally sung between two
themes and usually consists of the first subphrase of the
initial theme followed directly by the second subphrase of
the next theme. Theme transitions are discussed in detail by
Payne et al. (1983). On rare occasions, the units sung in
transitional phrases take on unusual forms, apparently
combining the characteristics of units from the two adjacent
themes. In the three seasons of this study, the only
occasion on which an apparently novel sound was recorded
was on 29 January 1979. The production of this sound was
not an isolated event. Rather, it was slowly incorporated
into the humpback song until, by the 1979-80 season, it was
sung in several song sessions (Payne and Guinee 1983) Even
this sound may not have been created *de novo* (e.g., a whale
from another population might have introduced it). But,
with the exception of transitional phrases and this one sound,
we have never heard units in songs that were not in their
usual themes.

Variation in phrase structure. Because units are yet
to be analyzed quantitatively, we cannot look closely at
variability within phrase structure. Nonetheless, we can get
a measure of variability in phrases of Theme 5. Although
there is a tendency for the number of units within subphrases
to change over time, the first subphrase of Theme 5 has been
fairly stable. At the beginning of the 1976-77 season, the
first subphrase of Theme 5 consisted of two units sung at a
similar frequency. In February 1977, each unit was replaced
by two units of a different frequency. The resulting
subphrase thus consisted of a pair of units of different
frequency followed by a repetition of the same pair (see
sample spectrograms of typical phrases of Theme 5 in Payne

Table 10. Variant phrase structure in Theme 5 for all song
sessions, 1976-77 - 1977-78.

Date Recorded	Total # Phrases	# Variant Phrases	% Variant Phrases
29 January 1977	16	5	31.3
1 February 1977	20	1	5.0
9 February 1977	17	1	5.9
14 February 1977	9	1	11.1
18 February 1977	30	1	3.3
19 February 1977	16	1	6.3
22 February 1977*	28	27	96.4
25 February 1977	44	2	4.5
17 May 1977	54	1	1.9
13 January 1978	21	1	4.8
7 February 1978*	14	1	7.1
23 March 1978*	180	1	0.6

*Aberrant song session

et al. 1983). Occasional exceptions[7] to this consistent structure are found in both typical and aberrant song sessions (see Table 10) and, as before, discussion is limited to the two fully analyzed singing seasons, 1976-77 and 1977-78.

In these two seasons, 9 of the 80 typical song sessions (12.4%) and three of the twelve aberrant song sessions (25%) contain at least one variant phrase of Theme 5. Variant phrases of Theme 5 were sung throughout the two analyzed seasons, but were most frequently recorded in February 1977 (7 of the 12 sessions which included variant phrases of Theme 5 were recorded during this month). Thus the variability was greatest when the number of units in subphrase one was evolving from two to four. Most commonly, a single variant phrase is sung amid several typical Theme 5 phrases; only 3 of the 12 sessions included two or more variant Theme 5 phrases. Most striking is the Type B aberrant session of 22 February 1977, in which 27 out of the 28 phrases of Theme 5 contain more or less than the typical number of units in the first subphase (see Figure 4). Variant phrase structure is not a clear characteristic of Type B sessions, however; the second Type B session, recorded on 23 March 1978, contains only a single variant phrase in 180 phrases of Theme 5. The inclusion of variant phrases in several sessions of otherwise typical humpback song indicates that there is much to learn about how and why singers vary the sequence of sounds that comprise each phrase.

Rhythm. Finally, the study of aberrant song sessions sheds light upon the role of rhythm in humpback song. Singing humpbacks often stop singing for a few moments when approached by another whale, a boat, or a diver. The singers of two of the Type C sessions, however, consistently incorporated pauses as integral components of their songs. Figure 5 presents a sound spectrogram from a Type C aberrant session recorded on 19 April 1978. It shows two phrases of a theme that was sung between Theme 1 and Theme 6. Each phrase consists of several seconds of silence followed by the low pulsive subphrase typically sung in Themes 6, 8, and 9. In addition, in the first of the two phrases, the whale sings two brief and very faint pulsive notes (they are most similar to the notes typically sung in the first subphrase of Theme 1) before becoming silent.

---

[7] Exceptions are defined as:
    1) Other than two or four units from Period I to Period III, 1976-77.
    2) Other than four units from Period IV, 1976-77 to Period VI, 1977-78.

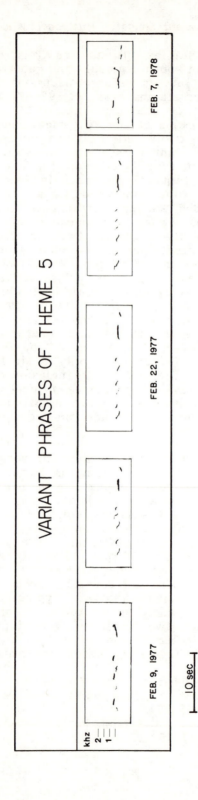

Figure 4. Variant phrases of Theme 5. Each phrase is variant because it contains more or less than four units in the first subphrase. The phrase recorded on 9 February 1977 is from a typical song session, the phrases recorded on 22 February 1977 are from a Type B aberrant session and the phrase recorded on 7 February 1978 is from a Type A aberrant session.

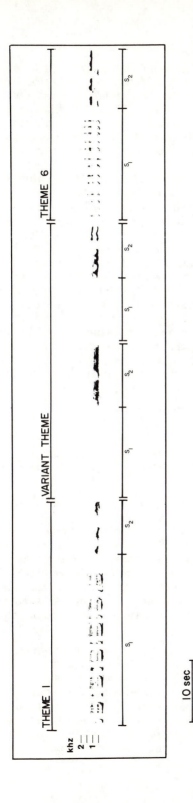

Figure 5. Spectrogram of an unusual theme consistently sung between Theme 1 and Theme 6 in a Type C aberrant song session recorded 19 April 1978. Each phrase consists of several seconds of silence followed by a low pulsive subphrase. Two units similar to those sung in subphrase 1 of Theme 1 are faintly heard before the initial period of silence.

Phrases of this unusual theme averaged 24.0 seconds, a phrase duration well within the range of variation for typical phrases of Theme 6 sung in the same period (within 1.53 standard deviations of the mean Theme 6 phrase duration). This singer thus maintained both a typical phrase duration and the typical rhythm in which phrases were sung while remaining silent during one of the two subphrases of each phrase.

The second incorporation of a pause in humpback song occurred in the 27 April 1978 Type C session. This singer consistently included a period of silence not as a component of a phrase, but rather like a theme, pausing for a mean of 94.5 seconds between the final phrase of Theme 5 and the first units of Theme 2. As a comparison, the duration of themes in typical song sessions of the same period ranged from 26.3 seconds for Theme 6 to 234.2 seconds for Theme 5.

Thus far, pauses within songs have been found only in these two Type C song sessions. It is unclear, however, whether within-song silence is an "option" available only to the singers of Type C sessions. Further, it is impossible to know how these two singers were keeping track of time in the absence of sound. Nonetheless, the maintenance of a rhythmic pattern without sound highlights the role of rhythm in humpback song.

Conclusion

Aberrant Song Categories

Each of the twelve aberrant song sessions of sufficient length for analysis falls into one of three distinct categories of aberrant song:

Type A. Each of four Type A sessions includes both typical and aberrant songs. Aberrant songs result when one or more fundamental themes are omitted from an otherwise typical sequence of themes. An analysis of song variability in three of the four Type A sessions indicates that the duration of individual themes is not altered; the aberrant songs in Type A sessions are thus structurally incomplete versions of typical humpback song. Type A sessions were recorded in three of the six 31-day periods and in two of the three study seasons.

Type B. The three Type B sessions are characterized by a highly variable theme order. Each song session includes repeated groups of themes in typical sequence, but no typical humpback song is sung. Song variability has been analyzed in two of the three Type B sessions and we find that the duration of individual themes is not significantly altered. Type B sessions were recorded in three of the six 31-day periods and in two of the three study seasons.

Type C. Each of the five Type C sessions is characterized by a consistent aberrant theme sequence or song. The sequence of themes is unique to each session. The songs in each session are of typical humpback song duration. In sessions in which more than a single fundamental theme is omitted from each song, the typical song durations result from the extremely long duration of individual themes. All Type C sessions in this sample were recorded during the final four weeks of the 1977-78 singing season.

Humpback Song Structure

The characteristic structure of humpback song is discussed by Payne and McVay (1971), Winn and Winn (1978), and Payne et al. (1983). The study of those humpback song sessions which deviate most strongly from the norm yields insight into the limits of humpback song variability. Several features of humpback song are unaltered even in the aberrant sessions here described. This suggests that there are inviolate "rules" of humpback song structure. The following "rules of structure" underlie both typical and aberrant song sessions analyzed in this study:

1) The introduction of novel material into humpback songs is very rare, if it happens at all.

Almost all units in humpback songs are characteristic elements of phrases of current themes. Over the three seasons of the present study, only one apparently novel sound has been incorporated into the song.

2) Theme order in humpback songs is never random.

While the theme order in Type B aberrant song sessions is variable, _all_ aberrant song sessions include repeated groups of themes, each derived from the theme sequence sung in typical song sessions.

3) The phrase duration of each theme shows remarkably little variability at any point in time.

Average phrase duration in all themes constantly changes, but at every point in time, all singers are singing phrases of very similar durations.

4) Humpback singers adhere more closely to a given song duration than to the duration of individual themes.

Comparison of the relative variability of song duration and theme duration in typical song sessions indicates that, at each point in time, the duration of songs is more consistent from one session to the next than is the duration of individual themes within the song. The fundamental nature of consistent song duration is highlighted in Type C aberrant song sessions in which singers omit several themes from each song yet sing songs of typical duration by greatly extending the durations of individual themes.

Discussion

One means of gaining insight into aberrant humpback songs is to look for parallels in the literature on bird song. Like humpback whales, several well-studied avian species sing a highly stereotypic song, including the Short-toed Tree Creeper *Certhia brachydactyla* (Thielcke 1969), White-crowned Sparrow *Zonotrichia leucophrys* (Marler and Tamura 1962; Baptista 1977), Chaffinch *Fringilla coelebs* (Marler 1952) and the Oregon Junco *Junco oregonus* (Konishi 1964). Within a species, slight dialect differences may distinguish the songs of one population from another, but I can find no reports of mature birds of these species whose songs deviate strongly from the typical species song. The songs of immature individuals, however, may be strikingly different from the typical species song. First-year male Chaffinches, for example, initially develop a characteristically quiet and variable vocalization termed "subsong" (Thorpe 1958; Nottebohm 1970). For subsong to develop into full adult song, the first-year Chaffinch must be exposed to the songs of adult conspecifics.

The complex and evolving nature of humpback song strongly indicates that, like some bird song and human language, its intricacies must be learned by individual "performers." The role of learning and memory in humpback song is supported by the finding of Guinee, Chu, and Dorsey's (1983) that the songs of individual whales change over time and that these changes tend to parallel the progressive song changes found in the entire population of singers. The variable theme sequences of Type B aberrant humpback song are suggestive of both avian subsong and the babbling of human infants. They lack, however, the inter-unit variability that is characteristic of young birds and young humans (D. Kroodsma pers. comm.). Further, the discrete aberrant song categories indicated by the present sample may not indicate different functional bases. It is possible, for example, that both Type A and Type B singers represent different points on a continuum of song acquisition. While there is accumulating evidence that humpback singers are primarily adult males (Winn, Bischoff, and Taruski 1973; Hudnall 1977; Tyack 1981; Darling, Gibson, and Silber 1983; Glockner 1983) there is no present information on the age, sex, or reproductive success of any singers of aberrant song. Thus, there is as yet no evidence for any functional basis for the three types of aberrant songs.

There is in the avian literature an interesting parallel to our finding that, in humpback songs, both phrase duration and song duration are more consistent than their components. The duration of the song of the Oregon Junco (Konishi 1964), which consists of a single repeated unit, is relatively constant between individuals of the same population, but the number of units within each song is highly variable. Konishi found that Oregon Juncos maintained a uniform song duration by varying the duration of individual units.

Hitherto, we have had to be content with definitions of what the salient features of humpback song are to human beings. The "rules of structure" listed above, however, suggest something of what the salient features of humpback song are to humpback whales. Their concept of song, if indeed they have one, must include the fact that it should last a specific length of time. Its themes are never to be given in a random order and units are rarely, if ever, to be introduced *de novo*. Rather, new material should be derived from elements already present in current or previous songs. The phrase is presumably the most fundamental block of which songs are composed. This is because its duration is accurately maintained at every point in time by all singers whether they are singing typical or aberrant songs. Although we have yet to learn how and why humpback song changes so rapidly, we have begun to understand the characteristics of song that are consistently unaltered.

If these "rules of structure" are truly characteristic of humpback song, then they should be characteristic of song sessions recorded both in other seasons in the Hawaii humpback population and from other populations of humpback whales. A comparative analysis of variability in song structure will thus provide a means of discriminating between features of humpback song that characterize a particular population and those that are indeed "universal" features of humpback song. If the categories of aberrant song described here are in fact characteristic of particular groups of individuals within humpback social organization (e.g., young humpbacks just beginning to sing) then they too should be found in other singing seasons and other humpback populations. An increased understanding of age, sex, and behavior of singers should shed light on the functional basis of both typical and aberrant song. As genetic mutations allow for insight into the structure and function of the genome, so too may aberrant songs allow new insights into the complex and beautiful songs of humpback whales.

Acknowledgements

The recordings used in this work were made by Jim Darling, Kim Gibson, Jan Heyman, Fred Levine, Katharine and Roger Payne, Peter Tyack, and Greg Silber. Analyses of most of the non-aberrant songs were made by Linda Guinee, Jan Heyman, Katharine Payne, Greg Silber, and Peter Tyack. I thank these people for making this material available to me. I benefited from useful discussions with and comments by the above people as well as Ronald Christensen, Eleanor Dorsey, Donald Kroodsma, and Cara Lieb. Jim Bird helped with the bibliography and final manuscript preparation and Linda Guinee and Vicky Rowntree helped with typing and made the final illustrations.

Appendix Table 1. Theme order in aberrant song sessions. In all examples given below, complete songs or repeated groups of themes are put on separate lines. This punctuation is provided only to highlight the repeated sequences and is not meant to indicate any break in singing between themes. Ellipses ("...") represent a break of a few seconds in our recording of the song session. An X in the Aberrant Category denotes a song session too short to categorize.

Date Recorded	Theme Order	Aberrant Theme Transitions	Fundamental Themes Omitted	Aberrant Category
22 Feb. 1977	1 5 6 7			
	5 6 7			
	6 7			
	6 7			
	1 5 6 7	1-5		
	1 5 6 7	7-5	4	B
	6 7	7-6		
	1 5 6 7...	2-5		
	..,5			
	1 2 6			
	5 6 7			
14 March 1977	...9			
	1 2 4 5 6 7 8	8-7	None	X
	7 8 9			
	1...			
11 May 1977	...6 8			
	1 2 5 6 8 9			
	1 6 8			
	1 2 5 6...9			
	1 2 5...			
	...5 6 8	2-5	4	B
	1 2 5 6 8 9	1-6		
	1 2 5 6 8 9...			
	...1 2 5 6 8 9			
	1 2 5 6 8 9			
	1 2 5...			

Appendix Table 1 (continued).

Date Recorded	Theme Order	Aberrant Theme Transitions	Fundamental Themes Omitted	Aberrant Category
4 Jan. 1978	...4			
	1 2 4			
	1 2 4			
	1 2...5			
	1 2 4			
	1...2 4			
	1 2 4			
	1 2 4			
	1 2...			
	...9	4-1	None	A
	1 2 4	5-1		
	1 2 4			
	1 2 4			
	1 2 4			
	1...			
	...5 6 7 8 9			
	1 2 4			
	1 2 4			
	1 2...5 6 7 8 9			
	1...			
29 Jan. 1978	...2 4 5 6 8			
	1 2 4 5 6 8 9			
	1 2 4 5 6 8 9			
	1 2 4 5 6 8 9			
	1 2 4 5 6 8 9			
	1 2 4 5 6 8			
	1...	4-1	None	A
	...2 4 5 6 8 9			
	1 2 4 5 6 8 9			
	1 2 4 5 6 8 9			
	1 2 4			
	1 2 4 5 6 8 9			
	1 2 4 5 6 8			
	1 2 4...			
7 Feb. 1978	...4 5 8 9			
	1 2 4			
	1 2 4...			
	...2 4 5 8 9			
	1 2 4	4-1	None	A
	1 2 4...			
	...4 5 8 9			
	1 2 4 5 8 9			
	1 2 4 5 8 9			
	1 2...			

Appendix Table 1 (continued).

Date Recorded	Theme Order	Aberrant Theme Transitions	Fundamental Themes Omitted	Aberrant Category
23 March 1978	...4			
	1 5			
	2 4 5			
	2 4			
	1...			
	...5			
	4			
	2 4 5			
	1...			
	...2 4 5			
	2	5-2		
	1 5	5-4		
	1 5	4-2		
	2 4...	1-5	8	B
	...5	2-1		
	1 5	5-1		
	1 5	4-1		
	2 4 5			
	4			
	2 4 5			
	2 4 5			
	1 5...			
	...4			
	2 4 5			
	2 4 5			
	1 2...			
	...5			
	1 5			
	1 5			
	2 4			
	2...			
1 April 1978	1 2...			
	...5			
	1 2 4 5			
	1 2 4...			
	...5			
	1 2 4 5			
	1 2 4 5...			
	...4 5	5-1	8	C
	1 2...			
	...5			
	1 2 4 5			
	1 2...			
	...1 2 4 5			
	1 2 4 5			
	1 2...			

Appendix Table 1 (continued).

Date Recorded	Theme Order	Aberrant Theme Transitions	Fundamental Themes Omitted	Aberrant Category
19 April 1978	1 Sil-6_2* 6 9			
	1 Sil-6_2 6 9			
	1...			
	...1 Sil-6_2 6 9			
	1 Sil-6_2 6 9	1-6	2,4,5	C
	1 Sil-6_2 6 9			
	1...			
	...1 Sil-6_2 6 9			
	1...			
22 April 1978	...4 9			
	1 2 4 9			
	1 2 4 9			
	1...			
	...1 2 4 9	4-9	5,8	C
	1 2 4 9			
	1 2...			
	...1 2 4 9			
	1 2 4 9			
	1 2...			
25 April 1978	...9			
	1 2 5 6 8 9			
	1 2 5...			
	...6 8 9	2-5	4	C
	1 2 5 6 8 9			
	1 2...			
27 April 1978	...2 4 5 (silence)			
	2 4 5 (silence)			
	2 4 5...			
	...2 4 5 (silence)	5-2	8,1	C
	2 4 5...			
	...5 (silence)			
	2 4 5...			

* Sil-6_2 represents two phrases sung between Theme 1 and Theme 6 in this session. Each phrase consists of a silence equivalent in length to the first subphrase of Theme 6 followed by two low pulsive units typically sung as the second subphrase of Theme 6.

Appendix Table 1 (continued).

Date Recorded	Theme Order	Aberrant Theme Transitions	Fundamental Themes Omitted	Aberrant Category
22 Feb. 1979	...1 2 4 5 1 2 4 5...	5-1	8	X
19 April 1979	...1 2 4 5 8 1 2 1 2 4 5 8 1 2 1 2 4 5... ...8 1 2 4 5 8 1 2 1... ...2 4 5 8 1 2 4 5 8	2-1	None	A

TYPE A: FEBRUARY 7, 1978

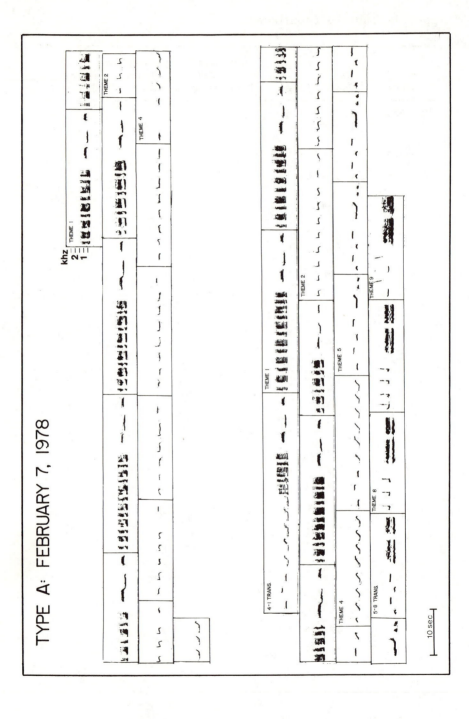

10 sec

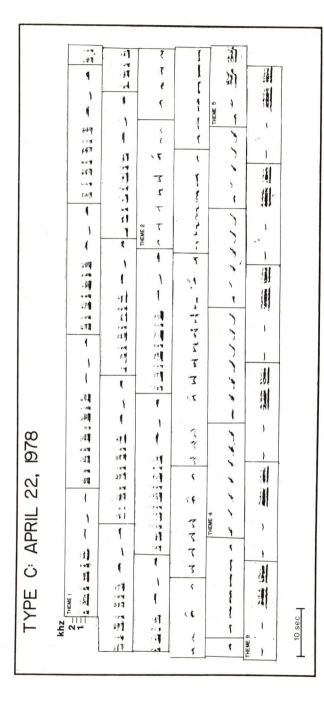

Appendix Figure 1. Sound spectrograms of aberrant humpback songs. Presented here are continuous sound spectrograms from song sessions of each of the three categories of aberrant humpback song. Opaquing fluid was used to remove backgrounds and the harmonics of pure tones.

TYPE B: MARCH 23, 1978

THEME DURATION

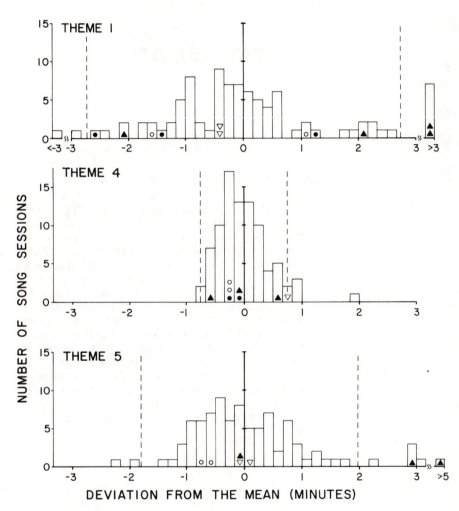

Appendix Figure 2.   See caption on page 122.

← Appendix Figure 1 (continued).   Sound spectrograms of
aberrant humpback songs.

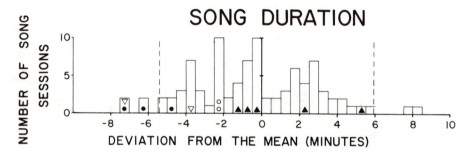

Appendix Figure 2. Variation in humpback song structure. Each histogram details the deviation from the mean value of a parameter of typical humpback song structure. This distribution is obtained by subtracting the mean of typical song sessions in a 31-day period from the mean of each song session within the same period. The distribution of individual aberrant song sessions is indicated as follows: ● = Type A (aberrant songs in a Type A session), O = Type A (typical songs in a Type A session), ▽ = Type B aberrant song sessions, ▲ = Type C aberrant song sessions. Dotted lines indicate 95% confidence intervals for typical song sessions.

PHRASE   DURATION

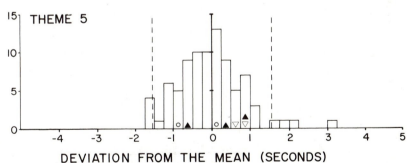

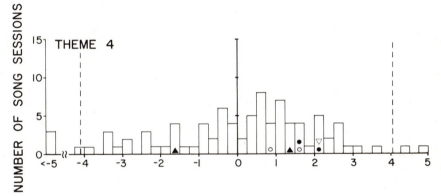
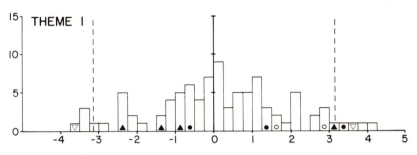

Appendix Figure 2 (continued).

PHRASES PER THEME

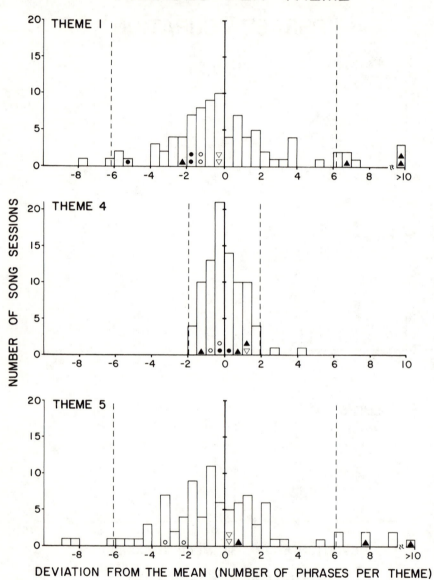

Appendix Figure 2 (continued).

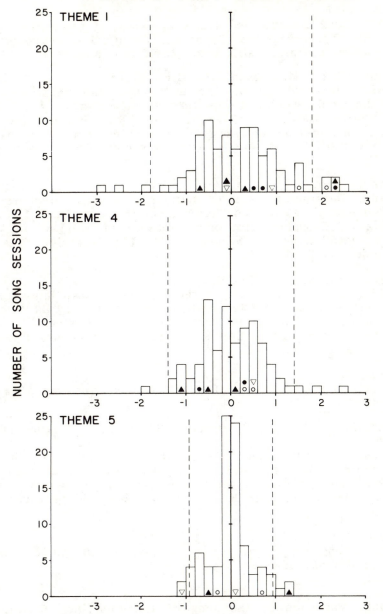

Appendix Figure 2 (continued).

Literature Cited

Baptista, L.F.
1975. Song dialects and demes in sedentary populations of the White-crowned Sparrow (*Zonotrichia leucophrys nuttalli*). Univ. Calif. Publ. Zool. 105, 53p.

Darling, J.D., K.M. Gibson, and G.K. Silber.
1983. Observations on the abundance and behavior of humpback whales (*Megaptera novaeangliae*) off West Maui, Hawaii, 1977-79. *In* R. Payne (ed.), Communication and behavior of whales. AAAS Selected Symposia Series, p. 201-222. Westview Press, Boulder, Colo.

Glockner, D.A.
1983. Determining sex of humpback whales (*Megaptera novaeangliae*) in their natural environment. *In* R. Payne (ed.), Communication and behavior of whales. AAAS Selected Symposia Series, p. 447-464. Westview Press, Boulder, Colo.

Guinee, L.N., K. Chu, and E.M. Dorsey.
1983. Changes over time in the songs of known individual humpback whales (*Megaptera novaeangliae*). *In* R. Payne (ed.), Communication and behavior of whales. AAAS Selected Symposia Series, p. 59-80. Westview Press, Boulder, Colo.

Hudnall, J.
1977. In the company of great whales. Audubon 79(3): 62-73.

Katona, S., B. Baxter, O. Brazier, S. Kraus, J. Perkins, and H. Whitehead.
1978. Identification of humpback whales by fluke photographs. *In* H.E. Winn and B.L. Olla (eds.), Behavior of marine animals - Current perspectives in research, Vol. 3: Cetaceans, p. 33-44. Plenum Press, N.Y.

Konishi, M.
1964. Song variation in a population of Oregon Juncos. Condor 66: 423-436.

Marler, P.
1952. Variation in the song of the Chaffinch *Fringilla coelebs*. Ibis 94: 458-472.

Marler, P. and M. Tamura.
1962. Song dialects in three populations of White-crowned Sparrows. Condor 64: 368-377.

Nottebohm, F.
1970. Ontogeny of bird song. Science (Wash. D.C.) 167: 950-956.

Payne, K. and R. Payne.
In prep. Large-scale changes over 17 years in songs of humpback whales in Bermuda.

Payne, K., P. Tyack, and R. Payne.
1983. Progressive changes in the songs of humpback whales (*Megaptera novaeangliae*): A detailed analysis of two seasons in Hawaii. *In* R. Payne (ed.), Communication and behavior of whales. AAAS Selected Symposia Series, p. 9-57. Westview Press, Boulder, Colo.

Payne, R.S. and S. McVay.
1971, Songs of humpback whales. Science (Wash. D.C.) 173: 585-597.

Payne, R. and L.N. Guinee..
1983. Humpback whale (*Megaptera novaeangliae*) songs as an indicator of "stocks". *In* R. Payne (ed.), Communication and behavior of whales. AAAS Selected Symposia Series, p. 333-358. Westview Press, Boulder, Colo.

Thielcke, G.
1969. Geographic variation in bird vocalizations. *In* R.A. Hinde (ed.), Bird vocalizations - their relation to current problems in biology and psychology, p. 311-339. Cambridge Univ. Press, London.

Thorpe, W.H.
1958. The learning of song patterns by birds with especial reference to the song of the Chaffinch *Fringilla coelebs*. Ibis 100: 535-570.

Tyack, P.
1981. Interactions between singing Hawaiian humpback whales and conspecifics nearby. Behav. Ecol. Sociobiol. 8: 105-116.

Winn, H.E., W.L. Bischoff, and A.G. Taruski.
1973. Cytological sexing of cetacea. Mar. Biol. (Berl.) 23: 343-345.

Winn, H.E. and L.K. Winn.
1978. The song of the humpback whale *Megaptera novaeangliae* in the West Indies. Mar. Biol. (Berl.) 47: 97-114.

John K. B. Ford, H. Dean Fisher

4. Group-Specific Dialects of Killer Whales *(Orcinus orca)* in British Columbia

Abstract

Underwater phonations recorded during 49 encounters with killer whales *(Orcinus orca)* in the coastal waters of British Columbia include clicks, whistles, and pulsed calls. The majority of sounds heard in killer whale pods (stable social groups) engaged in active behavior are repetitious pulsed calls which can be organized into discrete categories. From repeated encounters with photographically identified pods it is apparent that each pod has a limited repertoire of discrete pulsed calls which appears to be shared by most members of the group. There is evidence that pod repertoires of pulsed calls are highly stable for periods of at least 10 years and probably longer. There are considerable differences in the discrete pulsed call repertoires of certain pods. The extent of these vocal differences appears to reflect the degree of associations between pods. Group-specific calls, or 'dialects', may possibly serve as a mechanism for maintaining the cohesion and integrity of killer whale pods.

Introduction

Consistent geographically-related variation in the vocal patterns of conspecifics can occur between isolated populations or among contiguous groups. Both types of variation have been referred to as 'dialects', but it is important that a distinction be made between the two. Differences in the vocalizations of widely separated populations which do not normally mix can be referred to as 'geographic variation', while the term 'dialect' is best reserved for vocal differences on a local scale among neighboring groups or populations which can potentially intermix (Nottebohm 1969; Grimes 1974). Geographic variation in vocalization can result from genetic isolation of populations, differences in body size related to changes in altitude or latitude, or may even be a response to differences in acoustic environment (Green 1975, Krebs and Kroodsma 1980).

Although geographic variation in vocalization may often be a functionless by-product of isolation and genetic divergence, true dialects are considered to have some adaptive significance. Local dialects in birds have been suggested to serve as a means of reducing gene flow between breeding populations thereby allowing more efficient and rapid adaptation to locale-specific conditions (Nottebohm 1972; Baker and Mewaldt 1978). In the absence of conclusive evidence of this, several alternative hypotheses have been proposed (Krebs and Kroodsma 1980).

Dialects are known to occur in many species of birds (for review see Krebs and Kroodsma 1980), but they appear to be very rare in mammals. Geographic variation has been reported between isolated populations of langurs (*Presbytis entellus*, Vogel 1973), squirrel monkeys (*Saimiri sciureus*, Winter 1969), pikas (*Ochotona princeps*, Somers 1973), and prairie dogs (*Cynomys gunnisonni*, Slobodchikoff and Coast 1980). Dialects have been observed among three different troops of Japanese monkeys (*Macaca fuscata*), but only in certain calls which developed as a result of artificial feeding of the troops (Green 1975).

Among the marine mammals, dialects in the threat vocalizations of male northern elephant seals (*Mirounga angustirostris*) on different breeding rookeries off the coasts of California and Mexico have been described by Le Boeuf and Peterson (1969). More recently, it has been found that the magnitude of these differences has diminished with expansion of the population and the exchange of animals between rookeries (Le Boeuf and Petrinovich 1974). Geographic variation in vocalization has been suggested for several cetaceans (e.g., Taruski 1976; Thompson, Winn, and Perkins 1979), but has been demonstrated only in the humpback whale, *Megaptera novaeangliae*. The songs of humpbacks in the North Atlantic are significantly different from those in the North Pacific (Payne 1979; Winn, Thompson, Cummings, Hain, Hudnall, Hays, and Steiner 1981). However, songs recorded from different breeding grounds within each ocean are essentially identical. In an acoustic study of captive killer whales (*Orcinus orca*), Dahlheim (1980) noted vocal differences related to locations of capture, suggesting that geographic variation exists in this species.

In this paper, we describe the existence of group-specific dialects in the most common category of phonations, the pulsed calls, of free-ranging killer whales in British Columbia. This finding results from an ongoing study of the acoustic behavior of known pods of these animals. Pod identification was based on a photographic technique for identifying individual whales in the field. This technique was developed by M.A. Bigg and co-researchers of the Pacific Biological Station in Nanaimo, B. C., who have determined that killer whales in this area form discrete, long-term social units, or pods (Bigg, MacAskie, and Ellis 1976; Bigg, pers. comm.). We recorded the phonations of 12 of the 30 pods known to occur in the waters surrounding Vancouver Island, B.C., during 49 encounters in 1978-79. We present here preliminary results of analyses of these recordings, and describe the pulsed-call repertoires of five selected pods in order to illustrate the types of vocal differences which occur within the population. A complete description of the dialect system of killer whales in British Columbia will be published in a future report.

Methods

The Study Animals

The following is a summary of information on killer whale abundance, distribution, and social organization in the study area collected during 1973-80 by Bigg et al. (1976) and Bigg (pers. comm.), by Balcomb, Boran, Osborne, and Haenel (1980) during 1976-78, and by ourselves during 1978-80. As mentioned, these data result from a monitoring program based on the photographic identification of individual whales using naturally-occurring nicks and scars on the dorsal fin and the lightly-pigmented 'saddle patch' posterior to the fin.

Killer whales in British Columbia coastal waters live in highly stable groups or pods containing up to 50 individuals, although sizes of 5 to 20 are more typical. A population of 30 pods, containing a total of about 260 animals, occurs in the waters surrounding Vancouver Island. On average, pods are composed of 23% mature males, 34% mature females, 39% juveniles, and 4% calves or young of the year. Pods are made up of sets of related animals and may represent extended family or kin groups. Because of the very low birth and mortality rates in the population, pod composition may undergo little change over periods of several years. No permanent exchange of animals between pods has been observed, but pods often join and travel together for periods of up to a few days or weeks.

The killer whales off Vancouver Island are divided into three distinct communities. Two communities are each comprised of a number of 'resident' pods which are present year-round in two separate ranges. One resident community, referred to as the southern community, consists of three pods, identified as J, K, and L, containing a total of about 80 whales. The range of these pods includes the inshore waters of Vancouver Island south of Discovery Passage (Figure 1) and, for pods K and L, some offshore areas west of southern Vancouver Island as well. The second, or northern community of resident whales contains 12 pods with a total of approximately 135 animals. These pods range through inshore waters north of Discovery Passage, although one pod (A5) occasionally occurs south of this boundary (Figure 1). Although northern and southern community pods are not known to intermix, pods within each community regularly associate with one another.

The third community consists of 15 small pods, totalling about 45 individuals, which do not appear to have a well defined range. The movements of these 'transient' pods are unpredictable and some have been observed only rarely. Transients may be seen within the ranges of the two resident communities, but they do not associate with resident pods. However, transient pods periodically join and travel together.

Recording and Analysis

Killer whale pods were encountered and recorded acoustically in a number of locations near Vancouver Island in 1978-79. All observations and recordings were made from a 5 m, outboard powered boat. Our method of locating pods varied according to local conditions and availability of whales. At some locations, field camps were established and local waters were patrolled by boat. In other areas, we established a network of volunteer observers who notified us when whales were present. Upon receipt of a call, we would then attempt to intercept the group.

Whale identification photographs were taken on high-speed black and white film (ASA 1200) using a motor-driven 35 mm SLR camera with a 300 mm lens mounted on a shoulder brace. Individual whale identifications were later determined from the photographs by M.A. Bigg.

Our recording system consisted of a Nagra IV-SJ instrumentation recorder, Celesco BC-10 or BC-50 hydrophones, and a specially-designed preamplifier/filter unit. Frequency response was limited by the recorder and high-pass filter to 100 Hz - 35 kHz (+ 3 dB) at a tape speed of 38 cm/s. Some recordings were made at 19 or 9.5 cm/s, which reduced the upper frequency limit to 20 kHz or lower. Concurrent observations of the whales' behavior and spatial arrangement were recorded on a second channel of the recorder.

A simple shorthand technique, involving symbols which reflect the pitch contours of killer whale signals was used to

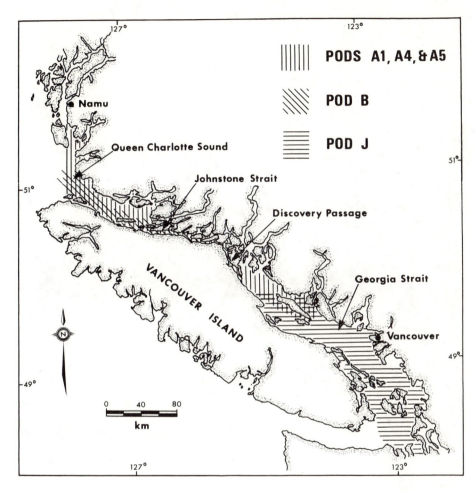

Figure 1. The known distribution of five killer whale pods discussed in this paper. Of the three A-pods, only A5 has been seen south of Discovery Passage. The range of pod J also extends south into Puget Sound, not shown in this map. Data from Bigg et al. (1976) and M.A. Bigg (pers. comm.).

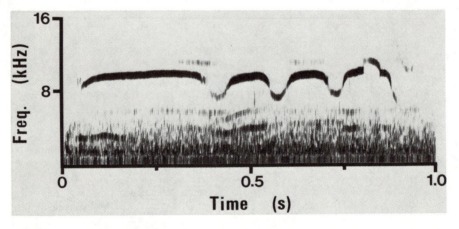

Figure 2. Spectrogram of a whistle recorded from a group of resting killer whales. Effective analyzer filter bandwidth 300 Hz.

transcribe the recorded phonations from the tapes. The phonations were then sampled and analyzed spectrographically using a Kay Elemetrics 7029A spectrum analyzer. With the exception of Figure 2, all spectrograms illustrated in this report have a linear frequency bandwidth of 80-8000 Hz and were produced using a narrowband (45 Hz) analyzing filter. Sound frequency and duration measurements were taken from the spectrograms with a calibrated acetate overlay.

Results

Characteristics and Contexts of Killer Whale Phonations

Our observations on the structural characteristics of killer whale phonations are consistent with descriptions given previously by other authors. Killer whales produce three basic types of sounds. The first type consists of short (0.8 to 25 ms) pulses or clicks, containing energy over a wide range of bandwidths, which are thought to function as echolocation signals (Schevill and Watkins 1966; Steiner, Hain, Winn, and Perkins 1979). We have recorded series of clicks having repetition rates of 1 or 2 to over 300 clicks/s, with frequencies as high as 35 kHz, the upper limit of our recording equipment. The second type of phonation is comprised of tonal signals having a continuous or non-pulsed waveform. Spectrographically, these 'whistles' are characterized by a dominant narrowband component with little or no harmonic or sideband structure (Figure 2). Whistles occur at frequencies of 1.5 to 18 kHz, although most are emitted between 6 and 12 kHz. Most whistles contain numerous rapid frequency modulations over their duration. Whistles range in length from 50 ms to 10-12 s.

The third and most characteristic type of killer whale vocalizations are intense sounds made up of high repetition rate pulses. Although the pulses incorporated in these sounds are similar in structure to clicks, pulse calls typically contain abrupt and patterned shifts in repetition rate which are not present in click series. These sudden shifts in repetition rate result in distinctive and often unusual aural qualities characteristic of pulse calls. It is these pulsed calls which are the main subject of this paper. Schevill and Watkins (1966) describe killer whale pulsed calls as being "scream" like, while Steiner et al. (1979) mention their "harsh, strident, or metallic" quality. We recorded a wide variety of pulsed calls, most of which showed complex physical structure. Spectrographically, pulsed calls with high repetition rates are represented by a number of sidebands at intervals on the frequency scale equivalent to the pulse repetition rate (Watkins 1967). Hence, the pulse repetition rate of a pulsed call (in pulses/s) can be read directly from the sideband interval (in Hz) shown on the spectrogram. Pulsed calls contain energy as high as 25 kHz in frequency and have pulse repetition rates of up to 5000/s. In most pulse calls, however, energy is concentrated between about 1 and 6 kHz and pulsing rates are typically between 250 and 2000/s. Durations of pulsed calls range from less than 50 ms to over 10 s, although most are between 0.5 and 1.5 s long.

In general, the rate of vocalization in killer whale pods is related to the level of activity in the group. Foraging is the most commonly observed group activity. At such times, pods break up into a number of subgroups which disperse over areas as large as 10 km^2. Each subgroup moves at roughly the same pace and in the same direction but each usually dives independently. Vocal activity among pod members during periods of foraging tends to be high. Within a single foraging pod, pulsed calls are typically given at rates of 5 to 30 or more per minute, either continuously or in bouts of varying duration. Click series are commonly heard, but whistles are given comparatively infrequently. Other behavioral contexts accompanied by many pulsed calls and increased whistling include beach-rubbing, where a whole pod may engage in vigorous rubbing of their entire bodies on pebble beaches and shelves, and pods traveling from one foraging area to another.

Periods of activity in killer whale pods are interspersed with periods of low activity or group-resting behavior, which may last for up to several hours at a time. While resting, pod members typically join together in a tightly-knit group, usually with animals lined up abreast. Movement slows and dives become synchronous and long in duration (4-8 min). Usually the animals travel less than 150 m during each resting dive. In most cases, pods become silent while group resting, although on occasion the animals will emit a variety of low-intensity whistles (Figure 2) and brief pulsed calls.

Pulsed Calls: Discrete and Variable

Most killer whale pulsed calls (hereafter referred to as 'calls') can be unambiguously arranged into discrete, non-overlapping categories on the basis of their distinctive physical structure. In most cases, the differences between discrete call types are readily apparent to the ear. In some instances, however, spectral analysis is required to differentiate between similar calls. A certain amount of variability of form exists among calls within each call category, but seldom does this variability mask the calls' distinguishing characteristics. The discrete nature of signals given by captive killer whales was noted by Spencer, Gornall, and Poulter (1967) and Singleton and Poulter (1967). The latter authors also mentioned that the calls can be recognized despite variations in structure and noisy backgrounds.

In addition to discrete calls, killer whales also emit a complex array of highly variable pulsed signals which cannot be placed in any clearly defined categories. These variable, or 'miscellaneous', calls cover a broad spectrum of sounds ranging from very short squeaks and trills to long, raucous squawks lasting up to 12 s in duration.

From repeated encounters with identified killer whale pods, it is apparent that each group emits a limited number of discrete call types. Certain calls in these repertoires are highly repetitious and dominate acoustic exchanges during periods of activity. Other call types are heard only occasionally. The common calls in a pod's repertoire often occur in repetitive series which appear to be exchanges among a number of individuals in the group. Intervals between calls in a series are generally less than a few seconds, and in rapid series calls may overlap each other. When such series are recorded from widely dispersed members of a pod, it is clear from the different amplitude levels and reverberation patterns of the calls that several individuals are involved in the exchange. Since this frequently occurs with many different discrete calls, it is apparent that at least the common calls in a pod's repertoire are produced by most animals in the group.

Discrete calls are by far the most common form of vocalization produced by active killer whales. While the animals are dispersed and foraging, often 90% or more of their calling is of this type (this excludes clicks). The remaining phonations consist of miscellaneous pulsed signals and occasional whistles. However, the whales often cease foraging and engage in beach-rubbing behavior or in activities involving much physical interaction among pod members and frequent aerial displays (tail- and flipper-lobbing, breaching, etc.). At such times the rate of phonation generally increases and the proportion of miscellaneous calls in exchanges increases substantially. As a typical example of this change in acoustic activity, a representative 10-min sample was taken from a recording of pods A1, A4, and A5 while widely dispersed and foraging in Johnstone Strait. This sample contained 188 calls, 176 (94%) of which were repetitions of 11 different discrete calls, while only 12 (6%) were variable or miscellaneous calls. The following day the same whales were observed in the same area, but the animals were grouped together and exhibiting intense social activity in the form of beach-rubbing and physical interaction at the surface. A 10-min sample from this occasion included a total of 460 calls, of which 350 (68%) were discrete and 110 (32%) were miscellaneous.

Discrete Call Repertoires of Selected Pods

A comparison of the discrete calls produced by different killer whale pods indicates that there are significant differences in the repertoires of most groups. In order to illustrate the types of similarities and differences in these vocal patterns, we have selected 5 out of 12 pods recorded to date. Since the highly variable, miscellaneous pulsed calls showed no consistent group-specific differences, only the discrete call repertoires of these pods will be described.

Four of the five pods to be considered here (pods A1, A4, A5, and B) are resident in the northern community while the fifth (pod J) resides in the southern community (Figure 1). Classification and identification of discrete calls were carried out separately for the northern and southern communities since the whales in these two areas had no discrete calls in common. Calls recorded from pods in the northern community are identified by the letter 'N', and those from the southern community by the letter 'S'. Discrete call categories from each community were assigned consecutive numbers in the order in which they were identified, regardless of the identity of the pod(s) responsible for their production (i.e., N1, N2, N3... and S1, S2, S3...). Thus, the call numbering system is largely arbitrary and a call's number does not necessarily reflect its rank in any pod's repertoire. When reference is made to the pod responsible for a particular call, the pod's name is included with the call's identification number (e.g., call N7 given by pod B is referred to as call N7/B). As mentioned previously, there is some degree of variability in the physical structure of calls within each category. This may be a result of many things, including variable motivation or arousal levels of phonating animals, possible individual differences in call production, etc. However, it is also apparent that different pods often emit unique, pod-specific renditions of the same general call format. Whenever these signals could be identified as being different versions of the same call type, they were given the same call number.

Pods A1, A4, and A5: Pods A1, A4, and A5, containing 14, 7, and 12 individuals, respectively, are very closely associated and often travel together in the northern community range. In our 34 encounters with the A-pods, all three were together on 12 occasions (35%), two were present on 8 (24%), and during the remaining 14 encounters (41%), only one of the three was present. Other northern-community resident pods associated with the A-pods on 13 of the 34 encounters. Pod A1 was encountered alone a total of 6 times, and pods A4 and A5 were observed either alone or accompanied only by non-A pods 2 and 3 times, respectively.

A total of 13 different discrete calls, N1 through N13 (Figure 3), was recorded during encounters when all three A-pods were together. Although this list may grow slightly with further recording and analysis, we expect any additional discrete call categories to be relatively uncommon. Recordings made of pods A1, A4, and A5 when encountered individually indicate that virtually the entire repertoire is produced by each of the three groups. The only exception is call N1, which is not emitted by pod A5. Although detailed analyses are not yet complete, there appear to be no obvious differences in the way pods A1, A4, and A5 render the calls shared by each or in the calls' frequency of occurrence in each group's repertoire.

There is some degree of variability in the physical structure of calls within each category. Certain calls, however, are more variable than others. Calls N7, N8, and N12, for example, showed relatively little variation, while others, such as calls N2, N5, and N9, showed significant fluctuations in both duration and structural details. Despite such variability, these calls retained their unique identifying qualities. Examples of the types of variation observed in N2 calls are shown in Figure 4.

Calls N1, N2, N4, N5, N7, N8, N9, and N12 were the most common of the A-pods' discrete signals, each being recorded during 30 or more of the 34 encounters. The remaining calls were heard during at least half of the encounters, with the exception of N13, which was represented in only 10 recording sessions. In a representative 4-min sample taken from a recording of pods A1, A4, and A5 while the animals were foraging, 80% of the 137 calls exchanged was comprised of calls N2, N4, N7, and N8 (Figure 5). The remaining 20% was made up of calls N1, N5, N9, N10, N12, and several variable or miscellaneous calls. This general hierarchy of call use by the A-pods was evident whenever the animals were actively foraging or traveling. The repertoire of the A-pods underwent no apparent change over the two summers the animals were recorded. Each call type showed the same structure and pattern of variability and was given at generally the same rate relative to other calls in the repertoire in both 1978 and 1979. Further evidence indicating a long-term stability in the call repertoire of these pods is discussed below.

An examination of the pattern of occurrence of the various calls in the 4-min excerpt shown in Figure 5 reveals that the more common calls in the repertoire were distributed throughout the sample and often occurred in series, while those calls used less often were given rather sporadically. Also apparent in Figure 5 is the close association of calls N7 and N8. In each encounter with the A-pods, N7 calls were often followed one or two seconds later by an N8 call. Both calls in any N7/N8 pair appeared to be given by the same

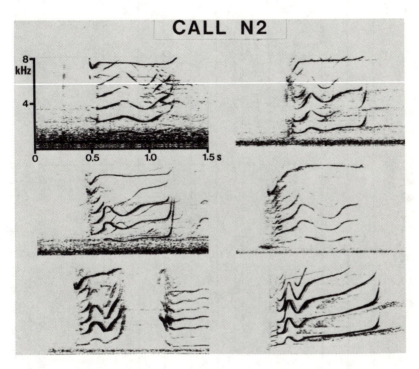

Figure 4. Examples showing the range of variability within the N2 call category.

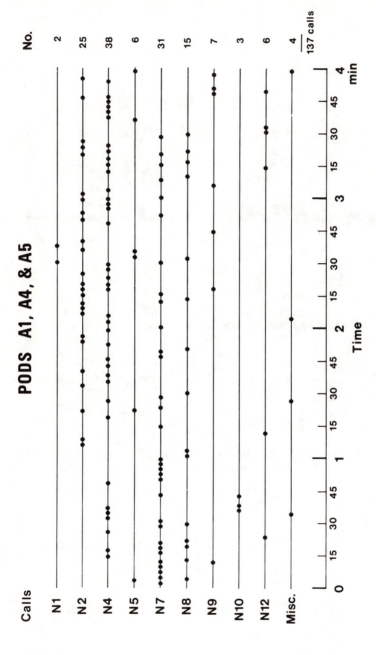

Figure 5. Calls present in a representative 4-minute sample from a recording of pods A1, A4, and 5. Each dot represents one call and the time it occurred in the sample. 'Misc.' refers to unclassifiable, variable pulsed signals.

individual since the relative signal strength of the two was generally consistent. In addition, on several occasions N7/N8 pairs were clearly observed to be emitted by single individuals as they passed close by the hydrophone. The N8 call is never heard without first being preceded by an N7. However, not all N7's are followed by an N8. In one 32-min sample from the three A-pods, 91 N7 calls were recorded but only 30 (33%) occurred in association with an N8. The mean time interval between the two components in the N7/N8 pairs in this sample was 2.1 s (range, 1.2 - 3.9 s).

A frequently-occurring variant of call N7 (identified as N7a) is shown in Figure 3, along with the more common form of this signal. This variant can be distinguished by the inclusion of a short (100-150 ms) upsweep in pitch (i.e., pulse rate) at the end of the call. Since this terminal upsweep was usually of a lower amplitude than the other portions of the call, it was often noticeable only upon spectral analysis. Twenty-four percent (14 of 58) of N7 calls analyzed were of the variant form. This variant is significant because it closely resembles the rendition of call N7 by pod B, discussed below.

As mentioned above, the repertoire of the A-pods did not exhibit any apparent change between 1978 and 1979, suggesting that stability exists in the discrete calls of these groups. Two early recordings provide further evidence that the pods' repertoire has considerable long-term stability. The first was made in northern Johnstone Strait on 29 August 1964, in the presence of an estimated 50 whales. Although there is no photographic confirmation of the A-pods' presence at the time, it can be considered highly probable. The recording contained the entire call repertoire of the A-pods, as well as a number of other discrete calls which were likely produced by other pods in the area. The second recording analyzed was obtained from pod A5 while the group was temporarily held captive in Pender Harbor, B.C., in December 1969. In a 48-min sample from these tapes, all discrete calls from pod A5's repertoire, with the exception of calls N10 and N13, were represented. In fact, of the 881 calls recorded in this sample, all but 3% were of the discrete call types presently given by the pod. The 28 exceptions were variable or miscellaneous calls. No difference was apparent in either the discrete calls' structure or importance in the repertoire. Examples of three of the most common A-calls as they appeared in 1964 and 1969 are shown in Figure 6. Thus, the call repertoire of the A-pods appears to have been essentially the same in 1969, as indicated by pod A5, and possibly as early as 1964.

Pod B: Pod B, containing 8 whales, was recorded on 4 occasions in the Johnstone Strait area of the northern community (Figure 1). On two encounters the pod was alone,

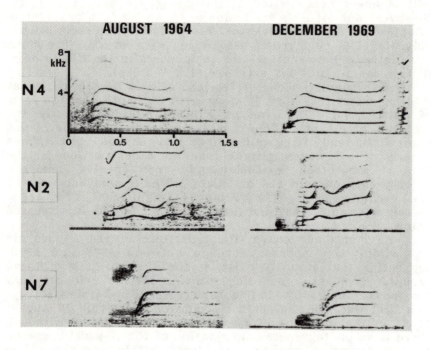

Figure 6. Three common discrete calls from the A-pods' repertoire recorded in August 1964 and December 1969.

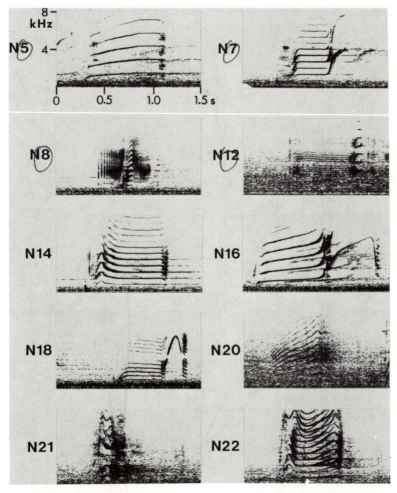

Figure 7. The discrete call repertoire of pod B.

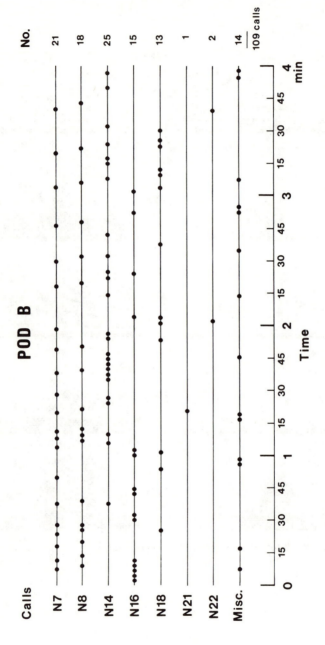

Figure 8. Calls present in a representative 4-minute sample from a recording of pod B. For explanation, see Figure 5.

once it was traveling with pod A4, and once it was present in a large aggregation with the three A-pods and pod C. Pod B was found to produce a repertoire of 10 discrete calls, N5, N7, N8, N12, N14, N16, N18, N20, N21, and N22, shown in Figure 7. Four of these calls, N5, N7, N8, and N12, were also given by the A-pods, but not all in the same form. None of the remaining pod-B calls showed any close resemblance to the discrete calls of the A-pods. Call N5, a rather broad signal category, included any calls having a relatively constant or slightly increasing pitch and often terminating with a sudden drop in pitch. The latter component was generally of a very short duration relative to the main portion of the call and variable in intensity. Pod B's renditions of this call (N5/B) were within the range of variation noted for the A-pods' version (N5/A).

Call N7/B, on the other hand, was consistently and obviously different in structure from N7/A in two ways. First, all pod B renditions of N7 included a terminal upsweep similar to that noted in a minority of N7/A calls (the variant N7a shown in Figure 3). This upsweep was emphasized in call N7/B whereas it was usually very weak in its counterpart given by the A-pods. Second, the middle 'pitch plateau', situated between the initial low pulse-rate burst (evident in the spectrograms as very closely spaced sidebands at the beginning of the calls) and the terminal upsweep, was of a much lower pitch in call N7/B. Structurally, this plateau maintained a pulse rate of 600 ($\pm$ 50) pulses/s in N7/B compared to rates of 1200-1500 pulses/s in N7/A. This difference can be seen in the closer spacing of sidebands in this component of N7/B (Figure 7) than in N7/A (Figure 3). It is noteworthy that calls N7/B and N8/B were also associated in the manner described for the A-pods' version of these signals (Figure 8). In a 14-min sample from a recording of pod B, 36 N7 calls were heard, 29 of which were followed by an N8. The average interval between N7 and N8 calls was 1.8 s, with a range of 1.0 to 4.3 s. As in the A-pods, no N8/B calls occurred without a preceding N7/B. Call N8/A and N8/B showed no significant differences in structure. Call N12, a signal also produced by the A-pods, was rendered in a consistently different manner by pod B. Comparing the two, N12/B (Figure 7) had a distinct, strongly-emphasized upsweep in pulse rate at its end, a component never seen in N12/A (Figure 3).

Figure 8 shows the frequency of occurrence and distribution of 109 calls in a representative 4-min sample of pod B phonations. Call N14 was the most common call (25% of all signals), followed by N7, N8, and N16. Calls N14 and N16 were not emitted by the A-pods. Calls N18, N21, and N22, all apparently unique to pod B, accounted for 16% of the calls in the sample. Although most calls in pod B's repertoire were quite stereotyped, call N16 showed considerable amounts of variation in structure and duration (Figure 9).

Our observations indicate that when pod B traveled in association with the A-pods, each group adhered to its own call repertoire (although it did not limit itself to group-specific sounds), and we have noted this with other pods as well. On many occasions, pods which traveled together remained some distance apart, allowing the recorded calls to be attributed to one pod or the other on the basis of relative signal intensity. At no time was one pod observed to emit calls characteristic of another pod with which it was traveling unless those pods shared the same repertoire, as in pods A1, A4, and A5. As an example, on one occasion pods B and A4 were traveling in association but separated by approximately 1 km. When recordings were made in close proximity to pod B, only the calls characteristic of that pod were heard at higher levels while the different calls of pod A4 were much weaker. The reverse was observed when recordings were made near pod A4. On another occasion, recording of pod A1 had been underway for approximately 1 hr when pod B approached the group. Faint pod-B calls were heard intermixed with A-calls about 15 min prior to pod B being sighted. As pod B drew near, the level of B-calls increased and finally matched that of the A-calls as the two groups joined. After interacting for a short time, pod A1 left pod B, and in the process, the A-calls were observed to grow weaker until finally they were inaudible.

Pod J: Pod J is a group of 19 whales which resides in the southern community range. Of the four northern-community pods described previously, pod A5 is the only group which occasionally visits northern Georgia Strait (Figure 1) and could come into contact with pod J. However, these two groups have never been seen together (M.A. Bigg, pers. comm.).

We encountered and recorded pod J on nine occasions in Georgia Strait; four times pod J was traveling with two other pods (K and L) and five times it was alone. A repertoire of 10 different discrete calls, numbered S1 through S10, was determined for pod J from analyses of the latter encounters. Examples of the calls forming this repertoire are shown in Figure 10. The most striking feature of pod J's repertoire was the predominance of calls ending in a pronounced drop in pitch. Calls S1, S3, S6, S7, and S9 all shared this trait. For this and other reasons, the vocalizations of pod J sound quite unlike those of the northern community pods, and this difference is readily apparent to an untrained ear. Call S1 was by far the most commonly heard signal in every encounter with pod J. It was also one of the most variable in structure (Figure 11). The prevalence of call S1 in pod J's

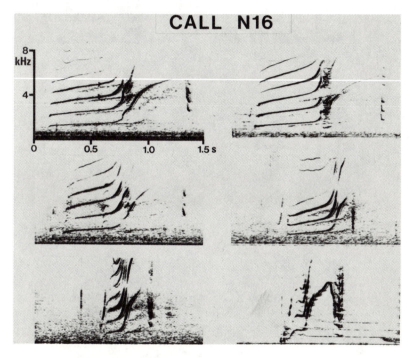

Figure 9. Examples showing the range of variability within the N16 call category.

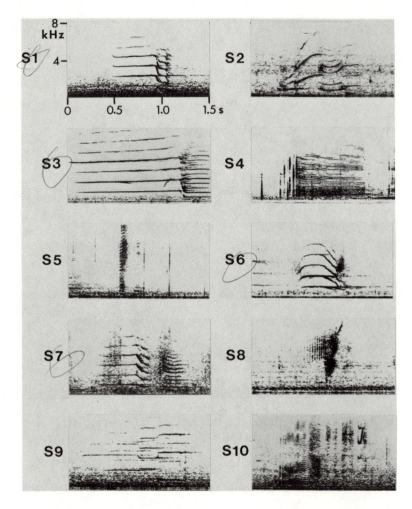

Figure 10. The discrete call repertoire of pod J.

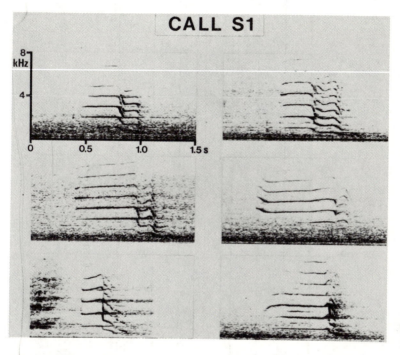

Figure 11. Examples showing the range of variability within the S1 call category.

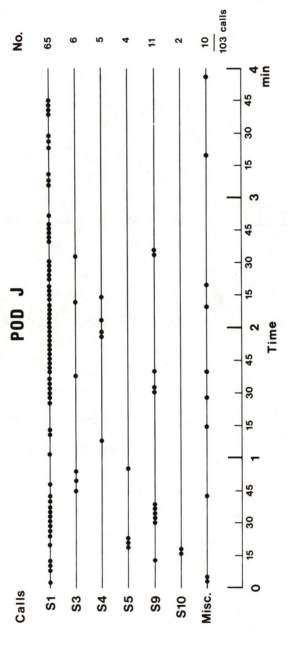

Figure 12. Calls present in a representative 4-minute sample from a recording of pod J. For explanation, see Figure 5.

repertoire can be seen from an analysis of the frequency distribution of the calls heard in a 4-min sample from a recording made while the pod was foraging (Figure 12). Of the 103 calls recorded, 65 (63%) were of the S1 type. The remaining 38 signals consisted of calls S3, S4, S5, S9, S10, and a number of variable pulsed calls. No calls in pod J's repertoire appeared to be associated in the manner of calls N7 and N8 in the A- and B-pods' repertoire.

The calls of pod J showed no apparent change over the 12 months spanned by our recordings. We also have some evidence suggesting that at least part of pod J's repertoire may be consistent over a period of many years. In the spring of 1961, the Canadian Navy made a tape recording of a pod of killer whales on the southeastern end of Vancouver Island. Although the group was unidentified, pod J is the most commonly encountered group in the area (M.A. Bigg, pers. comm.). Calls S1, S2, S4, S8, and S10 were represented on this tape, in addition to a number of unidentifiable signals.

Summary and Discussion

Killer whales generate three distinct sound types - tonal whistles, clicks, and pulsed calls. Many of the pulsed calls are unique to any given pod. In most cases, whistles are associated with low-activity levels, represented by group-resting behavior, while pulsed calls are heard mainly when the animals are active. Click series are recorded primarily when the whales are active but are also heard occasionally in inactive contexts. Periods of activity in killer whale pods most often involve foraging behavior, where whales travel in a highly-dispersed manner, on roughly parallel courses, moving at fairly constant speeds, but also include 'recreational' behaviors such as beach-rubbing. Vocal exchanges during foraging behavior and, to a lesser extent, 'recreational' behavior, are dominated by discrete pulsed signals which are often highly stereotyped. Each pod has a limited repertoire of these discrete pulsed calls. Certain calls in a pod's repertoire are emitted more frequently than others. The more common calls are often produced in rapid series, as if they are exchanged among several animals in the group. There is evidence that a pod's repertoire of discrete calls may remain stable over ten or more years.

A comparison of the discrete calls given by five killer whale pods reveals considerable differences in the repertoires of certain groups. The most marked differences exist between the calls of pod J, a resident of the southern community range, and the calls of pod A1, A4, A5, and B, all residing in the northern community range. The whales in the two areas have no discrete pulsed calls in common. Within the northern community, pods A1, A4, and A5 share essentially the same call repertoire, confirming observations that these three groups are very closely associated. Pod B, a group which occasionally associates with the A-pods, emits four of the call types given by the A-pods, in addition to a set of 6 different calls. Of the four calls shared by the A- and B-pods, two are rendered in a consistently different manner by pod B.

Group-Specific Calls of Killer Whales

Before attempting to interpret the differences in the vocal patterns of killer whale pods, some consideration must be given to the possible functions of an acoustic system based on discrete or stereotyped signals. As described earlier, killer whale pods spend much of their time traveling in a widely dispersed manner while foraging. At such times, pod members are out of sight of each other and any contact must be maintained acoustically. The exchanging of discrete calls would appear to be the most effective and reliable means of maintaining intra-group spacing and movement patterns. The strident nature and unique physical properties of discrete calls allowed us to hear and accurately identify them at ranges of up to 10 km even when partially masked by reverberation and background noise. The use of loud, discrete calls for maintaining intra- and inter-group spacing in forest-dwelling primates is well known (Marler 1972, 1977; Waser 1977). Like killer whales, these social species live in environments where visibility is severely restricted and must maintain contact by acoustic signals which are recognizable at considerable distances.

It is possible that much of the variation seen within discrete call categories of particular killer whale pods resulted from individual differences in call rendition. If these calls contain individual as well as group identity information, they would be even more effective cues for coordinating group deployment. Individuality has been observed in the long-range calls of chimpanzees (*Pan troglodytes*; Marler and Hobbett 1975) and mangabeys (*Cercocebus albigena*; Waser 1977).

It has been suggested from studies of two captive killer whales that there are sex-related differences in the calls of these animals (Singleton and Poulter 1967; Spencer et al. 1967). It is certainly possible that whales of different sexes emit consistently different renditions of discrete calls, or even that certain discrete calls are produced only by animals of one sex. However, the male killer whale studied by Singleton and Poulter (1967) and Spencer et al. (1967) was captured at Namu, B.C. (Figure 1), while the female was taken at Puget Sound, some 600 km south, and thus the two undoubtedly came from very different pods. We feel that the differences observed in their phonations would be better attributed to pod-specific rather than sex-specific vocal patterns.

No one has described close equivalents of the group-specific call system of killer whales in the phonations of other odontocete cetaceans which have been studied to date. The acoustic communication of captive dolphins is based predominantly on the exchange of complex whistles, although a variety of pulsed signals are also emitted (Caldwell and Caldwell 1967). A significant proportion of the phonations given by individuals of several delphinid species observed in captivity have been reported to be highly stereotyped whistles unique to each animal. These "signature" whistles are quite stable over time and are subject to only minor variations in structure with changes in behavioral situations (for review, see Caldwell and Caldwell 1977). Apparent individualized signalling has also been reported for several odontocetes recorded in the wild. Although pilot whales (*Globicephala melaena*) in the North Atlantic have been found to produce a complex array of continuously varying or graded whistles, they have also been observed to produce series of repetitious, discrete whistles, apparently given by single animals (Taruski 1976). Many of the discrete pulsed calls recorded from wild narwhals (*Monodon monoceros*) by Ford and Fisher (1978) were given in long, repetitive series which appeared to be correlated with the passage of different individuals past the hydrophone. Using a three-dimensional hydrophone array, Watkins and Schevill (1977) have determined that sperm whales (*Physeter catodon*) produce individually-unique pulse series, or codas.

Thus, the available information on the phonations of other odontocetes suggests that individual-specific signals, rather than group-specific signals as in the killer whale, may be an important component of the acoustic behavior of at least several species. However, detailed study of the vocal patterns of known social groups of these species must be carried out before the existence or significance of such apparent differences can be determined.

Dialects in Killer Whale Calls

The differences in the vocal patterns of killer whale pods described here represent true dialects since they exist in groups which are not geographically isolated and can potentially interbreed. Killer whale dialects appear to be unusual, however, since they involve group-specific variation in vocalization which is not necessarily geographically-related. In killer whales, pods with different calls may frequent the same waters and regularly associate together. This is unlike the song dialects of birds, which are consistently related to geography (Krebs and Kroodsma 1980).

The discrete call repertoires of pods appear to be consistent over a period of years, which is important if the dialects are to have any functional significance (Le Boeuf and Petrinovich 1974). At this point, we are unable to offer any confident interpretation of the possible role of dialects in the social organization of killer whales. However, it is possible to make some suggestions. As described earlier, killer whale pods are highly stable social groupings of related animals (Bigg et al. 1976). It is probable that there are important selective advantages in the maintenance of long-term kin associations. Distribution patterns of killer whales in the waters surrounding Vancouver Island suggest that pods have well-defined home ranges and foraging routes that are likely to have developed over consecutive generations. Such pod or family traditions may be of considerable importance to the survival of the group, and hence the inclusive fitness of pod members would be enhanced by maintaining close familial ties and passing such traditions on to subsequent generations.

This being the case, dialects in killer whale groups could be viewed as having developed under the same selective pressures responsible for the long-term kin associations within such pods. The use of group-specific signals may provide an effective means of preserving pod identity and cohesion, which would allow kin units to be kept intact. The importance of having a reliable system for maintaining pod integrity can be appreciated when one considers the amount of time killer whales spend in association with other pods. Of the 49 encounters with killer whales in this study, 21 (43%) involved groups containing more than one pod (the closely-related pods A1, A4, and A5 are considered here to be a single group). Up to four different pods were observed together at one time. Under such conditions, cohesion of kin groups could be easily broken down without an effective means of familial recognition. Although this hypothesis may be the most reasonable, it does not account for the mechanism by which the three A-pods, which share essentially the same discrete call repertoire, maintain their discreteness. It may be that these three pods represent one large kin group which frequently but temporarily breaks into consistent subgroups. Cohesion of these subgroups could be maintained through recognition of possible individualized renditions of shared discrete calls. Unfortunately, the mating system of killer whales is unknown, so it is impossible to say whether the breeding is carried out within or between these groups. It is also unknown whether the killer whale dialects described here are genetically encoded (i.e., result from reproductive isolation of dialect groups) or result from behavioral traditions which are passed from parent to offspring through cultural transmission. However, the latter would appear more likely since the ability to mimic and learn new sounds, while rare in most mammals, is well known among dolphins (Caldwell and Caldwell 1972, 1979). Furthermore, dialects in birds have consistently been found to involve vocal learning (Nottebohm 1972).

There are several possible reasons for the similarities observed in the repertoires of different pods. One is that pods simply incorporate into their own repertoires certain calls given by the pods with which they most often travel. Another possibility, perhaps more likely, is that pods with similar call repertoires originated from the same ancestral kin group which split into smaller groups as it grew. As these subgroups became more independent, their acoustic repertoires changed as a result of behavioral or genetic isolation and drift. Old calls gradually became modified or were lost and replaced with entirely new signals. Perhaps the three A-pods, which have very similar repertoires yet spend considerable amounts of time separated from each other, represent an early stage in such a process.

Whatever their function, killer whale dialects are useful as a means of providing insight into the inter-relationships of different killer whale pods. Whether dialects are controlled by genetic or behavioral factors, the degree of similarity or difference in the call repertoires of different pods can be viewed as an index of group association. As our study continues, we will hopefully gain a more complete understanding of the function of the dialect system of killer whales through further correlations between call types and behaviors and through further encounters with the pods described here and with other groups not included in this report.

Acknowledgements

The generous assistance and support of many people helped make this study possible. In particular, we would like to thank Michael A. Bigg (Pacific Biological Station, Nanaimo, B.C.) for his continuing support and close involvement in our study. In addition to making all the whale identifications from our photographs, making several recordings, and helping logistically in a variety of ways, Dr. Bigg made his extensive knowledge of the killer whales of B.C. freely available and offered many helpful suggestions throughout this project. Special thanks are extended to Deborah M. Cavanagh for able assistance in all aspects of the work, and to Graeme Ellis for help and advice. We thank Mervin Black (Defense Research Establishment Pacific) for loaning us hydrophones and killer whale tapes, and also Paul Spong and William A. Watkins for providing tapes. We appreciate the critical comments offered by Michael Bigg, Helene Marsh, William E. Schevill, James N.M. Smith, and W. John Smith on an early draft of this paper. Finally, we thank the many persons who reported whale sightings to us and others who helped in different ways. This research was supported by a contract with the Canadian Department of Supply and Services and Department of Fisheries and Oceans.

Literature Cited

Baker, M.C. and L.R. Mewaldt.
1978. Song dialects as barriers to dispersal in White-crowned Sparrows, *Zonotrichia leucophrys nuttali.* Evolution 32: 712–722.

Balcomb, K.C., J.R. Boran, R.W. Osborne, and N.J. Haenel.
1980. Observations of killer whales (*Orcinus orca*) in greater Puget Sound, state of Washington. U.S. Dep. Commer., NTIS PB80-224728, 42p.

Bigg, M.A., I.B. MacAskie, and G. Ellis.
1976. Abundance and movements of killer whales off eastern and southern Vancouver Island with comments on management. Unpubl. MS, Arctic Biological Station, Ste. Anne de Bellevue, Quebec, Canada. 20p.

Caldwell, D.K. and M.C. Caldwell.
1977. Cetaceans. *In* T.A. Sebeok (ed.), How animals communicate, p. 794–808. Indiana Univ. Press, Bloomington.

Caldwell, M.C. and D.K. Caldwell.
1967. Intraspecific transfer of information via the pulsed sound in captive odontocete cetaceans. *In* R.G. Busnel (ed.), Les systemes sonars animaux, biologie et bionique, vol. 2, p. 879–936. Nato Advanced Study Institute.

1972. Vocal mimicry in the whistle made by an Atlantic bottlenosed dolphin. Cetology 9: 1–8.

1979. The whistle of the Atlantic bottlenosed dolphin (*Tursiops truncatus*) - ontogeny. *In* H.E. Winn and B.L. Olla (eds.), Behavior of marine animals, vol. 3: Cetaceans, p. 369–401. Plenum Press, N.Y.

Dahlheim, M.E.
1980. A classification and comparison of vocalizations of captive killer whales (*Orcinus orca*). Unpubl. MSc. thesis, San Diego State Univ., 76p.

Ford, J.K.B. and H.D. Fisher.
1978. Underwater acoustic signals of the narwhal (*Monodon monoceros*). Can. J. Zool. 56: 552–560.

Green, S.
 1975. Dialects in Japanese monkeys: Vocal learning and cultural transmission of locale-specific vocal behavior? Z. Tierpsychol. 38: 304-314.

Grimes, L.G.
 1974. Dialects and geographical variation in the song of the Splendid Sunbird *Nectarinia coccinigaster*. Ibis 116: 314-329.

Krebs, J.R. and D.E. Kroodsma.
 1980. Repertoires and geographical variation in bird song. Adv. Study Behav. 11: 143-177.

LeBoeuf, B.J. and R.S. Peterson.
 1969. Dialects in elephant seals. Science (Wash. D.C.) 166: 1654-1656.

LeBoeuf, B.J. and L.F. Petrinovich.
 1974. Dialects of northern elephant seals: Origin and reliability. Anim. Behav. 22: 656-663.

Marler, P.
 1972. Vocalizations of east African monkeys. II: Black and white colobus. Behaviour 62: 175-197.

 1977. The evolution of communication. *In* T.A. Sebeok (ed.), How animals communicate, p. 45-70. Indiana Univ. Press, Bloomington.

Marler, P. and L. Hobbett.
 1975. Individuality in a long-range vocalization of wild chimpanzees. Z. Tierpsychol. 38: 97-109.

Nottebohm, F.
 1969. The song of the Chingolo, *Zonotrichia capensis*, in Argentina: Description and evaluation of a system of dialects. Condor 71: 299-315.

 1972. The origins of vocal learning. Am. Nat. 106: 116-140.

Payne, R.
 1979. Humpback whale songs as an indicator of "stocks." Seattle: Abstracts from presentations of the third biennial conference on the biology of marine mammals, October 7-11, p. 46.

Schevill, W.E. and W.A. Watkins.
 1966. Sound structure and directionality in *Orcinus* (killer whale). Zoologica (N.Y.) 51: 70-76.

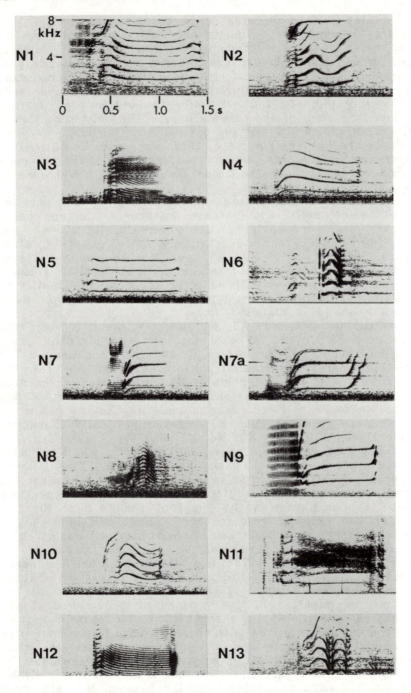

Figure 3. The discrete call repertoire of pods A1, A4, and A5.

Singleton, R.C. and T.C. Poulter.
   1967. Spectral analysis of the call of the male killer
   whale. IEEE Trans. Audio Electroacous. AU-15: 104-
   113.

Slobodchikoff, C.N. and R. Coast.
   1980. Dialects in the alarm calls of prairie dogs. Behav.
   Ecol. Sociobiol. 7: 49-53.

Somers, P.
   1973. Dialects in southern Rocky Mountain pikas, *Ochotona
   princeps* (Lagomorpha). Anim. Behav. 21: 124-137.

Spencer, M.T., T.A. Gornall III, and T.C. Poulter.
   1967. Respiratory and cardiac activity of killer whales. J.
   Appl. Physiol. 22: 974-981.

Steiner, W.W., J.H. Hain, H.E. Winn, and P.J. Perkins.
   1979. Vocalizations and feeding behavior of the killer
   whale (*Orcinus orca*). J. Mammal. 60: 823-827.

Taruski, A.G.
   1976. Whistles of the pilot whale (*Globicephala* spp.):
   Variations in whistling related to behavioral/en-
   vironmental context, broadcast of underwater sound, and
   geographic location. Unpubl. Ph.D. thesis, Univ. of
   Rhode Island. 86p.

Thompson, T.J., H.E. Winn, and P.J. Perkins.
   1979. Mysticete sounds. *In* H.E. Winn and B.L. Olla (eds.),
   Behavior of marine animals, vol. 3: Cetaceans, p. 403-
   431. Plenum Press, N.Y.

Vogel, C.
   1973. Acoustical communication among free-ranging
   common Indian langurs (*Presbytis entellus*) in two
   different habitats of north India. Am. J. Phys.
   Anthropol. 38: 469-480.

Waser, P.M.
   1977. Individual recognition, intragroup cohesion and
   intergroup spacing: Evidence from sound playback to
   forest monkeys. Behaviour 60: 28-74.

Watkins, W.A.
   1967. Harmonic interval: Fact or artifact in spectral
   analysis of pulse trains. *In* W.N. Travolga (ed.), Marine
   bio-acoustics, vol 2, p. 15-42. Pergamon Press, Oxford.

Watkins, W.A. and W.E. Schevill.
1977. Sperm whale codas. J. Acoust. Soc. Am. 62: 1485-1490.

Winn, H.E., T.J. Thompson, W.C. Cummings, J. Hain, J. Hudnall, H. Hays, and W.W. Steiner.
1981. Song of the humpback whale - Population comparisons. Behav. Ecol. Sociobiol. 8: 41-46.

Winter, P.
1969. Dialects in squirrel monkeys: Vocalization of the Roman Arch type. Folia Primatol. 10: 216-229.

Christopher W. Clark

5. Acoustic Communication and Behavior of the Southern Right Whale *(Eubalaena australis)*

Abstract

The sounds from most of the great whales have been recorded, but little is known about the biological function or adaptive significance of these sounds. This report is an investigation of the acoustic communication system of a free-ranging population of southern right whales. Animals were studied intensively for two field seasons between July 1976 and January 1978. Results indicate that activity, size, and sexual composition of groups of whales were correlated with the types of sounds produced. Sounds with the simplest and most predictable structure were associated with long distance contact situations while the highly variable, acoustically complex sound types were associated with groups of socially active whales. The complexity of the social context was directly related to the complexity of the sounds made. This suggests not only that the whales are communicating acoustically but that they are using sounds to communicate a complex array of messages.

Introduction

It is generally assumed that the majority of vocal sounds made by animals serve some communicative function (Thorpe 1961; Busnel 1968; Sebeok 1977). Demonstration of acoustic communication has relied primarily on the technique of correlating observable changes in the behavior of an animal with the acoustic signals produced or received (Marler 1967; Emlen 1972). The function of sound types is often deduced by an interpretation of these results and the social contexts in which they were made (Struhsaker 1967; LeBoeuf and Peterson 1969; Green 1975; Owings and Virginia 1978).

Sound playback experiments are often used to test these deductions concerning a sound's "meaning" and its dependence on the social context (Smith 1965). It is often implied that the degree of complexity in the signal repertoire is indicative of the complexity of the social system (Marler 1970; Smith 1977).

Mysticete whales are known to make a variety of sounds (Schevill and Watkins 1962; Cummings, Fish, and Thompson 1971; Payne and McVay 1971; Thompson, Winn, and Perkins 1979) but there is little scientific evidence to demonstrate that these sounds are communicative. Until recently, observations on the behavior of mysticete whales associated with sounds could best be described as anecdotal and most have been restricted to identifying the species responsible for the sounds (see for example Schevill, Watkins, and Backus 1964; Cummings and Thompson 1971; Watkins and Schevill 1972; Thompson et al. 1979). Efforts to associate social behavior with sounds have usually failed to reveal correlations (Best 1970; Payne and Payne 1971).

This state of affairs is not surprising in the light of the obvious difficulties encountered when working with these large, pelagic mammals. They are hard to observe and follow. There are obvious difficulties in reidentification of individuals from one day to the next and in determining which animal under observation is vocalizing. Recently, through a combination of efforts, solutions to these problems have been found. R. Payne has developed a technique for accurately tracking the movements of whales with a surveyor's theodolite. This technique has been used by Würsig and Würsig (1979a, 1979b), Clark and Clark (1980) and Tyack (1981) to determine swimming patterns for cetaceans. Methods for identifying individual whales using photographs of natural markings have now been applied to humpback whales (Katona, Baxter, Brazier, Kraus, Perkins, and Whitehead 1979) and southern right whales (Payne, Brazier, Dorsey, Perkins, Rowntree, and Titus 1983). Acoustic tracking of individuals from the sounds they make has been reported by Walker (1963, though he was not certain that whales were the source), Watkins and Schevill (1972), Watkins (1977), Clark and Clark (1980), and Clark (1982, in press) using hydrophone array systems.

The data reported here are the results of an eighteen month field study on a free-ranging population of southern right whales. The aims were to determine the acoustic repertoire of the whales, to correlate the sounds with other activities, and to demonstrate the possible communicative function of the sounds. The evidence shows that there are correlations between sounds and activities, that the sounds are communicative, and that the function of the sounds can be interpreted from the activities of the whales and the contexts in which the sounds were made.

Materials and Methods

Location

Data are based on the results of observations on free-ranging southern right whales in the Golfo San José, Argentina from 16 July 1976 to 21 December 1977. A major portion of the 1976 season was spent arranging electronic equipment and devising methods for systematically gathering data. For this reason, the majority of the data presented here are from the 1977 season, when data collection was standardized and consistent.

Eight hundred and fifty hours of observations were made from an observation hut situated on the edge of a 46 m cliff overlooking the southeast corner of the gulf. Additional observations (80 hours) were made from other nearby locations when sun glare obscured a group or additional identification photographs were needed. No observations were made from boats.

Group

The term "group" refers to one or more whales seen within close proximity of one another (approximately 15 m or one adult whale length). Animals more than one whale length apart were also considered a group if their behavior appeared to be coordinated. Both the theodolite tracking data and the qualitative impressions of the whales' behaviors were used to decide when a group had split up. For example, if one group member moved several lengths away from the others and then maintained that distance, it was still considered a group member. If the distance continued to increase, it was considered as a separate group. In general, if two members of a group could be tracked and the distance between them was greater than three or four whale lengths for more than five minutes, they were considered separate groups. The point of separation was defined as the point at which we began tracking both groups separately.

Identification

The head callosity patterns of southern right whales can be used to reliably distinguish between individuals and reidentify individuals over long periods of time (Payne et al. 1983). Whales were photographed whenever possible using cameras (Nikon F) with 50 mm to 1000 mm lenses. Photography concentrated on the head callosity patterns,

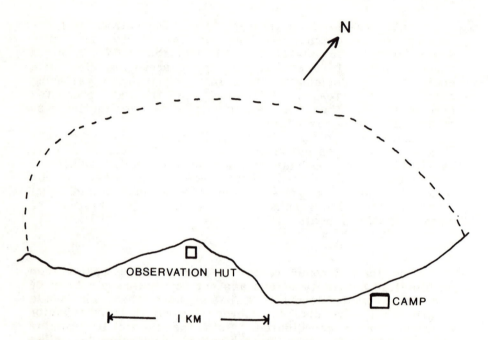

Figure 1. Schematic representation of the observation area. The dotted boundary outlines the area within which whale observations were made. The observation hut is located at $42^{\circ}25'30''$S latitude and $64^{\circ}9'00''$W longitude.

white back and belly patches, the back edges of flukes, and the anogenital region. An effort was also made to photographically document behavior, particularly that of active groups.

The sex of right whales can be determined by the morphology of the anogenital region (See Matthews 1938 p. 178). Sexing these whales on behavioral criteria is risky and not definitive (Best 1970; Saayman and Tayler 1973; Payne and Dorsey 1983) except in the case of a lactating mother and calf. A whale was sexed as female if it was seen alone with the same calf on at least three different days, or it had mammary slits on either side of the genital slit, or no anus was seen separate from the ventral slit during a complete view of the ventral area between the genital slit and the flukes. A whale was determined to be male if an anus widely separate from the ventral slit or a penis was seen.

Visual Observations

Whales were observed on 159 days between 14 May and 9 December 1977 with 2 to 5 observers watching for an average of 5.6 hours per day. All whale groups seen within the approximate area bordered by the dotted line in Figure 1 were studied by ad. lib. sampling (Altmann 1974). All observers strived to determine how many groups were in this area and to sample a group whenever its activity or location changed. A surveyor's theodolite (Kern model DMK 1) was used to record the location of a group with a maximum error of $\pm$ 1 m at 1 km. A pair of binocular spotting scopes (Bausch and Lomb with X15 eye pieces) mounted on a heavy tripod in an observation hut was used for behavioral observations. The data collected for each group sampled included time, group identification, description of activity, number of animals, compass headings of group members, and location. The procedure was designed so that the number of observers would always be sufficient to log data consistently, independent of the total number of whales in the area or their activities. The sampling method assumes that the rate at which a group is sampled is independent of the number or identity of observers taking the samples. It does not assume that the sampling rate is independent of the type or level of activity. For example, a group that was swimming through the area would get sampled more often than a group that remained in the same spot doing the same activity. Of course, this sampling procedure could not be maintained if the number of groups and the movement of individuals between groups became high. In these cases, we would stop ad. lib. sampling and scan repeatedly and sequentially from one group to another or concentrate on the activities of the most vocal groups.

Acoustic Sampling

Tape recordings were made during all periods of observation. The receiving system consisted of a hydrophone array (Clark 1980) and a Sony TC880B tape recorder (15/16 ips) or Nagra IV-S stereo tape recorder (3 3/4 ips). This system was flat ± 3 db from 50 Hz to 5000 Hz. The hydrophones were bottom mounted in relatively shallow water (5 m at mid tide) and were not used during low tides (<2 m). A total of 1102 hours of tape recordings were made during the daylight hours in 1977 for an average of 4.5 ± 1.5 (SD) hours of recordings per day.

Sound Direction Sampling

A portable real-time sound direction finding system for underwater sounds (see Clark 1980 for details) was used for a total of 231 hours during 32 days of observation. This device computed and displayed the direction to the sound source whenever a whale made a sound with a fundamental frequency between 50 and 500 Hz, that was 3 dB above the ambient noise level. The bearing was either dictated onto a stereo recorder or written down with the time and a general description of the sound. Bearings could be redetermined by replaying the stereo tapes into the direction finder at a later date. The system was not capable of determining which whale within a group of two or more whales was responsible for the sound unless the group was so close to the receiving hydrophones that the bearing angles between members of the group exceeded considerably the heading error of the system (± 6–12°).

Definitions of Activities

Details of one whale's behavior within a group of active whales were difficult to record. Often a familiar posture or movement would be seen but assigning the behavior to an individual was problematical. During observations on active groups that remained just below the surface, an observer would only see the constant churning of the water and the edges of flukes and flippers knifing through the foam. In order to avoid defining a large set of behaviors which could not be reliably scored, all references to the behaviors of groups were reduced to a set of five activities; resting, swimming, mild activity, full activity, and sexual activity. These activities are not necessarily mutually exclusive.

Resting. Whales that remained in the same location without any evidence of physical exertion were considered as a resting group. Most resting groups drifted at the surface with their nares and a portion of their backs above the water. Sometimes a resting whale would remain underwater in the same spot and surface occasionally to breathe.

Swimming. Whales moving from one location to another at a fairly constant speed were considered as a swimming group. In shallow water, whales would swim at the surface or alternate between swimming at and just below the surface. In deeper water, whales would usually swim long distances underwater, surface, take a quick series of breaths and submerge for another long, underwater swim. Swimming whales that were approaching an active group would typically submerge when 0.5 to 1 km away and then swim the remaining distance entirely underwater.

Mild Activity. A group was mildly active if there was enough activity to disturb the surface to produce white water, but the activity was frequently interrupted by brief periods of resting or swimming.

Full Activity. A group was fully active if its movements produced constant white water.

Sexual Activity. A group was sexually active if it was known to contain both males and females and a male was seen with his penis extended.

The most distinctive feature of active groups was white water. White water could be produced by movements such as tail or head lashing, flipper slapping, lobtailing, breaching, spyhopping, belly sinking, flipper stroking, clasping, etc. (Donnelly 1967; Best 1970). The movements of mildly active groups were generally slower and more deliberate than those of fully active groups whose actions tended to be rapid and pronounced. The amount of white water associated with sexually active groups varied considerably and seemed directly related to group size and the number of males in the group.

The use of white water as an indication of activity is somewhat confounded by environmental factors. During rough sea conditions, waves would break over the whales (like waves over a reef) exaggerating the amount of white water generated by their activities. Observers were well aware of these conditions and always took them into account when judging activity levels, particularly in distinguishing mildly active from fully active groups. Cues such as the constancy of the waves breaking over a back or the steady direct progress of the group were usually enough to tell that the whales were resting or swimming.

Description of Sound Types

The sounds of southern right whales have been generally described by Payne and Payne (1971) and Cummings et al. (1971, 1972, 1974). Recently, Clark (in press) has described and classified southern right whale sounds using principal components analysis (Rohlf, Kishpaugh, and Kirk 1980). His sound sample was based on over 1500 hours of recordings made throughout the 1977 season during all times of day. The sounds were categorized into three general classes; blow sounds, slaps, and calls, and each class was subdivided into types. There were three types of blow sounds: normal, tonal, and growl blow sounds. There were four types of slaps: flipper, breach, lobtail, and underwater slaps. There were six types of calls: up calls, down calls, constant calls, high calls, hybrid calls, and pulsive calls. All blow sounds and all slap sounds, except underwater slaps, are audible both in air and underwater. Table 1 gives a general description and spectrographic illustration of a normal blow sound, flipper slap, and each of the six call types. In general, the call repertoire is best described as a continuum with certain types more common than others (Clark, in press).

Group Data

A total of 358 groups were investigated with the sound direction-finding equipment when there were no ambiguities as to which group was responsible for the sounds. Each group was described by the following variables: total number of whales (a single whale was considered as a group of size one), number of females, number of males, activity (including its duration), whether it was with or without porpoises or sea lions, and rates of sound production for blow sounds, slaps, and all six call types. Although differences in the types of blow sounds and slaps are recognized, sound rates were computed for these two classes and not for their various types. Differences between the types of blow sounds and slaps will be incorporated into the discussion on communicative function. Sound rates were computed by dividing the number of sounds per hour by the number of whales in the group. It seems highly probable that some whales in the group were actually producing sounds at higher rates.

The groups were sorted into three sets: the Silent Groups, the Acoustic Groups, and the Interspecific Groups. The Silent Groups never made a sound and were not with porpoises or sea lions. The Acoustic Groups made at least one sound and were not seen with porpoises or sea lions. The Interspecific Groups were seen with porpoises (*Lagenorhynchus obscurus* or *Tursiops truncatus*) or sea lions (*Otaria flavescens*).   Interspecific Groups were separated from the others since porpoises and sea lions tended to affect the whales' behavior.

Statistical Tests

The following statistical tests were applied to data in relation to group size, activity, sexual composition, and rates of sound production for blow sounds, slaps, and all six types of calls.   The Kruskal-Wallis (KW) test (Sokal and Rohlf 1969) was used to determine the significance of the effect of activities, sizes, and sexual compositions on sound rate for each of the eight sound types.   The Mann-Whitney U (MWU) test (Sokal and Rohlf 1969) was used to determine the significance of differences in the sound rates between pairs of activities, sizes, and sexual compositions.

Results

Individual Sighting Data

By comparing the whales in over 10,000 photographs taken throughout the 1977 season both with themselves and with the identified whale heads in an identification catalog (Payne et al. 1983), we were able to identify 135 different whales.   Because we saw many of these individuals repeatedly, the total number of times we made an identity was 700.   Seventy-five identified whales were sexed, 72 by us and three females by others.   Forty-six were females and 29 were males (this does not necessarily reflect the sex ratio in the gulf).   Six females were sexed only on the basis of having a newborn calf.   Otherwise the sex of an individually identified whale was based on the positive association between the sexually distinctive features of its anogenital region and its head callosity pattern.   Forty-eight of the 60 whales of unknown sex were relatively small animals (estimated body length <12 m).   They were sighted on only one or two days and were rarely seen in fully active or sexually active groups.   This suggests that these animals were subadults and sexually immature.

The Activity, Size, and Sexual Composition of the Silent, Acoustic, and Interspecific Groups

The Silent Groups were less active and tended to be smaller than the Acoustic Groups or the Interspecific Groups (see Table 2).   The activity and size of the Silent Groups and Acoustic Groups were different depending on the sexual composition of the group.   All female and all male groups were smaller and less active than groups of mixed sex (this

Table 1. Sound types in the acoustic repertoire of the southern right whale. The power spectral density functions (PSDF) were computed by averaging the spectra from ten random samples for each type.

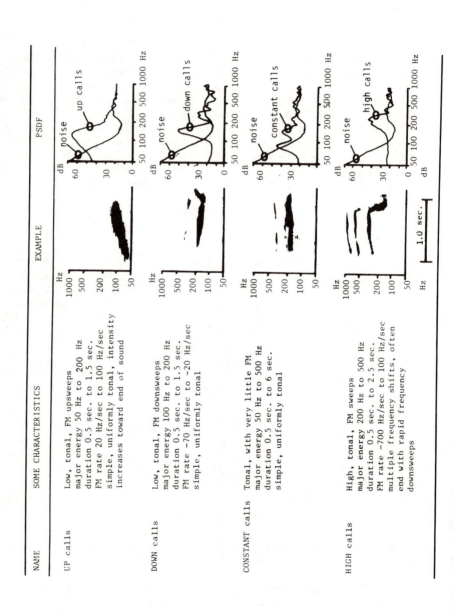

NAME	SOME CHARACTERISTICS	EXAMPLE	PSDF
UP calls	Low, tonal, FM upsweeps major energy 50 Hz to 200 Hz duration 0.5 sec. to 1.5 sec. FM rate 20 Hz/sec to 100 Hz/sec simple, uniformly tonal, intensity increases toward end of sound		
DOWN calls	Low, tonal, FM downsweeps major energy 100 Hz to 200 Hz duration 0.5 sec. to 1.5 sec. FM rate −70 Hz/sec to −20 Hz/sec simple, uniformly tonal		
CONSTANT calls	Tonal, with very little FM major energy 50 Hz to 500 Hz duration 0.5 sec. to 6 sec. simple, uniformly tonal		
HIGH calls	High, tonal, FM sweeps major energy 200 Hz to 500 Hz duration 0.5 sec. to 2.5 sec. FM rate −700 Hz/sec to 100 Hz/sec multiple frequency shifts, often end with rapid frequency downsweeps		

Table 1 (continued).

NAME	SOME CHARACTERISTICS	EXAMPLE	PSDF
HYBRID calls	Complex mixtures of FM sweeps and amplitude modulation major energy 50 Hz to 500 Hz duration 0.5 sec. to 2.5 sec. usually begin like a high call but become pulsive at the end		
PULSIVE calls	Complex mixtures with amplitude modulation of noise and/or an FM signal major energy 50 Hz to 200 Hz duration 0.5 sec. to 3.5 sec. usually very harsh, strident or growlly		
BLOWS	Noisy, broadband major energy 100 Hz to 400 Hz duration 0.5 sec. to 26 sec. sometimes tonal like a long moan sometimes noisy and pulsive		
SLAPS	Noisy, broadband sharp onset major energy 50 Hz to 1000 Hz duration 0.2 sec. when produced underwater very intense and painful		

(From Clark, in press, reproduced by permission of Animal Behavior Society)

ACTIVITY						
	Total	Resting	Swimming	Mild	Full	Sexual
INTERSPECIFIC						
Ave. Size	1.8+3.5	–	1.5+0.9	2.1+0.9	2.4+1.1	–
Size Range	1-4	–	1-3	1-4	1-4	–
N	42	0	15	20	7	0
ACOUSTIC						
Ave. Size	2.3+4.0	1.5+0.8	1.6+1.0	2.6+1.1	3.6+1.9	3.9+1.2
Size Range	1-10	1-3	1-5	1-6	2-10	2-6
N	187	10	87	59	23	8
SILENT						
Ave. Size	1.6±1.8	1.6±0.7	1.6±0.6	2.0±1.2	–	–
Size Range	1-4	1-3	1-3	1-4	–	–
N	129	11	101	17	0	0

SIZE					
	Total	Single	Diad	Triad	3
INTERSPECIFIC					
Ave. Act.	3.0±0.5	2.6±0.5	2.9±0.3	2.9±0.5	3.5±0.5
Act. Range	2-4	2-4	2-4	2-4	3-4
N	42	19	11	10	2
ACOUSTIC					
Ave. Act.	2.7±0.7	2.0±0.2	2.8±0.7	2.8±0.7	3.1±0.8
Act. Range	1-5	1-3	1-5	1-5	1-5
N	187	74	41	47	25
SILENT					
Ave. Act.	2.0±0.2	2.0±0.2	2.0±0.2	2.2±0.4	3.0±0.0
Act. Range	1-3	1-3	1-3	1-3	3
N	129	59	56	12	2

SEXUAL COMPOSITION			
♀♀	♂♂	♀♂	??
INTERSPECIFIC			
Ave. Size 1.3±0.5	1.3±0.7	2.7±0.4	2.0±0.4
Size Range 1-3	1-2	2-3	1-3
Ave. Act. 2.7±0.5	2.7±0.7	3.2±0.6	3.1±0.5
Act. Range 2-4	2-4	2-4	2-4
N 15	6	6	15
ACOUSTIC			
Ave. Size 1.6±0.9	1.1±0.4	3.3±1.1	2.2±0.9
Size Range 1-4	1-2	2-6	1-10
Ave. Act. 2.3±0.5	2.3±0.4	3.7±0.9	2.5±0.6
Act. Range 1-3	1-2	2-5	1-4
N 41	22	40	84
SILENT			
Ave. Size 1.4±2.0	1.0±0.0	2.8±0.7	1.9±0.3
Size Range 1-3	1	2-4	1-3
Ave. Act. 2.1±0.4	1.8±0.3	2.6±0.6	2.0±0.1
Act. Range 1-3	1-2	1-3	1-3
N 56	12	8	51

Table 2. Average sizes (± SD) and activities for Interspecific, Silent, and Acoustic groups depending on the activity, size, and sexual composition of the group (see text for definitions). Size = number of whales in the group; activities were coded as follows: resting = 1, swimming = 2, mild = 3, full = 4, sexual = 5. Nomenclature for sexual composition was as follows: ♀♀ = all female, ♂♂ = all male, ♀♂ = at least one female and one male in the group, ?? = sexual composition considered unknown since it does not fall into any of the three previous categories.

does not include mother and calf pairs). In general, most of the Silent Groups were single or pairs of swimming whales, while most of the Interspecific Groups were single, mildly active whales.

Correlation Between Activity, Size, and Sexual Composition and the Sound Rates of Acoustic Groups

The following results are derived from an analysis of the data from 187 groups that made sounds but were not with porpoises or sea lions. These will constitute the major data by which the activity, size, and sexual composition of groups will be correlated with rates of sound production. Throughout these results, the activity, size and sexual composition will be considered as the independent variables to which the sound rates are referred.

Activity vs. sound rates. The activity of a group differentially affected rates of sound production (KW, p<0.01) for six of the eight sound types: up calls, high calls, hybrid calls, pulsive calls, blow sounds, and slaps. Significant differences between the sound rates per whale for different activities are noted in Figure 2.

Size vs. sound rates. The size of a group differentially affected rates of sound production (KW, p<0.01) for five of the eight sound types: up calls, high calls, hybrid calls, pulsive calls, and blow sounds (Figure 3). In a few cases, the size of the group engaged in a particular activity had significantly different effects on sound rates (see Figure 4). Swimming whales that were alone had higher rates of up calling than larger swimming groups. Single, mildly active whales had higher rates of up calling, down calling, and slapping than mildly active groups that contained more than a single individual. Fully active groups containing three whales had significantly higher rates for hybrid calls than fully active pairs.

Sexual composition vs. sound rates. The sexual composition of a group differentially affected rates of sound production for four of the eight sound types: up calls, hybrid calls, pulsive calls, and blow sounds. The significant differences between the sound rates per whale for different sexual compositions are noted in Figure 5. In a few cases, depending on the type of activity, the sexual composition of a group had significantly different effects on the sound rates (Figure 6). Swimming or mildly active unisexual groups had significantly higher up call and slap rates than swimming or mildly active groups of mixed sex. These differences are a reflection of the fact that almost all of the swimming and mildly active unisexual groups were single animals.

176   *Christopher W. Clark*

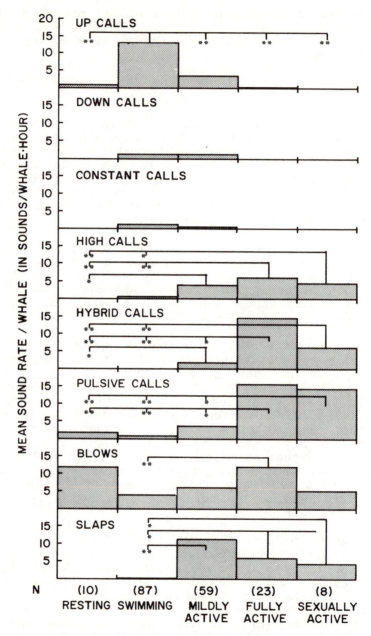

Figure 2. Group activity versus average sound rates per whale (in sounds per whale-hour). Asterisks (** for p<0.01, * for p<0.05) indicate those cases in which groups engaged in different activities had significant differences in their call rates per whale.

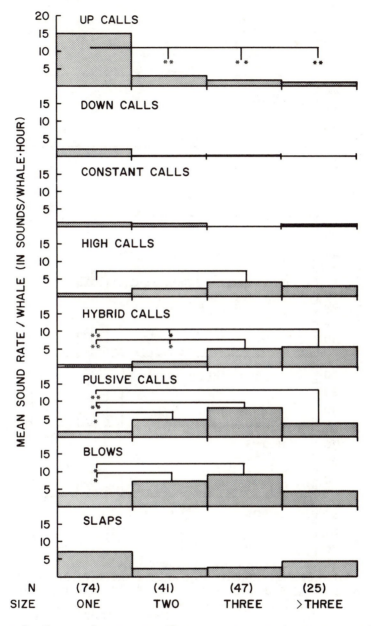

Figure 3. Group size versus the average sound rates per whale
(in sounds per whale-hour). Asterisks (** for
p<0.01, * for p<0.05) indicate those cases in which
groups of different sizes had significant
differences in their call rates per whale.

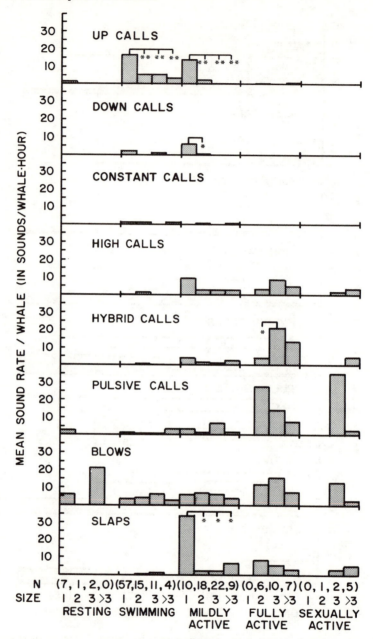

Figure 4. Group activity by size of group versus average sound rates per whale (in sounds per whale-hour). Asterisks (** for p<0.01, * for p<0.05) indicate those cases in which groups of different sizes but similar activity had significant differences in their call rate per whale.

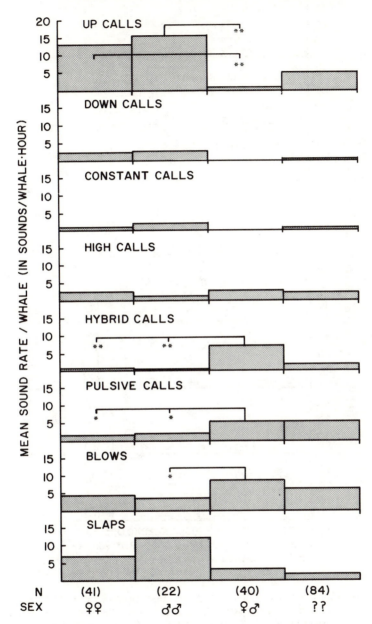

Figure 5. Group sexual composition versus average sound rates per whale (in sounds per whale-hour). Asterisks (** for p<0.01, * for p<0.05) indicate those cases in which groups of different sexual compositions had significant differences in their call rates per whale.

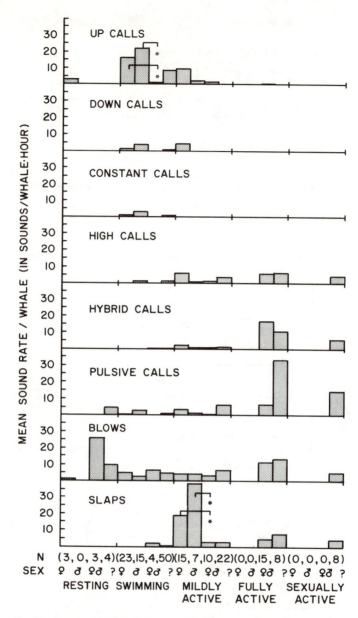

Figure 6. Group activity by sexual composition versus average
sound rates per whale (in sounds per whale-hour).
Asterisks (** for p<0.01, * for p<0.05) indicate
those cases in which groups of different sexual
composition but similar activity had significant
differences in their call rates per whale. ♀ = all
females; ♂ = all males, ♀ ♂ = at least one female
and one male in the group, ? = sexual composition
considered unknown since it does not fall into any
of the three previous categories.

The Contexts in Which Sounds Were Made

One of the major purposes of this study was to correlate the whales' behaviors with the sounds they made. The eventual aim is to understand what, if any, communicative function is served by each of these sounds. The results just presented demonstrate that there are correlations between behaviors and sounds. In the sections which follow, these results will be summarized and each sound type placed in its behavioral context. This evidence will then be used to interpret the communicative functions of the sounds.

Resting groups (N=10). Eleven of the 21 resting groups (see Table 2) were silent. For those groups that did make sounds, resting groups made significantly fewer up calls than swimming groups, significantly fewer high calls and hybrid calls than any of the active groups, and significantly fewer pulsive calls than fully active or sexually active groups (see Figure 2). Resting groups were smaller than fully active or sexually active groups (see Table 2). There was a tendency for smaller resting groups to have higher sound rates than larger resting groups (see Figure 4).

Only three of the resting groups made calls and all three were single animals. Two of these groups made a few up calls just prior to changing their behavior and swimming off. The third made a number of calls that sounded very much like up calls but these were distinctly pulsive during the upsweep. This whale did not move but continued to drift backward near the observation hut for over an hour. Its pulsive calls were produced at about the same tempo as its breathing rate and some of the calls were correlated with the opening of the nostrils. This suggests that by opening its nostrils, the whale modified an up call into a pulsive call.

Most of the sounds from resting groups were blow sounds (see Table 3). Many of these blow sounds (65%) were heard both in air and in water by the observers and were longer than 2 seconds (43%). All of the resting groups that made long blow sounds were single whales or a mother and calf pair. The single whales would typically stay at the surface in the same location (sometimes for hours at a time) and at regular intervals of several minutes make long (up to 26 s.) sounds accompanying their exhalation. Because this sound was heard in the air, such whales were referred to as moaners. No other whales were ever seen joining moaners but one moaner did drift into a fully active dyad. When it joined them, and this was probably unintentional, it stopped moaning and swam away from the active group. A mother accompanied by her calf was heard on four different days making long blow sounds that usually sounded like a harsh growl. She made these sounds when a third whale swam over

Table 3. Numbers of sound types produced by the 187 groups of whales engaged in the five different activities.

	# OF GROUPS	OBS. TIME (WHALE-HRS)	SOUND TYPE						BLOWS	SLAPS
			CALLS							
			UP	DOWN	CONSTANT	HIGH	HYBRID	PULSIVE		
RESTING	10	11.2	9	0	0	0	0	19	121	0
SWIMMING	87	74.8	1122	113	104	61	17	87	348	36
MILD ACTIVITY	59	80.0	207	77	30	230	112	218	360	667
FULL ACTIVITY	23	54.9	9	0	5	143	336	363	283	136
SEXUAL ACTIVITY	8	27.8	0	0	0	36	49	116	117	34

and joined them (N=2) or when her calf had moved several hundred meters away (N=7). On four of these seven occasions when her calf swam away and she growled, the calf quickly returned to her. When he didn't, she swam and retrieved him.

In two cases, a single resting whale made several loud, short tonal blow sounds (<2 s.). Both these animals had been resting quietly when they were suddenly joined by a whale who swam in silently underwater. After the rester made several loud tonal blow sounds, the intruder swam away.

Swimming groups (N=87). One hundred and one of the 188 swimming groups were silent. For those groups that did make sounds, swimming groups had significantly higher up call rates than all other groups, but significantly lower high call, hybrid call, pulsive call, blow sound, and slap rates than any of the active groups (see Figure 2). Swimming groups were smaller than fully active or sexually active groups (see Figure 2). Single, swimming whales had significantly higher up call rates than swimming groups containing two or more whales (see Table 4).

Most of the swimming group sounds were up calls (see Table 3). Seven of the swimming groups began up calling after they split away from a group. Forty-two up called as they swam toward another swimming group that was also up calling. As the two swimming groups approached each other, their rates of up calling tended to accelerate, but once the groups were together all calling stopped. Only 7 of the 101 swimming groups that never made a sound were observed joining another silent swimming group.

Once, late in the 1977 season, we observed a small calf making up calls as it swam alone without its mother. In 1976, there was another occasion when a mother and calf became separated and up calls were produced. The two were swimming silently (which was typical of mother and calf pairs) in shallow water that was murky due to a recent storm. As they rounded an underwater outcropping of the cliff, the calf swam about 75 m away from its mother and out of underwater visual range. She then made an up call which was answered almost immediately with an up call by the calf. The calf's up call was quickly followed by a second up call from the mother and the two swam to each other and continued on their way. This was one of a few instances when bearings to individual members of a group were sufficiently different to enable us to distinguish acoustically between members of the same group.

On several occasions, we heard swimming groups make several up calls and down calls as they approached a group of active whales. We never heard these active groups make either an up call or down call in response.

Only three swimming groups made only down calls. All three were single whales that made a series of three to five calls. One was a small female who later began slapping and rolling about by herself. The other two were males who were swimming in from deep water.

Two swimming groups made only constant calls. They were both solo pregnant females that gave birth within two weeks after the observations. Their low constant calls sounded similar to the long moan blow sounds made by solitary resting whales, but these calls were shorter (<2 s.).

Swimming groups were heard making blow sounds but usually only when the whales were within 100 m of the hydrophones. These blow sounds always sounded normal and, judging from the rates at which the whales were breathing, only a small fraction of their expirations were audible at the hydrophones.

On four different occasions, a pair of whales was seen swimming together with one whale swimming belly up underneath the second whale that was swimming at the surface. The back of the surface whale would hump up about a meter out of the water and as the back returned to a normal posture, an extremely loud, sharp slap would be heard. This humping behavior took several seconds and was repeated six to eleven times in succession. A slap was heard each time the humping was seen.

On one occasion, the behavior commenced with the underwater whale approaching the surface whale while swimming upside down. In this instance, the belly of the underwater whale, a female, had a large white mark which aided our observations on this unique behavior. When this belly marked whale was underneath the surface whale, she lifted it up on her belly and carried it along as she swam. The surface whale then humped up its lower back and thrust it down onto the belly of the underwater whale. Each time, the result was a sharp, intense slapping sound. In this and one other belly slapping incident, the observers could see the blubber of the surface whale vibrate from the impact of the collision. Three of the four groups seen belly slapping were pairs of females and the fourth was a mother and her male calf. The larger whale in a pair was always underneath except in one case when the two were of approximately equal size. This pair switched positions between their two belly slapping bouts.

Mildly active groups (N=59). Seventeen of the 76 mildly active groups were silent. Those mildly active groups that did make sounds had significantly lower up call rates than swimming groups; significantly lower hybrid call, pulsive call, and blow sound rates than fully active groups; and significantly higher slap rates than resting or swimming groups (see Figure 2). The sizes of mildly active groups were intermediate to the sizes of swimming and fully active groups (see Table 2). Single mildly active whales had significantly higher up call, down call, and slap rates than mildly active groups containing two or more whales (see Figure 4).

In general, mildly active groups made all types of sounds at rates which were intermediate to those for resting or swimming groups and fully active or sexually active groups.

All mildly active whales that were alone made either up calls, down calls, high calls, or slaps. Four of the six whales that up called were joined by a swimming whale that also up called. In one of these four instances, the whale also made a series of three down calls and several isolated high calls before swimming quietly away.

Three of the mildly active groups were small females that made a series of 3 to 11 down calls before or after a series of flipper slaps. One of these females also made several high calls during her period of activity. The remaining three groups made between two and eight isolated high calls and no other sounds except a few blow sounds and slaps. On only one occasion did a mildly active whale make up calls, down calls, high calls, and slaps.

There were four mildly active, all-female groups that made high calls exclusively. All of these were groups of two or three females that were observed early in the season between June and August when there were few whales in the gulf. The high calls were usually heard within several minutes after the group first formed. Activity consisted mostly of pectoral stroking, rolling slowly about together and draping flukes over each other's backs.

The one occasion when a mildly active group made a series of pulsive calls was when a mother and her three-day-old calf were separated by another female (who was pregnant). The mother and calf were swimming when they were overtaken by the pregnant female. She swam between the pair and was able to move the calf about 50 m away from its mother. As this was happening, a series of pulsive calls was produced. Several minutes later, when the mother retrieved her calf, another series of pulsive calls was made. The pregnant female then swam in a circle around the area of the incident, made one up call and swam slowly away as the mother and calf moved rapidly off in another direction.

The blow sounds from mildly active groups were sometimes forced and loud. Many of these blow sounds were audible in air and in water. These blows sounded so distinctive that they consistently alerted us to the presence of the group or to the moment when a resting or swimming group became a mildly active group. For example, whales that were resting or swimming would suddenly make tonal or growl blow sounds if they were joined by porpoises or sea lions. Usually a few of these blow sounds would be made just before the whale became physically active.

In seven of the 14 mildly active groups seen with porpoises (and therefore included only in the Table 2 results), a series of sharp slaps was heard but we could not see any behavior coincident with the sounds. During one observation of a single, mildly active female who was accompanied by *Tursiops truncatus*, a series of slaps was made as the whale moved the anterior half of her body back and forth in a sweeping motion. Her head was almost entirely out of the water during this behavior and the slaps did not coincide with the rocking motion of her body or her exhalations. Of course it could have been the porpoises that made the slapping sounds, but the sounds sounded identical to the intense slap sounds made by whale groups at the same range when porpoises were absent.

In two other mildly active groups, a whale made a long series of slaps (>20). These slapping bouts included several different behaviors that produce slaps (breaching, lobtailing, or flippering), interrupted by brief silent periods of resting or swimming. None of the five single whales that slapped were joined by another whale although three did eventually begin swimming quietly toward an active group.

Fully active groups (N=23). None of the 23 fully active groups of whales were silent. Fully active groups had significantly lower rates for up calls than swimming groups, significantly higher rates for high calls, hybrid calls, pulsive calls, and slaps than resting or swimming groups; and significantly higher hybrid call and pulsive call rates than mildly active groups (see Figure 2). Fully active groups were larger than resting or swimming groups (see Table 2). Fully active groups of three whales had significantly higher rates of hybrid calls than fully active groups that contained only two whales (see Figure 4). No fully active groups contained only males or only females (see Figure 6).

Fully active groups made mostly high calls, hybrid calls, and pulsive calls. In general, the rates for high calls and hybrid calls tended to increase while the rate for pulsive calls tended to decrease with larger group sizes (see Figure 4). The significant difference between hybrid call rates for fully active groups containing three or more animals is explained by the following observations: In 16 of the fully active groups nearly two-thirds of the calls (68%) were produced in a series containing a combination of 3 to 16 high, hybrid, and pulsive calls. Typically the number of hybrid calls in a series was greater than the number of <u>high</u> calls and much greater than the number of <u>pulsive</u> calls. Only two of the six fully active dyads, but 14 of the 17 larger fully active groups produced calls in a series. The 2 dyads produced a total of 13 calls in 3 series and 6 of the calls were <u>hybrids</u>. The 14 larger groups produced a total of 696 calls in 107 series and 317 of these calls were <u>hybrids</u>.

Three of the fully active groups made <u>pulsive</u> calls almost exclusively. One of these groups was initially a mildly active pair that appeared to be courting. The whale at the surface was a female but the sex of her companion was not known. She was not obviously avoiding the advances of this whale who spent most of its time underwater stroking and clasping her. Most of their sounds were blow sounds and a few isolated <u>high</u> and <u>hybrid</u> calls. When a large male joined them the situation changed. The whales suddenly became very animated and several series of <u>pulsive</u> calls were immediately heard. The two other fully active groups heard making almost exclusively <u>pulsive</u> calls were composed of a mother and calf and a large male. On both occasions, <u>pulsive</u> calls were produced when the male separated the calf from its mother by lifting it up onto his belly, clasping it between his pectorals and rolling about with the tiny whale. The time interval between <u>pulsive</u> calls seemed less regular than in most series and instead many of the calls were either isolated or produced in short series of only two to four calls.

Occasionally, two separate fully active groups would make <u>high</u> calls, <u>hybrid</u> calls, <u>pulsive</u> calls, or a series of these types at the same time. But we never heard just one fully active group produce two of these call types or a call series simultaneously.

Fully active groups almost always produced blow sounds that were short (<2 s.), tonal, or growl types. Both females and males made these loud, forced blow sounds, although females in fully active groups of mixed sex usually stayed at the surface while the males pushed and rolled beneath them. Often a member of the group would blow out just below the surface or as it lifted its head rapidly out of the water. These underwater blow sounds tended to sound pulsive while the surface blow sounds were harsh and growly.

Whales in fully active groups would usually only slap once or twice with a pectoral and sometimes the slapper would actually hit one of the other whales in the group. Whales were never seen breaching or lobtailing in these groups. A whale would begin breaching or lobtailing only after it swam away from a fully active group. Slaps were often heard that did not coincide with any obvious behavior at the surface. These slaps were usually very loud and sharp sounding and were presumably produced underwater.

Sexually active groups (N=8). None of the sexually active groups were silent. Sexually active groups were never heard up calling. They had significantly higher rates for high calls, hybrid calls, pulsive calls, and slaps than resting or swimming groups and significantly higher rates for hybrid calls and pulsive calls than mildly active groups (see Figure 2). Sexually active groups were larger than either resting or swimming groups (see Table 2).

Sexually active groups made only high calls, hybrid calls, and pulsive calls. In two of these sexually active groups, many of these calls were produced in a series. In one of the groups, long (>2 sec.), isolated pulsive calls were heard each time that a female was forcefully pushed by one or several of the males. The female was lying upside down at the surface and each time the male pushed at her midsection with his rostrum, she lashed with her tail and the long pulsive call was made.

Sexually active groups made mostly blow sounds that were short tonal or growl types, but normal blow sounds were also heard. The types of blow sounds produced seemed to depend on the general level of activity. In a pair of mating whales, most blow sounds were normal while in the larger more active sexual groups, the blow sounds were almost always forced and loud.

Like the whales in fully active groups, whales in sexually active groups would usually only slap once or twice with a pectoral and never breached or lobtailed when in the group. Other slap sounds were heard which did not coincide with any surface behavior. These slaps were intense, sharp sounds which were presumably produced underwater.

Discussion

The evidence presented here indicates that the sounds made by a southern right whale are correlated with the activity of the animal and its social context. Whales that were resting did not make many calls but did sometimes make blow sounds which were exceptionally long and relatively constant in frequency. When a resting whale did make calls, they were usually produced just before the whale began to swim. Swimming whales that were alone usually produced up calls and often joined other swimming whales that also up called back. Their blow sounds were normal. Slaps were produced when two swimming whales hit their bellies together. Whales in mildly active groups made all types of sounds depending on the number of whales in the group. Single, mildly active whales made mostly up calls, down calls, and slaps, while larger mildly active groups made mostly high calls, hybrid calls, and pulsive calls. The blow sounds from mildly active groups were sometimes forced and loud, and both surface and underwater slaps were heard. Whales in fully active and sexually active groups made almost exclusively high, hybrid, and pulsive calls. Their blow sounds were almost always short, loud, and either tonal or pulsive sounding, quite unlike the less intense blow sounds made by swimming whales. Flipper slapping and underwater slaps were common in both fully active and sexually active groups. In the following discussion, I will interpret these results and attempt to derive communicative functions for the sound types.

Communicative Function

Blow sounds. Some variation in blow sounds may be directly associated with different levels of activity. Whales that were resting or swimming tended to be exerting themselves less than whales in active groups and therefore their breathing might be expected to be shallower and less pronounced.

There were several observations which seem to require some other explanation. The mother who growled through her blowhole whenever her calf left her was always resting and not exerting herself. She did not growl when the calf was by her side, which suggests that the change in the blow sound was associated with the proximity of her calf and not with physical exertion. Whales that were resting or swimming would make loud, harsh blow sounds if they were joined by porpoises or sea lions. This was also true for single, resting whales that were joined by a whale who swam in silently. In the cases when the porpoises or sea lions did not leave, the whale became mildly active. When the intruder was another whale, it left after the disturbed animal made several loud, tonal, or growl blow sounds. Females in fully active or sexually active groups would make these same distinctive blow sounds even though they themselves were not particularly active.

This evidence suggests that blow sounds which are loud, tonal, or pulsive serve some communicative purpose. In all of the contexts in which these distinctive blow sounds were heard, the animal producing them was disturbed. This disturbance could be a calf leaving its mother, the sudden spooking of a whale by an unannounced animal, or the aggressive advances of several amorous males.

I therefore propose that some blow sounds have a function that is communicative. In general, these sounds are threats which serve to warn other animals that the whale is disturbed. Possibly the level of disturbance is encoded by grading in the intensity and harshness of the sound. If these blow sounds are threats, this would possibly suggest that the long moan blow sounds from resting whales serve to indicate that the animal does not want to be disturbed. Another explanation could be that these moans are simply the snores of a sleeping whale (Payne, in press).

Slap Sounds. In general, the number of flipper slaps made by single, mildly active whales was greater than the number made by whales in fully active or sexually active groups. This would suggest that slapping is indicative of an aroused animal but that the signal is not specific. In the context of a single individual that is rolling about and slapping repeatedly, the sound generated by the behavior may have little or no communicative value. The whale is excited as evidenced by its behavior but the sound is inconsequential. In fully active or sexually active groups, the slapping behavior is somewhat different. Whales in these groups usually slapped only once or twice and sometimes the slapping whale would actually strike another member of the group. The observation that no whale was ever seen lobtailing or breaching in the midst of an active group would suggest that whales are aware of the damage they can cause each other by a direct hit from their tail or body.

The sound produced by flipper slapping in the context of an active group of several whales probably functions as a threat with the possible consequence being a physical strike. How the loud, sharp underwater slaps were produced is unknown. These slaps were painful to an unsuspecting listener and might also have been painful to the other whales, porpoises, or sea lions that were closer. I believe that these sounds are threats and because of their intensity are stronger threats than flipper slaps. The function of belly slapping by pairs of whales is not clear.

Up call. The up call was the most common sound made by the whales. Calves, subadults, adults, males, and females all made this call. Up calls were usually associated with swimming whales or in a few cases with single resting or single mildly active animals that called just before they started swimming. In about half the cases when a swimmer made an up call, another whale called back, the two swam to each other as they called and then stopped calling once they were together. This evidence strongly suggests that up calls are "contact" calls. They function as long distance signals which aid in bringing the whales together.

Down call. The down call was not common. It was made by lone males and lone females when they were swimming or mildly active and when they were in situations similar to those in which up calls were heard. In fact, if a swimmer did make down calls it also made up calls. Some single, mildly active whales made only down calls and these animals were usually young females that appeared agitated. They would roll and twist about at the surface while thrashing with their heads, flippers, and tail. Swimmers that mixed <u>down</u> calls with <u>up</u> calls were seen joining other swimming whales that <u>up</u> called back (i.e. which alternated their <u>up</u> calls with <u>up</u> calls made by the swimmer) or active groups that did not. But no swimming whales which made <u>down</u> calls joined other whales soon thereafter. For mildly active whales that <u>down</u> called and were eventually joined by another whale, so much time elapsed between the last <u>down</u> call and the appearance of the second whale, it seemed very unlikely that <u>down</u> calling had mediated the meeting.

This evidence suggests that the <u>down</u> call is a type of "contact" signal made by a whale that is moderately excited. It functions to keep whales in acoustic contact but does not serve to bring whales into physical contact.

Constant call. The <u>constant</u> call was quite rare and was made only occasionally by swimming, mildly active or fully active groups of whales. The two times when a whale made exclusively <u>constant</u> calls, the animal was a single, pregnant female. Since, in general, <u>constant</u> calls were made so infrequently and almost always when other call types were prevalent, it is not possible to infer the function of this call type.

High call. A very few isolated <u>high</u> calls were made by swimming whales but, in general, the <u>high</u> call was made by active groups and the number of calls made by the group increased as the level of activity increased. Single mildly active whales would make a few isolated <u>high</u> calls sometimes along with a <u>down</u> call series. Groups of two or three females would make a few <u>high</u> calls and no other calls after joining and becoming mildly active together. Fully active and sexually active whales would make both isolated <u>high</u> calls and <u>high</u> calls arranged in a series with <u>hybrid</u> and <u>pulsive</u> calls. The rate of <u>high</u> calling in these fully active and sexually active groups increased in proportion to the number of whales in the group. Since rates were measured in calls/hour/whale, this would mean that a whale in a group of four would tend to make more <u>high</u> calls than a whale in a group of two or three. As a result of this evidence, I suggest that <u>high</u> calls are indicative of a general level of excitement but their true function remains to be determined.

Hybrid call. The <u>hybrid</u> call was made almost exclusively by active groups and mostly by fully active and sexually active groups containing males and females. <u>Hybrid</u> calls would sometimes be made as isolated calls but usually were produced in a series with <u>high</u> calls and <u>pulsive</u> calls.

Groups of two whales made fewer hybrid calls per whale than larger groups and the hybrid calls from pairs would usually be isolated while the hybrid calls in the larger groups would almost always be included in a series.

Pulsive call.   The pulsive call, like the hybrid call, was made almost exclusively by active groups containing males and females. The rate of pulsive calling decreased as the number of whales in the group increased, suggesting that group size and a whale's pulsive call rate were negatively correlated. In fact, this is probably not the case: if rates are computed in calls/hour for the various group sizes, there are no differences in rates as group size increases. This would suggest that perhaps most of the pulsive sounds are made by only one or two individuals in fully active or sexually active groups and not by all members of the group. This suggestion is supported by the observations that two pulsive calls were never produced simultaneously by an active group.

Pulsive calls would either be produced as isolated calls or in a series, and in most cases the series contained both high calls and hybrid calls. The one time that a mildly active group made a series containing only pulsive calls was when a mother was separated from her calf by a pregnant female. The other times that only pulsive calls were produced occurred when a calf was being jostled and lifted up on the belly of a male and when a male interrupted what appeared to be a courting pair.

There is a general association between hybrid and pulsive calls given in a series, and large, active groups of mixed sexes. Although the data are still too limited to infer a definite communicative function for these two call types, my distinct impression is that hybrid and pulsive calls are aggressive signals directed at other members of the group. A further impression is that the amount of pulsiveness in a call and the number of calls in a series are indicative of the level of aggression within the group.

These results and conclusions concerning the function of sound types in southern right whales are supported by the results of Clark and Clark (1980) and Clark (in press). In sound playback experiments, Clark and Clark (1980) demonstrated that whales responded to conspecific sounds but not to a variety of other sounds. In their first experiments, one of the authors (CWC) imitated up calls and the one whale in the area responded by making several up calls while swimming toward the loudspeaker. In most of the later experiments a tape containing a typical assortment of calls from a fully active group was played back. Whales responded by swimming toward the loudspeaker and making mostly up calls. A few down calls and high calls were also produced, but usually when the whale was circling in the vicinity of the loudspeaker.

My other results (Clark, in press) and those presented here are quite similar but the approaches are different. Here, I assumed that the types of sounds produced are dependent on activity. In separate analyses of the whales' acoustic repertoire, I was able to show that sounds and behavior were associated by including sounds in the sample that were "labelled" by activity (see Clark in press for details). I concluded that the whales' repertoire could be divided into two functional subdivisions. One subdivision contained calls that were discrete: up calls. The second subdivision contained calls that were highly variable: high calls, hybrid calls, and pulsive calls. Discrete calls were associated with resting or swimming whales while the variable calls were associated with fully active or sexually active groups. The calls that were intermediate to the two subdivisions, down calls, were made by mildly active groups, suggesting that these groups were in some way intermediate to resting or swimming groups and fully active or sexually active groups. I was not able to associate the constant calls with any particular type of activity.

The results showing that the up call is a "contact" signal are supported by my conclusion (Clark, in press) that this is a long distance call. It is an intense, low, frequency-modulated upsweep. These physical characteristics are all features which increase the signal's range of detectability and contrast with the ambient noise conditions (Watkins and Schevill 1979b). The frequency upsweep might also degrade predictability with range and permit a receiving whale to estimate its distance from the caller (Tolstoy and Clay 1966; Wiley and Richards 1978; Richards 1981).

Many of the sounds produced by other mysticete whales have similar acoustic properties. Most of the *Balaenoptera* species (blue whale, fin whale, etc.) have been reported to make low frequency-modulated sounds (Schevill et al. 1964; Cummings and Thompson 1971; Thompson et al. 1979) that probably function as long distance calls (Payne and Webb 1971). Northern right whales, *Eubalaena glacialis*, and bowhead whales, *Balaena mysticetus*, make calls that are essentially identical to the up call of the southern right whale (Schevill and Watkins 1962; Clark and Johnson in prep.; D. Spero pers. comm.), which suggests that these calls might serve as contact calls in these species as well.

The suggestion that high calls indicate excitement and hybrid calls and pulsive calls function as aggressive signals are in agreement with Morton's (1977) generalized conclusion that high frequency sounds are indicative of less hostile situations while pulsive sounds are typically made in agonistic contexts.

Summary

The sounds made by southern right whales are not random but are intimately related to the social context and activity of the animals. A resting whale does not call very often but sometimes makes long moans while exhaling through its nostrils. A swimming whale that is alone and seeking out other whales makes up calls. A whale that is alone and moderately excited makes up calls, down calls, flipper slaps, and loud blow sounds. A female that is excited and with other females makes high calls. A whale that is excited and involved in a group of sexually active animals makes high calls, hybrid calls, pulsive calls, flipper slaps, and loud forceful blow sounds. It remains to be determined whether some variable in the contact call encodes for the identity of the caller and whether the more complex associations between variables in the sounds from active whales encode for some subtle parameters of the social context.

Acknowledgements

I am endebted to G. Blaylock, J.M. Clark, J. Crawford III, and A. Mcfarland for their invaluable assistance with the collection of field data, A. Bindman, G. Blaylock, C. Breen, J.M. Clark, S.J. Clark, R. DiOrio, and D. Munafo for their dedicated help with the data reduction. J. Perkins, V. Rowntree, and E.M. Dorsey assisted in the process of identifying whales from photographs. Helpful criticism of the manuscript was provided by R. Payne, R.J. Rohlf, W.E. Schevill, D.G. Smith, P. Tyack, C. Walcott, and G. Williams. This work was supported in part by a grant from the National Geographic Society, U.S. Public Health Service Biomedical Research grant to the State University of New York at Stony Brook, and facilities and equipment from the New York Zoological Society.

Literature Cited

Altmann, J.
   1974. Observational study of behavior: sampling methods. Behavior 44: 227-267.

Best, P.
   1970. Exploitation and recovery of right whales *Eubalaena australis* off the Cape Province. Investl. Dir. Sea Fish S. Afr. 80: 1-20.

Busnel, R.G.
   1968. Acoustic communication. *In* R.G. Busnel (editor), Animal communication: Techniques of study and results of research, p. 127-153. Indiana Univ. Press, Bloomington.

Clark, C.W.
  1980. A real-time direction finding device for determining
  the bearing to the underwater sounds of southern right
  whales, *Eubalaena australis*. J. Acoust. Soc. Am. 68: 508-
  511.

Clark, C.W. and J.M. Clark.
  1980. Sound playback experiments with southern right
  whales (*Eubalaena australis*). Science (Wash. D.C.) 207:
  663-665.

Clark, C.W.
  In press. The acoustic repertoire of the southern right
  whale: a quantitative analysis. Anim. Behav.

Clark, C.W. and J. Johnson.
  In prep. Bowhead whale, *Balaena mysticetus*, sounds during
  the spring migrations of 1979 and 1980.

Cummings, W.C., J.F. Fish, and P.O. Thompson.
  1971. Bio-acoustics of marine mammals of Argentina: R/V
  Hero cruise 71-3. Antarct. J. U.S. 6: 266-268.

  1972. Sound production and other behavior of southern right
  whales, *Eubalaena glacialis*. San Diego Nat. Hist. Trans.
  17: 1-14.

Cummings, W.C. and P.O. Thompson.
  1971. Underwater sounds from the blue whale, *Balaenoptera
  musculus*. J. Acoust. Soc. Am. 50: 1193-1198.

Cummings, W.C., P.O. Thompson, and J.F. Fish.
  1974. Behavior of southern right whales: R/V Hero cruise
  72-3. Antarct. J. U.S. 9: 33-38.

Donnelly, B.G.
  1967. Observations on the mating behavior of the southern
  right whale, *Eubalaena australis*. S. Afr. J. Sci. 63: 176-
  181.

Emlen, S.T.
  1972. An experimental analysis of the parameters of bird
  song eliciting species recognition. Behavior 41: 130-171.

Green, S.
  1975. Variation of vocal pattern with social situation in
  the Japanese monkey (*Macaca fuscata*): a field study.
  Prim. Beh. 4: 1-102.

Katona, S., B. Baxter, O. Brazier, S. Kraus, J. Perkins, and
  H. Whitehead.
  1979. Identification of humpback whales by fluke

photographs. *In* H.E. Winn and B.L. Olla (editors), Behavior of marine animals - current perspectives in research, vol. 3: Cetaceans, p. 33-44. Plenum Press, N.Y.

LeBeouf, B.J. and R.S. Peterson.
1969. Dialects in elephant seals. Science (Wash. D.C.) 166: 1654-1656.

Marler, P.
1967. Animal communication signals. Science (Wash. D.C.) 157: 769-774.

1970. Vocalizations of East African Monkeys I: Red Colobus. Folia Primatol. 13: 81-91.

Matthews, L.H.
1938. Notes on the southern right whale *Eubalaena australis*. Discovery Rep. 42: 169-182.

Morton, E.S.
1977. On the occurrence and significance of motivational-structural rules in some bird and mammal sounds. Am. Nat. 111: 855-869.

Owings, D.H. and R.A. Virginia.
1978. Alarm calls of California ground squirrels (*Spermophilus beecheyi*). Z. Tierpsychol. 46: 58-70.

Payne, R.
In press. Behavior of southern right whales (*Eubalaena australis*). University of Chicago Press, Chicago, Ill.

Payne, R., O. Brazier, E. Dorsey, J. Perkins, V. Rowntree, and A. Titus.
1983. External features in southern right whales (*Eubalaena australis*) and their use in identifying individuals. *In* R. Payne (editor), Communication and behavior of whales. AAAS Selected Symposia Series, p. 371-445. Westview Press, Boulder, Colo.

Payne, R. and E.M. Dorsey.
1983. Sexual dimorphism and aggressive use of callosities in right whales (*Eubalaena australis*). *In* R. Payne (editor), Communication and behavior of whales. AAAS Selected Symposia Series, p. 295-329. Westview Press, Boulder, Colo.

Payne, R. and S. McVay.
1971. Songs of humpback whales. Science (Wash. D.C.) 173: 585-597.

Payne, R. and K. Payne.
1971. Underwater sounds of southern right whales. Zoologica (N.Y.) 58: 159–165.

Payne, R. and D. Webb.
1971. Orientation by means of long range acoustic signalling in baleen whales. Ann. N.Y. Acad. Sci. 188: 110–141.

Richards, D.G.
1981. Estimation of distance of singing conspecifics by the Carolina Wren. Auk 98: 127–133.

Rohlf, F.J., J. Kishpaugh, and D. Kirk.
1980. NTSYS. Numerical taxonomy system of multivariate statistical programs. State Univ. N.Y. at Stony Brook.

Saayman, G.S. and C.K. Tayler.
1973. Some behavior patterns of the southern right whale *Eubalaena australis*. Z. Saeugetierk. D. 38: 172–183.

Schevill, W.E. and W.A. Watkins.
1962. Whale and porpoise voices, a phonograph record. Woods Hole Oceanogr. Inst., Woods Hole, Mass., 24p.

Schevill, W.E., W.A. Watkins, and R.H. Backus.
1964. The 20-cycle signals and *Balaenoptera* (fin whales). *In* W.N. Tavolga (editor), Marine bio-acoustics, p. 147–152. Pergamon Press, N.Y.

Sebeok, T.A.
1977. How animals communicate. Indiana Univ. Press, Bloomington, 1128p.

Smith, W.J.
1965. Message, meaning and context in ethology. Am. Nat. 49: 405–409.

1977. The behavior of communicating. Harvard Univ. Press, Cambridge, Mass., 545p.

Sokal, R.R. and R.J. Rohlf.
1969. Biometry. W.H. Freeman, San Francisco, 776p.

Struhsaker, T.T.
1967. Auditory communication among vervet monkeys (*Cercopithicus aethiops*). *In* S.A. Altmann (editor), Social communication among primates, p. 281–324. Univ. Chicago Press, Chicago, Ill.

Thompson, T.J., H.E. Winn, and P.J. Perkins.
1979. Mysticete sounds. *In* H.E. Winn and B.L. Olla

(editors), Behavior of marine animals - current perspectives in research, vol. 3: Cetaceans, p. 403-431. Plenum Press, N.Y.

Thorpe, W.H.
1961. Bird song; the biology of vocal communication and expression in birds. Camb. Monogr. Exp. Biol. 12: 142p.

Tolstoy, I. and C.S. Clay.
1966. Ocean acoustics. McGraw Hill, N.Y., 293p.

Tyack, P.
1981. Interactions between singing humpback whales and conspecifics nearby. Behav. Ecol. Sociobiol. 8: 105-116.

Walker, R.A.
1963. Some intense, low-frequency, underwater sounds of wide geographic distribution, apparently of biological origin. J. Acoust. Soc. Am. 35: 1816-1824.

Watkins, W.A.
1977. Acoustic behavior of sperm whales. Oceanus 20(2): 50-58.

Watkins, W.A. and W.E. Schevill.
1972. Sound source location by arrival-time on a non-rigid three-dimensional hydrophone array. Deep-Sea Res. 19: 691-706.

1979a. Aerial observations of feeding behavior in four baleen whales: Eubalaena glacialis, Balaenoptera borealis, Megaptera novaeangliae and Balaenoptera physalus. J. Mammal. 60: 155-163.

1979b. Distinctive characteristics of underwater calls of the harp seal, Phoca groenlandica, during the breeding season. J. Acoust. Soc. Am. 66: 983-988.

Wiley, R.H. and D.G. Richards.
1978. Physical constraints on acoustic communication in the atmosphere: Implications for the evolution of animal vocalizations. Behav. Ecol. Sociobiol. 3: 69-94.

Würsig, B. and M. Würsig.
1979a. Behavior and ecology of the bottlenose dolphin, Tursiops truncatus, in the South Atlantic. Fish. Bull., U.S. 77: 399-412.

1979b. Behavior and ecology of the dusky dolphin, Lagenorhynchus obscurus, in the South Atlantic. Fish. Bull., U.S. 77: 871-890.

## Part 2

Behavior on Breeding Grounds

James D. Darling, Kimberly M. Gibson,
Gregory K. Silber

6. Observations on the Abundance and Behavior of Humpback Whales (*Megaptera novaeangliae*) off West Maui, Hawaii, 1977-79

Abstract

Observations were made of humpback whales in their winter assembly area off West Maui, Hawaii, from December to May 1977-79. Whales were individually identified by photographs of the black and white skin patterns on the underside of their flukes. A total of 264 individuals were identified giving a minimum population estimate. A continued high rate of discovery of previously unidentified whales at the end of this period (on the average 7 out of every 10 whales were "new") indicates that there are still many more whales to identify. A mark-recapture test, performed on the 1979 repeat sightings of whales identified in 1978, gives a population estimate of 895 (95% confidence limits of 592, 1837). Repeat sightings of the same individuals show that at least a percentage of the population returns to Hawaii each year and that individuals may stay in the area for as long as eleven weeks, February to April. There are large fluctuations in whale numbers seen near West Maui, even in mid-season, suggesting that whales are moving in and out of the study area. During the winter in Hawaii the whales can be divided into a number of behavioral sets. They are: a cow with a calf, an adult (or "escort") accompanying a cow and calf, a singer (usually a lone adult), a single non-singing adult, a pair, a trio, a larger group containing only adults, or a larger group containing adults with a cow and calf. In our sample, the only sets between which whales did not switch were singers or escorts and cows with a calf. This may indicate that whales which were identified as singing or as escorts are male. We observed a rivalry, including aggressive interactions, between whales competing for the role of escort. "Escorts" may be males

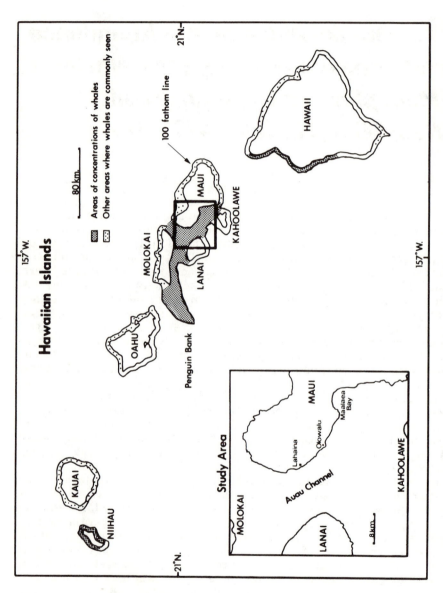

Figure 1. The general winter distribution of humpback whales in Hawaii (after Rice and Wolman 1978), and the study area.

competing for that position in order to eventually mate with the cow.

Introduction

Each winter humpback whales assemble in waters surrounding the main Hawaiian Islands (Rice 1978). They arrive as early as November and peak in numbers between February and April. A few may stay through late May and early June (Herman and Antinoja 1977). These whales have only been studied in a systematic way since 1975-76. Basic population parameters are not known, although some estimates exist, and behaviors have not been described in any depth.

Surveys to determine distribution and/or numbers were conducted by Shallenberger (1976), Wolman and Jurasz (1976), Herman and Antinoja (1977), and Rice and Wolman (1978). The distribution of the whales is summarized in Figure 1. All of the authors found a concentration of whales in the Four Island Group (Maui, Molokai, Lanai, and Kahoolawe) - Penguin Bank area; Rice and Wolman (1978) also define concentrations off northwest Hawaii and in the Kauai - Niihau area. Population estimates derived from these surveys are 200-250 (Herman and Antinoja 1977), 373 (Wolman and Jurasz 1976), and 500 + 90 (Rice and Wolman 1978).

Some attempts were made to define specific nursery areas within the Four Island Group. Herman and Antinoja (1977), although concluding there is no unique nursery area, state that the north coast of Lanai is clearly the most densely concentrated nursery region. Hudnall (1978) suggested that Maalaea Bay, Maui is a major nursery area. However, even though their study area did not include Maalaea Bay, Glockner and Venus (1983) found more cows and calves along the west Maui coast than Herman and Antinoja (1977) reported in the same year for the entire Four Island region. In fact, Herman and Antinoja found West Maui to be the area least populated by cows and calves. Thus the delineation of specific areas as nursery areas may not be warranted.

As cows with very young calves are a common sight through the winter, calving and nursing are apparently important behaviors in the area. Other behaviors involving adults strongly suggest that courtship and mating occur. These include singing, pairing, and consorting in groups. Chittleborough (1965) found that winter assemblages in the Southern Hemisphere correspond with a peak of spermatogenic and oestrus activity. Studies now underway by Tyack (1981) should elucidate this behavior.

From December to May 1977-79 we worked with Dr. Roger Payne on his research on humpback songs conducted off West Maui. Our study area is shown in Figure 1. To differentiate animals in behavior studies we identified individuals by photographs of the black and white skin patterns on the underside of their flukes. This technique is described by Katona, Baxter, Brazier, Kraus, Perkins, and Whitehead (1979).

Individual identification allowed estimates of abundance as well as behavioral analyses. We identified 264 whales in 1977-79 and our rate of discovery of new whales is still high. This, combined with a mark-recapture test between years, suggests that the highest previous estimate, 500 + 90 by Rice and Wolman (1978), may be low for the numbers of whales that are utilizing the area.

Repeated observations of individuals indicate that some whales may be present in Hawaii for most of the winter and that whales may circulate throughout the islands. Comparison of the behaviors of individuals which were seen repeatedly suggests that whales which are found singing or escorting a cow and calf are males. Observations were made on two groups consisting of a cow, calf, escort and other adults which indicates a rivalry for the escort role.

Methods and Results

Photographs of the undersides of tails were taken from small craft, no further than 100 m from the whale, with a 200 or 300 mm lens. As Katona et al. (1979) describe for the western North Atlantic, and as shown in Figure 2, the flukes range from mostly white to all black and most are patterned with both. Figure 3 illustrates that the markings are constant over at least 3 years in adults. In example 'b' even the thinnest lines and dots seen on the left fluke remained unchanged over the three years.

As illustrated in Figure 4, the identification effort increased from year to year. Over 6,000 photographs of whale tails were made on a total of 163 days. Sampling was best in mid-season. Samples were made before January or after April on only four of the 163 days.

The photographs were sorted and graded as to the sureness that a photograph could identify an individual. This resulted in 317 photographs of C+ grade or better, 43 from 1977, 74 from 1978, and 200 from 1979.[1] Figure 5 shows

---

[1]Photographs were given a C+ grade or better if we were sure that an individual could be identified from the picture. Higher grades were given to better quality photographs.

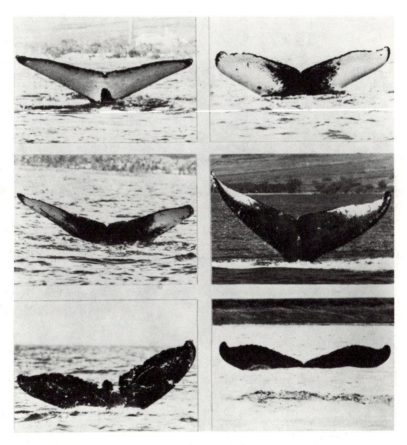

Figure 2. Examples of the black and white skin patterns on the underside of humpback flukes.

Figure 3. Photographs of the same whales taken in 1977 and
        again in 1979. In example 'b' notice that the
        thinnest lines and small dots did not change
        (example 'a' photographs by Sylvia Earle and
        Deborah Glockner.

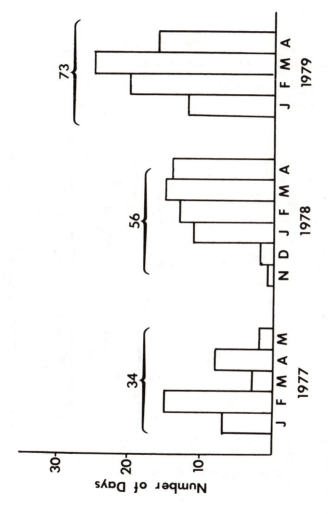

Figure 4. The number of days in each month (for all three study years) on which whales were photographed.

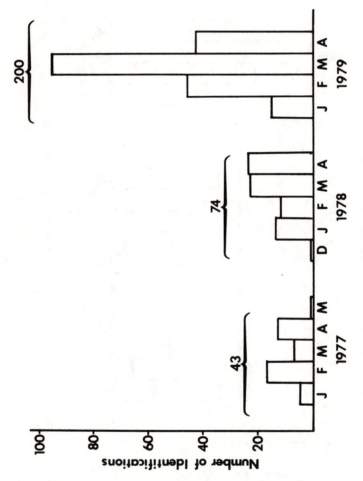

Figure 5. When photographs used in analysis were taken.

when the photographs above C+ grade were taken. These, along with one photograph taken in 1975, were used in our analyses.

The repeats within a year were subtracted from the 318 photographs. There were 3 in 1977, 2 in 1978, and 27 in 1979. Thus 40 individuals were identified in 1977, 72 in 1978, and 172 in 1979. There were also repeat observations between years: 1 between 1977 and 1978, 8 between 1977 and 1979, and 13 between 1978 and 1979. These too were subtracted, resulting in a 3-year total of 264 individuals. This is the absolute minimum number of whales present in the three seasons.

The cumulative rate of discovery of new whales is graphed in Figure 6. Over the three seasons, plus one useful photo taken in 1975, 318 photographs identified 264 individuals. Each time the graph moves horizontally it means the whale was previously identified and when all the whales are identified it should level off. The rate of discovery remained high throughout the study, however more repeats do occur near the top of the graph, indicating that we are at least making progress. For the first 155 photographs analyzed, 9 out of every 10 whales were new. For the last 163, 7 out of every 10 had not previously been seen by us in the area.

To approximate where the graph may level off we used a mark-recapture test between the 1978-79 seasons. Essentially we were 'marking' whales by photograph and 'recapturing' them each time we got a matching photograph. Unfortunately, we cannot be positive that we met the random sampling assumption necessary for the test to be accurate. That is, we do not know whether the chances of 'recapturing' in 1979 a whale 'marked' in 1978 are random. In right whales (*Eubalaena australis*) (Payne, in press), some sex and age classes do not return every year to the same area. However, under the theory that it seems better to make a guess with this test than without, we used it. We employed a Peterson test, modified by Baily to reduce the positive bias if the assumptions are not correct (Overton and Davis 1969; Tanner 1978). Of 72 whales 'marked' in the 1978 season, 13 were 'recaptured' in the 1979 season out of a total of 173 identified that year. The estimate of population size is 895 with 95% confidence limits of 592 and 1837. (We do not attach much importance to the precision of this estimate.)

A possible random sampling problem arises from one whale which spent one winter in Hawaii and a subsequent winter in the Mexican assembly area (Darling and Jurasz 1983). If this is a common occurrence, the chances for matches from year to year may be less than random, making the estimate high for the Hawaiian population. The work of Dawbin (1966) and Chittleborough (1965) in the southwest Pacific-Australia area with Discovery tags suggests that,

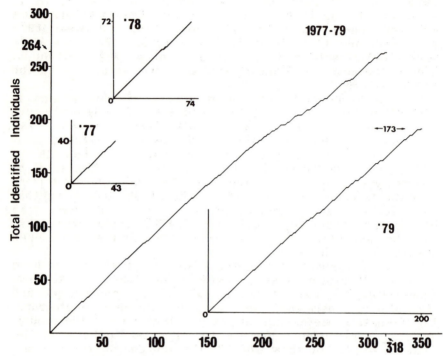

Number of photographs in which a whale can be
identified, C+ grade or better, arrayed in the
order in which they were taken.

Figure 6. The rate of discovery of previously unidentified
whales, for each year, and cumulatively.

although switching of wintering areas does occur, it is not common.

This should be one case in which the accuracy of a population estimate based on the mark-recapture test will be checked, inasmuch as we expect to identify most of the whales in the population over the next few years.

The repeat sightings within each year, with the number of days between the first and last sightings, are plotted in Figure 7. These indicate the minimum time each whale spent in Hawaii. Although 26 out of the 28 repeats were seen on dates less than 6 weeks apart, there were 2 that were identified over spans of 8 and 11 weeks. The longest interval between sightings, 74 days, was from 11 February to 25 April, which indicates that at least some whales may stay in Hawaiian waters for most of the season.

There are large fluctuations in numbers of whales off West Maui, even in mid-season. For a week or more the area may be heavily populated, then the next day it is difficult to find one animal. It is apparent they are moving in and out of that specific area. In 1977 and 1978 our rate of discovery of new whales was so high (see Figure 6) we wondered if the whales were streaming through the area, with assemblages staying for a few weeks then moving on, creating a lull in activity until the next bunch arrived. The data just presented indicate that this is probably not so simple, at least a few whales were resighted several times over the season in the same specific area. Perhaps there is a social cohesiveness amongst whales such that their distribution is more likely to be clumped than spread evenly around the islands. A loose congregation of whales, composed of individuals involved in different behaviors, may move in and out of a specific area. The lulls in any one area could be the result of the majority of whales being together elsewhere.

For the 22 between-year repeats we compared the dates they were seen in each year to see if there was any correlation from year to year between dates on which they were present (Table 1). There seems to be no such correlation in mid-season (15 January - late April) when we took most of our photos. For example, one whale (#50) was photographed on 22 April 1977 and on 27 January 1979. Another (#62) was present on 15 January 1978 and on 24 March 1979. Thus, if different individuals have differing preferences for what point in the season they visit Hawaii, our data have not yet demonstrated it.

During the winter in Hawaii the whales can be divided into a number of social groupings or sets. They are: a cow with a calf, an adult (the "escort") escorting a cow and calf, a singer (usually a lone adult), a single (non-singing) adult, a pair, a trio, a larger group containing only adults, or

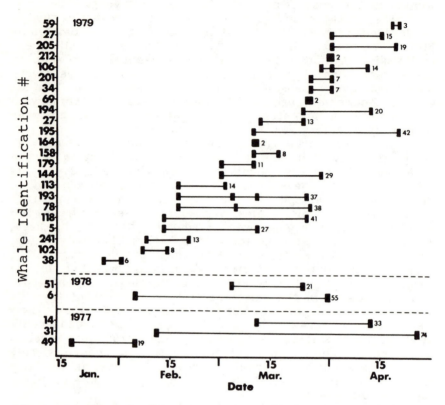

Figure 7. Repeat sightings of individuals within each year. The numbers beside the last sighting are the days between the first and last sighting.

Table 1. Whales resighted in Hawaiian waters, 1977-79.

Whale #	Date Identified		
	1977	1978	1979
8	Feb. 13		Mar. 12-24
50	Apr. 22		Jan. 27
31	Feb. 11		Feb. 27
40	Jan.		Mar. 11
18	Mar. 11		Apr. 18
28	Mar. 11	Mar. 4-24	Feb. 17-Mar. 2
32	Jan.		Mar. 2
10	Mar. 5		Mar. 11
5	Mar. 8		Apr. 5
6	Feb. 5-Mar. 31		Mar. 22
7		Mar. 8	Mar. 24
13		Feb. 7	Feb. 13
73		Jan. 22	Mar. 6
14		Mar. 3	Mar. 24
36		Mar. 22	Mar. 29, Apr. 1-11
56		Mar. 22	Apr. 3
59		Apr. 8	Mar. 22
65		Apr. 9	Mar. 24, Apr. 12
62		Jan. 15	Mar. 24
64		Apr. 8	Mar. 5

a larger group containing adults with a cow with calf.[2] We
took 60 repeat sightings and noted in which set the
individual was on each sighting. We then noted in which
pairs of sets any individual could be, as shown in Figure 8.
Some of the sets are combined. A cow+ indicates a whale
which had a calf on one of the sightings but not necessarily
on another; that is, was a known—reproducing female. The
only sets between which individuals did not switch were
known-reproducing females and singers or escorts. That is,
singers can be escorts and vice versa, but in our sample of
repeat sightings, neither was resighted as an obvious mother
with calf, and cows were not resighted singing or as escorts.
The reason for this may be that only males sing or act as
escorts. However, our sample of known females which have
been seen repeatedly is small (6), not allowing this
conclusion. In our sample, known-reproducing females
without a calf were always encountered in the company of
one or more other adults. It may indicate that reproducing
females are rarely left alone by other adults during this
season.

Larger groups of whales can be made up of whales
which were seen in smaller groups, pairs or trios, or which
were singers, escorts, or cows.

The comparison of the associations between whales in
the 60 repeated sightings shows no evidence that humpback
whales form permanent groups. Whales which were alone or
in smaller groups were also seen in larger groups.

Observations of Escort Behavior

On 11 and 18 April 1979 we were able to make close
observations of the interaction between whales in groups
containing a cow, calf, escort, and other adults.
Underwater observations were made by an observer (G.S.)
wearing a snorkle and hanging head down over the side of a
boat.

On 11 April the group contained a cow, calf, escort,
and two other adults. The cow and calf swam in front, with
the escort ranging to the sides or just behind them. One of
the other adults followed more or less directly behind the
cow, calf, and escort; the other was behind and over to one
side. The procession was moving very slowly northwest, and
was often close enough to the boat to be visible to Silber.
All were individually identified from photographs. The
escort adult spent most of its time apparently keeping the
other adults away from the cow and calf. Its behavior
included blowing bubble streams on the approach of other

---

[2] Whales which were staying close together, within two
whale lengths, were considered a group.

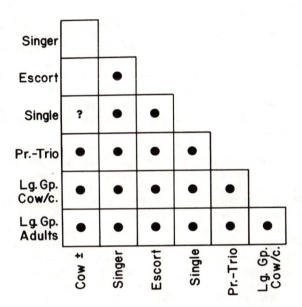

Figure 8. Humpback behavioral sets are matched with each other to show which ones individuals move between.

whales, directing tail lashes at them (a sideways swipe of the tail towards the other whale), and blocking any move they made to get near the cow and calf. At one point, as the escort dove after blowing, it came face to face with one of the other adults. The escort blew a burst of bubbles from its blowholes, stopped short, then dove under the other whale and returned to the cow and calf. Shortly after, the escort dove away from the cow and calf, swam straight to the closest following whale, dove underneath it, rolled on its side and directed a tail lash at it. Then the escort quickly returned to the cow and calf. Similar activities continued for approximately two hours with no change in the status quo. The cow and calf were swimming slowly in front of the other whales the whole time and did not seem hurried by the activities. The escort had red (we presume bloody or raw places) on the trailing edge of its dorsal fin and on several of the knobs on its head.

On 18 April we watched a different group which contained a cow, calf, escort, and, initially, one other adult. All the whales were identified photographically. The cow and calf swam in front of the escort and other adult, which were 15-20 m behind. The escort, easily recognizable by a large white scar in front of its dorsal hump, was between the cow and calf and the other adult. The pace was much faster, and the behaviors more aggressive than on 11 April, making underwater observation difficult. However, what was going on was clear enough this time from the surface. The escort repeatedly lashed the other adult with its tail while they were both at the surface. There were several high-speed chases with the dorsal fins inches below the surface, the other adult apparently attempting to outflank the escort. The group moved southwest, with the cow and calf swimming steadily in front of the action. Eventually another adult joined in, and this was followed by some very aggressive charging around and tail lashing in which we temporarily lost track of individuals. Then suddenly all calmed down and the group disbanded. One whale of the three that had been accompanying the cow and calf blew about 50 m to the south, another breached and moved off to the east, leaving the cow, calf, and an escort travelling slowly northwest up the coast of Lanai. The escort was swimming beside the cow and calf. On identification we found the whale filling the escort role had changed. The initial escort with the white scar had been displaced by the adult which was first challenging it. The knobs on the head of the new escort were rubbed raw. The escort had changed with no apparent change in the behavior of the cow and calf. This entire observation lasted approximately one hour.

There is apparently an advantage to being an escort and consequently there is rivalry for that position, at least in some cases.

In 1979 we noted 14 examples of cow-calf interaction with other adults, other than a single escort. Two of these, on 24 and 26 February, were apparently short-lived and details were not clear. In the first, on 24 February, a singer we were recording stopped singing, moved towards and joined a nearby cow and calf. This was immediately followed by thrashing and splashing for a few minutes. Then the group apparently broke up and the cow and calf moved quickly out of the area. In the second, on 26 February, two adults appeared to be chasing a cow and calf which were moving quickly in front of them. The pursuing adults were side by side, lunging and splashing, and at times in contact with each other. It is unknown if one of them had been an escort prior to the chase, and we were unable to follow in order to determine the result.

In the other 12 examples we saw in 1979 (which include the 11 and 18 April observations described above), the group we were watching contained a cow and calf and up to 9 other adults. The groups swam at varying speeds from slow to quite fast. In some cases the cow and calf were in the middle of the group. The interactions described for 11 and 18 April were not noticed, however most observations were not as extended, or as clear, as on those days. On 4 March, after watching one such group including 7-9 adults plus the cow and calf, we noted that "they were all acting like escorts in a sense." On a 1 March sighting of this type of group it was noted that the apparent escort had a bloody dorsal hump.

We have noticed that singers and escorts were often more marked and scarred than cows. The physical interaction in these observations, with bloody dorsal humps and head knobs, may be part of the reason.

Discussion

Although our estimate of abundance is preliminary, it indicates that there may be more whales using the Hawaiian waters than previously estimated. At this time it seems safe to consider Rice and Wolman's (1978) 500 $\pm$ 90 as minimum, keeping in mind there may be double that number. If this is true, it may affect current estimates of the North Pacific humpback population, now at 800-1,000 (Rice 1978). Estimates will become more precise as more whales are identified.

It is not known if all the whales that are in Hawaii one winter return the next, a crucial point for understanding the significance of Hawaiian population estimates. This also raises the question of a definition of the Hawaiian population. There may be in fact no such thing as "the Hawaiian population." The smallest meaningful unit may be "the North Pacific population."

It seems that counting individually identified whales is a more reliable way of estimating population size than vessel and aerial surveys. These have led to misleading population estimates. We have already identified more whales (and still have a continued high rate of discovery of new whales) than have been estimated for the entire Hawaiian population by Herman and Antinoja's (1977) aerial-vessel survey. Also, subsequent vessel surveys by Wolman and Jurasz (1976) and Rice and Wolman (1977) came up with significantly higher counts. (The accuracy of all of these will be determined with further individual identification.) Herman and Antinoja (1977), basing estimates of birth rate on their survey results, concluded that the Hawaiian population had a low birth rate and went on to wonder why, suggesting such reasons as human harassment. We feel that population estimates and recruitment rates have not yet been determined accurately enough and that speculation on the effects, if any, of human harassment are unwarranted at this time.

Chittleborough (1965) and Dawbin (1966) describe a partial age and sex segregation during humpback migrations in the southwest Pacific. One wonders if there might be, a similar parade of different age and sex classes through Hawaii in the winter. We do not yet have enough data to investigate this question. There is no indication from our data that this is true throughout the main part of the season. The data indicate that individuals for which we have repeat sightings from one year to the next can be seen at any time between January and late April. This, combined with the knowledge that some whales stay for at least 11 weeks, suggests that in mid-season any temporal segregation in the migration has been eliminated.

Herman and Antinoja (1977) found that 76% of all the adult groups they saw were "swimming on determined courses suggesting considerable local migration but without any obvious pattern." If there is some degree of cohesion amongst whales in a general area, these movements would tend to lead whales in and out of a specific area. If the travelling groups disbanded once out of the area in which they began, the result would be a shift in locality of most of the whales. Observers in one area would notice an increase in activity while those in another would record a lull in activity.

Glockner (1983) determined that 14 escorts were males. Winn, Bischoff, and Taruski (1973) showed that a humpback singing in the western Atlantic was male. Using Glockner's (1983) sexing technique, we determined a singer to be male. These, combined with our observations that an individual can be both a singer and an escort (in fact Glockner and Venus 1983, report two escorts singing), and that we never found a cow to be a singer or escort or vice versa, strongly suggest that whales found singing or as escorts are males. If this is true, the large groups of whales seen during the winter in Hawaii can be composed of males and reproducing females. This supports the assumption made by Herman and Antinoja (1977) that they are mating groups.

Herman and Antinoja (1977) cite Chittleborough's (1953) report of an escort beating off killer whales (*Orcinus orca*) with its flukes near Australia, and suggest that escorts serve a protective function for the cow and calf. Glockner and Venus (1983) describe an escort apparently guarding the calf until it could station itself with its mother, further suggesting a protective role. Our observations of a rivalry for the escort position, along with the likelihood that escorts are males, raises the possibility that protecting the cow or calf is not their sole purpose in being there. The fact that nothing happened to the cow or calf after an escort was replaced by another adult which it had initially held off by behaviors such as blocking, tail lashing, and bubbling, suggests that an escort . may expend as much energy protecting its own position as in protecting the cow and calf. The "protective behavior" which is at times exhibited by an escort on approach of a boat or diver, may be the result of the escort reacting to the intrusion as a threat to its position. This might be similar to a harem bull elephant seal's propriety over his females. It is his position, not the physical well-being of the females, that is more likely to be damaged by an intruder. The result may be the protection of the cow and calf.

It seems likely that the escort is primarily interested in mating with the cow. Cases in which we saw a number of adults accompanying a cow and calf may have represented groups of males following her. Also our observation of a single whale (probably a male) which stopped singing and momentarily joined a cow with calf, causing much thrashing around and finally fleeing of the cow and calf, suggests an attempt at mating.

In 1979 we noticed no cow and calf interactions with other adults, other than a single escort, until late February. Large groups including a cow and calf were not seen until March. The aggressive behavior between adults in the cow/calf group was not obvious until April. It may be that the cow doesn't come into oestrus until later in the season, thus causing an increase in competition for her. Chittleborough (1965) found that oestrus can follow parturition while the female is suckling a calf, but notes that postpartum ovulation is probably not a regular occurrence in this species. Our observation that males (escorts) competing for access to a female, often get minor injuries in the process, strongly suggests that the female is in oestrus. Perhaps this indicates that at least some females mate while suckling, and have calves in consecutive years.

Payne and Dorsey (1983) determined that marks and scars were more common on male southern right whales than females, and were probably caused by physical interaction between males, particularly in mating groups. Our observation that singing or escorting humpback whales (probably all males) are often more marked and scarred than cows fits well with their findings.

Acknowledgements

For contributing identification photographs we thank Scott Anderson (National Marine Fisheries Service), Bemi Debus, Ellie Dorsey, Sylvia Earle, Consuelo Gaza, Deborah Glockner, Ted Goodspeed, Chuck Flaherty, Jan Heyman-Levine, Maureen Hoskyn, Fred Levine, Jim Moore, Roy Nickerson, Katharine Payne, Ronda Luther-Raithaus, Peter Tyack, and Leigh Wilkes.

We gratefully acknowledge the use of Dan McSweeney's and Rick Chandler's photographs from the island of Hawaii.

Special thanks are due Maureen Hoskyn, Nancy Cison, and Victoria Rowntree for darkroom and other photographic work.

We are indebted to numerous Maui residents who helped us throughout this study. Jim Luckey, manager, and Shirley M. Herpick, secretary, of the Lahaina Restoration Foundation gave never-ending support, without which this work would not be at this stage. Drake and Maureen Thomas, Ron Roos, and Diane Fells repeatedly solved accommodation problems for us. Whale-watching boat operators Chuck Clark of Vida Mia Viajero Cruises and Rusty Nall of Windjammer Cruises allowed us many free rides to continue identification work when we could not use our own boats. Chuck Sutherland and the 1978 Maui Chapter of the American Cetacean Society gave a grant for us to begin the identification catalog.

We thank Roger Payne, William Schevill, and Susan Shane for reading and making suggestings on the manuscript.

This work is the result of the research programs of Roger Payne and the New York Zoological Society. It was funded by New York Zoological Society, World Wildlife Fund, and National Geographic Society grants to Roger Payne and a 1978 Vancouver Public Aquarium grant to Jim Darling.

Literature Cited

Chittleborough, R.G.
    1953. Aerial observations on the humpback whale, *Megaptera nodosa* (Bonnaterre), with notes on other species. Aust. J. Mar. Freshw. Res. 8: 219-226.

1965. Dynamics of two populations of the humpback whale, *Megaptera novaeangliae* (Borowski). Aust. J. Mar. Freshw. Res. 16: 33-128.

Darling, J.D. and C.M. Jurasz.
1983. Migratory destinations of North Pacific humpback whales (*Megaptera novaeangliae*). *In* R. Payne (editor), Communication and behavior of whales. AAAS Selected Symposia Series, p. 359-368. Westview Press, Boulder, Colo.

Dawbin, W.H.
1966. The seasonal migratory cycle of humpback whales. *In* K.S. Norris (editor), Whales, dolphins and porpoises, p. 145-170. Univ. Calif. Press, Berkeley.

Glockner, D.A. and S.C. Venus.
1983. Identification, growth rate and behavior of humpback whale, *Megaptera novaeangliae*, cows and calves in the waters off Maui, Hawaii, 1977-79. *In* R. Payne (editor), Communication and behavior of whales. AAAS Selected Symposia Series, p. 223-258. Westview Press, Boulder, Colo.

Glockner, D.A.
1983. Determining sex of humpback whales, *Megaptera novaeangliae*, in their natural environment. *In* R. Payne (editor), Communication and behavior of whales. AAAS Selected Symposia Series, p. 447-464. Westview Press, Boulder, Colo.

Herman, L.M. and R.C. Antinoja.
1977. Humpback whales in the Hawaiian breeding waters: Population and pod characteristics. Sci. Rep. Whales Res. Inst. Tokyo 29: 59-85.

Hudnall, J.
1978. Report on the general behavior of humpback whales near Hawaii, and the need for the creation of a whale park. Oceans Mag. 11(3): 8-15.

Katona, S., B. Baxter, O. Brazier, S. Kraus, J. Perkins, and H. Whitehead.
1979. Identification of humpback whales by fluke photographs. *In* H.E. Winn and B.L. Olla (editors), Behavior of marine animals - current perspectives in research, vol. 3: Cetaceans, p. 33-44. Plenum Press, N.Y.

Overton, W.S. and D.E. Davis.
1969. Estimating the numbers of animals in wildlife

populations. *In* R.H. Gildes, Jr. (editor), Wildlife management techniques, p. 403–455. Wildlife Management Soc., Wash. D.C.

Payne, R.
In press. Behavior of southern right whales, *Eubalaena australis*. University of Chicago Press, Chicago, Ill.

Payne, R. and E.M. Dorsey.
1983. Sexual dimorphism and aggressive use of callosities in right whales (*Eubalaena australis*). *In* R. Payne (editor), Communication and behavior of whales. AAAS Selected Symposia Series, p. 295–329. Westview Press, Boulder, Colo.

Rice, D.W.
1978. The humpback whale in the North Pacific: Distribution, exploitation, and numbers. *In* K.S. Norris and R.R. Reeves (editors), Report on a workshop on problems related to humpback whales (*Megaptera novaeangliae*) in Hawaii, p. 29–44. U.S. Dept. Commer., NTIS PB280 794.

Rice, D.W. and A.A. Wolman.
1978. Humpback whale census in Hawaiian waters - February 1977. *In* K.S. Norris and R.R. Reeves (editors), Report on a workshop on problems related to humpback whales (*Megaptera novaeangliae*) in Hawaii, p. 45–53. U.S. Dep. Commer., NTIS PB280 794.

Shallenberger, E.W.
1976. Report to Seaflight and Sea Grant on the population and distribution of humpback whales in Hawaii, 13p.

Tanner, J.T.
1978. Guide to the study of animal populations. Univ. Tenn. Press, Knoxville, 186 p.

Tyack, P.
1981. Interactions between singing Hawaiian humpback whales and conspecifics nearby. Behav. Ecol. Sociobiol. 8: 105–116.

Winn, H.E., W.L. Bischoff, and A.G. Taruski.
1973. Cytological sexing of cetacea. Mar. Biol. (Berl.) 23: 343–346.

Wolman, A.A. and C.M. Jurasz.
1977. Humpback whales in Hawaii: Vessel census 1976. U.S. Natl. Mar. Fish Serv. Mar. Fish Rev. 39(7): 1–5.

Deborah A. Glockner, Spearous C. Venus

7. Identification, Growth Rate, and Behavior of Humpback Whale (*Megaptera novaeangliae*) Cows and Calves in the Waters off Maui, Hawaii, 1977–79

Abstract

Humpback whale cows, calves, and escorts were studied in the ocean waters off West Maui, Hawaii, January through May, 1977 to 1979. Through surface and underwater photographs, 49 individual cows and their calves and 23 escort whales were identified. Each whale was classified according to the pigment pattern of its flippers, flukes, and flanks, the numerical spatial pattern of its lip grooves, and the shape of its dorsal fin. The pigment pattern of the flippers of calves was found to darken with age. Twenty-six individuals were resighted. Three yearlings were observed. The majority of cows and calves were located within 0.8 km of shore. The change from a predominantly southward to a predominantly northward movement of cows and calves occurred in March. The lengths of calves were measured as a percentage of their mother's length and the initial growth rate for the calves was determined to be 47.5 cm/month. An escort whale accompanied the cow and calf in 69% of the sightings. Escorts were observed to exhibit protective behavior of the cows and calves. The presence of bottlenose dolphins, remoras, and leatherback runners were noted with the whales. The techniques used to study these humpback whales in their natural environment can be adapted to studying other species of endangered whales.

Introduction

The humpback whale (*Megaptera novaeangliae*) has been protected by the International Whaling Commission since 1966. Prior to this time, most biological information on this species came from examination of corpses at whaling stations and through data supplied by the Bureau of International Whaling Statistics (Omura 1955; Nishiwaki 1959, 1962; Chittleborough 1965; Tomilin 1967). Because the whaling industry has in modern times been forbidden to kill calves and/or their mothers, basic parameters for this species such as the growth rate of calves were often only guessed. In 1975, Whitehead and Payne (1983) introduced a new technique for identifying and measuring individual right whales (*Eubalaena australis*) through aerial photographs. They were able to determine the growth rates and length frequencies of right whales off the coast of Argentina. Their study confirmed that identification and observation of individual whales in their natural environment is an effective way to obtain basic information about the life history and social interactions of baleen whales.

Humpback whales frequent the coastal waters of the Hawaiian Islands during the winter and spring months (Dawbin 1966; Rice 1974; Herman and Antinoja 1977; Wolman and Jurasz 1977). Because of their preference for warm coastal waters during the breeding season (Dawbin 1966) and the clarity and relative accessibility of the protected waters which they inhabit, we have found it possible to study individual whales over prolonged periods. Kraus and Katona (1977) reported on a technique to identify individual North Atlantic humpback whales through photographs of the pigmentation patterns on the undersurfaces of their flukes. We have adopted some of these techniques, adding new ones of our own in a three-year study of humpback whales off the West Coast of Maui, Hawaii. In this paper, we will discuss the morphological characteristics we used to identify each whale. These include the pigmentation pattern on the flippers, flanks, and undersurfaces of the flukes; the distinctive spacing and number of lip grooves; the shape of the dorsal fins; and the presence of scars and unusual markings. We will also detail resightings of individual whales, discuss the movements and behavior patterns of cows, calves, and "escorts" (see below for definition), and present the first evidence for the growth rate of the calves during their first two years.

Methods

During the months of January through May 1977, 1978, and 1979, we spent 776 hours observing humpback whales from land and 554 hours observing on the ocean. We took 1337 surface photographs and 1706 underwater photographs. From these photographs, we identified 49 cows, 49 calves, and 23 "escort" whales. We defined the "escort" as a whale accompanying a cow and calf usually seen below them. In cases in which there is more than one whale accompanying a cow and calf, the escort is that whale which stays closest to the cow and calf. We were able to make observations six days a week but only during the morning hours on four of these days. Figure 1 shows our study area in the waters off the west coast of Maui. We began each day by observing whales along the west coast of Maui from various lookout points between Mahinahina and McGregor Point, a distance of approximately 32 km. Upon spotting a pod of whales, we noted the number of individuals and the presence, if any, of a calf as well as its relative position in the pod. We recorded the various behaviors of the whales and the direction they were moving. Depending on the proximity and approach- ability of the whales and on the winds and wave heights, we then proceeded to carry out our open ocean work.

Using a 10'6" inflatable Avon powered by a 9.9 hp Mercury engine, we would launch from shore, motor the correct distance out, cut the engine, and wait for the whales to approach. We often used currents and oars to guide us into position. A diver then entered the water with mask and snorkel to take underwater photographs while an observer remained aboard the boat to carry out surface observations and take surface photographs. We used a Nikonos II 35 mm camera with a 28 mm lens for taking underwater photographs and a Nikonos III 35 mm camera with an 80 mm lens for taking topside photographs. In 1979, one of us joined J. Darling and G. Silber on a 17' Boston Whaler on seven additional days to obtain photographs of cows and calves throughout the Auau Channel and off the northeast coast of Lanai.

Results

Through detailed analysis of our photographs, we catalogued each whale according to the pigment patterns on its flippers, flanks, and flukes, and according to the numerical pattern of its lip grooves. Appendix Table 1 lists the classification of each cow, calf, and escort we photographed. We also noted the shape of the dorsal fin and the presence of scars and any unusual markings.

Pigmentation Patterns

The pigment patterns of humpback whales have been described by numerous biologists (True 1904; Lillie 1915; Matthews 1937; Omura 1953; Pike 1953; Chittleborough 1965; Tomilin 1967). In general terms, the back and flanks are charcoal black dappled with areas of grey. The throat, belly, and genital region vary from being almost entirely black to

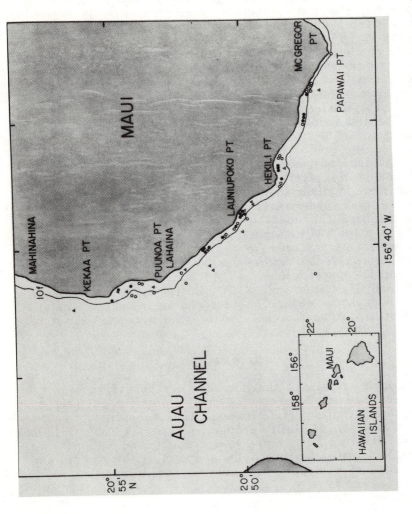

Figure 1. Location of cows, calves, and escorts photographed in 1977-79 in the waters off the west coast of Maui, Hawaii. Triangles (▲) = 1977, closed circles (●) = 1978, open circles (○) = 1979. Note: In 1979, two sets of cows and calves were photographed 0.4 km from the northeast shore of Lanai. These are not depicted on the map.

having extensive areas of white. Rawitz (1897) suggested that whales lighten in color with age. Angot (1951), studying humpback whales caught off the coast of Madagascar, suggested that young whales are lighter in coloration and turn darker with age. Chittleborough (1953) stated that calves off the western coast of Australia sighted late in the season were darker and larger than those sighted earlier. He found that near-term fetuses were lightly pigmented and suggested that this characteristic lasts for a short time after birth. The smallest calf we photographed, calf #3207C9, was the lightest calf we observed. It had a light cinnamon pigmentation. Furthermore, in our studies, we found that calves do indeed darken with age. We found that the pigment pattern of the flippers of two different calves #1006C8 and #2203C8 had darkened when we resighted them at a later date (discussed in detail in the following section).

Pigment Pattern of the Flippers

We have classified the pigment pattern on the dorsal and ventral surfaces of the flippers (pectoral fins) into five categories:

Type 1: The flipper is entirely white.

Type 2: The flipper is predominantly white. A large black area extends from the base of the flipper outward as far as the first node on the anterior edge of the flipper. Small islets, streaks, rings, and dots of black may occur on the distal half of the flipper. A rim of black may occur along the anterior or posterior edge.

Type 3: The flipper is neither predominantly white nor black. The black area extends from the base of the flipper towards the region of the fourth node. A secondary black area may occur in the distal portion of the flipper. Smaller black marks may be scattered between the two areas.

Type 4: The pattern is predominantly black. The black region extends from the base of the flipper outward past the fourth node. A white margin may be present along the anterior and posterior edges. A white area usually occurs at the distal end of the flipper but may be present proximally or at the mid-section.

Type 5: The flipper is black, frequently dappled with areas of grey. Streaks, rings, or specks of white may occur along the anterior or posterior edges.

Figure 2 shows examples of three of the types of flippers. We found that although they usually have the same type of pigment pattern, the left and right flippers occasionally have different types. The detail of the pigment pattern is never identical on both of the flippers.

Table 1 shows that the majority of the whales we photographed had flippers whose dorsal surface were in the two darkest types of pigment patterns: 75.9% of the total cows, calves, and escorts had flipper patterns of type 4 or 5 while 20.8% were in the type 2 category. The ventral surfaces of the flippers fell into lighter categories, 90.6% of the total being in type 2, 9.4% in type 3. Pike (1953) examined 184 humpback whales at a whaling station in British Columbia. He found that 73% of the females and 72% of the males had flipper patterns of predominantly black on the outside and white on the inside. However, he also found that 16% of the females and 11% of the males had predominantly black ventral surfaces.

Herman and Antinoja (1977) obtained pigmentation data on 189 humpback whales in the Hawaiian Islands from the air. They found that only 52.4% of the whales had flippers with predominantly black dorsal surfaces and 37.0% of the whales had flippers with extensive white dorsal surfaces. They did not find significant differences between the calves, cows, and other adults. Herman and Antinoja state that of the 189 whales, many may have been the same whale seen on separate occasions. This, or perhaps the fact that white is highly reflective, could account for their finding an unusually high percentage of whales with white flippers.

Through resighting five of the same individuals over successive years, we found that the pigment pattern of the flippers remains the same in adults. However, we found that the pigment pattern on the flippers of calves does change over a period of time as illustrated in Figure 3. On 30 January 1978, we photographed calf #2203C8. It had type 3 pigment pattern on its right flipper and type 4 on its left flipper. Each of the flippers had a black region extending from its base. This region was surrounded by a grey area and then by white. On 26 February 1978, we photographed the same calf. The right flipper had changed to the darker type 4 category. The left flipper had also darkened. The grey area had changed to black. The black filled in much of the white area. A very small area of the black region had lightened to grey. Calf #1006C8 was photographed as a calf and as a yearling. On both occasions, its flippers were of type 2, a predominantly white pigment pattern. However, the black region on the dorsal surface of the yearling's flipper became blacker and increased in size.

Figure 2. Examples of the various types of flippers, flukes, lip grooves, and fins.   ⟶

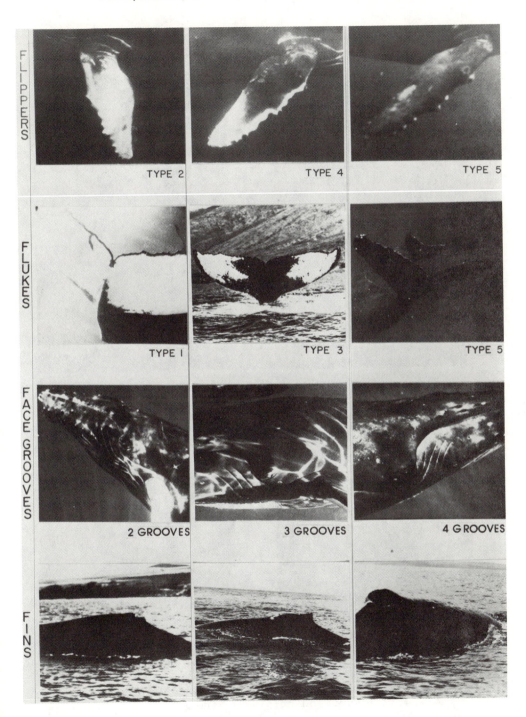

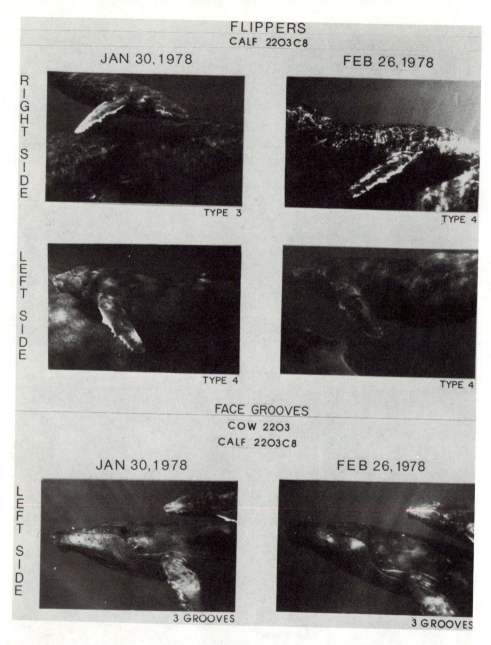

Figure 3. A) Pigment pattern change of the flippers of a calf
with age.
B) Lip groove spatial-numerical pattern retention
of a cow and calf over time.

Table 1. Pigment patterns of Flippers.

Dorsal Surface

Category	Cows N	%	Calves N	%	Escorts N	%	Total N	%
Type 5	29	60.4	10	20.4	11	47.8	5u	41.7
Type 4	7	14.6	25	51.0	9	39.1	41	34.2
Type 3	2	4.2	1	2.1	1	4.4	4	3.3
Type 2	10	20.8	13	26.5	2	8.7	25	20.8
Type 1	0	0.0	0	0.0	0	0.0	0	0.0
Total	48	100.0	49	100.0	23	100.0	120	100.0

Ventral Surface

Category	Cows N	%	Calves N	%	Escorts N	%	Total N	%
Type 2	33	91.7	37	94.9	7	70.0	77	90.6
Type 3	3	8.3	2	5.1	3	30.0	8	9.4
Total	36	100.0	39	100.0	10	100.0	85	100.0

Table 2. Pigment patterns of flukes.

Category	Cows N	%	Calves N	%	Escorts N	%	Total N	%
Type 5	11	33.4	10	31.3	3	27.3	24	31.6
Type 4	14	42.4	7	21.9	2	18.2	23	30.3
Type 3	3	9.1	5	15.6	4	36.3	12	15.8
Type 2	4	12.1	8	25.0	2	18.2	14	18.4
Type 1	1	3.0	2	6.2	0	0.0	3	3.9
Total	33	100.0	32	100.0	11	100.0	76	100.0

Pigmentation Pattern of the Flukes

We classified the pigment patterns on the ventral surface of the flukes into five categories:

Type 1: The flukes are predominantly white. The white region may be edged with a black border.

Type 2: The flukes are again predominantly white. The black border extends inward, fusing at the midline of the flukes to divide the white region into two distinct areas.

Type 3: The flukes are neither predominantly white nor black. A large black area exists in the mid-region and along the borders. Two distinct areas of white are present.

Type 4: The flukes are predominantly black. Two small areas of white occur, each at the distal corners of the flukes.

Type 5: The flukes are entirely black. Streaks and circles of white may be present.

Figure 2 illustrates types 1, 3, and 5.

Table 2 shows that the pigment pattern of the flukes varies widely amongst the cows, calves, and escorts: 75.8% of the cows' flukes occurred in the two darkest categories while the calves' and escorts' flukes were distributed throughout each of the categories. The higher percentage of the flukes of calves in the light categories could reflect a darkening with age. Pike (1953) found that 73% of the flukes of British Columbian females were predominantly black while only 33% of the males' flukes were predominantly black.

The general pattern of the adults' flukes remains the same over successive years. However, we discovered additional black streaks in a white region on a female's flukes 12 days after we originally photographed the whale. These streaks were apparently superficial scratches. Whether they persisted is not known.

Pigment Pattern of Flanks and Belly

Lillie (1915) classified 30 whales killed at a New Zealand whaling station into seven categories according to the pigment pattern of the ventral surface and flanks. In Lillie's type 1, the flanks and ventral side are white. In type 2, three points of black occur along the flanks. In type 3, the points of black form bands around the abdomen. In type 4, the black area covers the ventral surface except for an area of white below the mandible and two areas of white below the dorsal fin. Lillie also defined three categories occurring between each of these types. Lillie found that the majority of the whales occurred in the intermediate categories with only a few in the extreme types. Matthews (1937) found that the majority (60.2%) of South Georgian and South African whales occurred in the darkest category with only 1.8% in the lightest category. Omura (1953) found that the South Atlantic humpback whales were spread throughout the categories, the majority occurring in the intermediate type. Nishiwaki (1959) examined 217 humpbacks from Ryukyu and found that 92.2% occurred in type 4, 6.4% in type 3, and 1.4% in type 2. Pike (1953) found that all the humpbacks from British Columbia fell into type 4 category except for one female in the type 3-4 category. We found that out of 121 humpbacks from Hawaii 96.7% occurred in the darkest type 4 category, 1.65% (1 cow and 1 calf) in type 3, and 1.65% (2 calves) in type 3-4. Thus, humpback whales in the North Pacific appear to be much darker than humpbacks in the Southern Hemisphere.

Numerical-Spatial Pattern of the Lip Grooves

Several grooves occur on the lower jaw of each whale near the corner of its mouth and in the area above its flippers (True 1904; Tomilin 1967). The pattern of these grooves is unique to each whale, the number varying from 1 to 6. The pattern is not symmetrical on the right and left sides. True (1904) noted of five Newfoundland humpback whales that the groove pattern was distinct to each whale. We photographed the lip grooves of one adult #1015, on several occasions and found they remained the same over a period of a year. We photographed the lip grooves of seven other adults on several occasions separated by a range of 9 to 27 days. In each instance, their lip grooves had remained the same. In the case of three different calves, we found that their lip grooves had not changed over a period of 10, 12, and 27 days, respectively. Figure 3 shows the same pattern of lip grooves of cow #2203 and her calf #2203C8 on 30 January 1978 and 26 February 1978. The number and pattern of grooves of each whale was the same on both days. The lip groove pattern is especially useful in identifying individual whales whose black flippers appear similar.

Dorsal Fin

The shape of its dorsal fin is distinctive to each humpback. Schevill and Backus (1960) used the shape of the dorsal fin along with the pigment pattern of the flukes to track a single humpback off the coast of Maine over a 10-day period. Balcomb and Nichols (1978) also used this feature to

Table 3. Direction of Migration.

Month	Southbound		Northbound	
	n	%	n	%
January	5	100.0	0	0.0
February	8	72.7	3	27.3
March	9	47.4	10	52.6
April	9	39.1	14	60.9
May	0	0.0	2	100.0

identify humpbacks in the North Atlantic seen in different locations over a year's time. Figure 2 gives illustrations of various shapes of dorsal fins.

Scars and Other Markings

Many of the whales we have photographed carry scars which make them easily recognizable. Numerous white oval markings occur along the back and flanks of the whales. Ivashin and Golubovsky (1978) have suggested that these markings are caused by the parasite *Penella*. Pike (1951) suggested they are caused by lampreys, Shevchenko (1970) by sharks. The presence of such markings has been helpful in confirming the identification of whales sighted over several days. How long these markings are retained is unknown.

Position and Migratory Pattern

All of the cows and calves we photographed were located within approximately 1.6 km of shore, 90.2% within 0.8 km of shore as demonstrated in Figure 1. We photographed one cow-calf-escort pod 6.5 km from shore 20 days following their first sighting 0.4 km from shore. These distances are visual approximations. The daily movement pattern of the cows and calves consisted of a general northward or southward movement following the contour of the shore, usually within the 10 fathom line. Figure 4 and Table 3 show the proportions of cows and calves moving north and south for each month from January through May. In January, all of the cows and calves we photographed were moving southward. In March, approximately equal numbers were moving north and south. By April, the trend was predominantly northward. Chittleborough (1953) showed that the change from a predominantly northward to a predominantly southward movement for humpbacks off western Australia occurs in late August. In the Northern Hemisphere, where the seasons are six months later, August

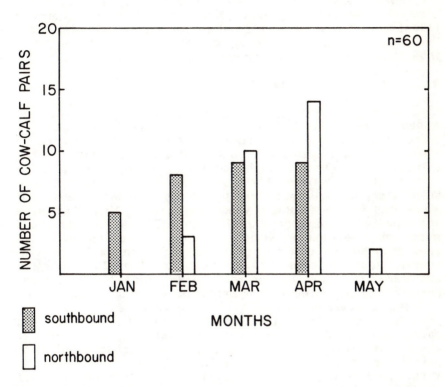

Figure 4. Migration of cows and calves traveling north and
south.

would correspond to late February. Cows and calves, being last in sequence of migration of humpbacks (Dawbin 1966), begin moving northward slightly later than the rest of the population.

The average rate of travel of the cows and calves was 1.1 knots. Dawbin (1956) reported the average rate of 1.3 knots for humpbacks passing New Zealand. Chittleborough (1953) determined an average speed of 4.3 knots. He stated that whales with very young calves travelled more slowly than those without young calves.

Resightings of Individual Whales

All sightings of recognized individuals are presented in Appendix Table 2. We were able to identify two whales in three successive years and six other whales in two successive years. We first identified whale #1006 on 14 February 1977 travelling in a pod of eight adults. The next year, on 29 January, we identified this whale with a calf. Exactly one year later, on 29 January 1979, we again photographed this whale and her calf, now a yearling. At this time they were accompanied by an escort whale. In 1978, the cow and calf were in the exact same location in which we had photographed the cow the previous year. In 1979, the cow and its yearling were 22.6 km north of their 1978 location but were travelling southward. Dawbin (1960) and Chittleborough (1958) noted the presence of yearlings accompanied by their mothers in Southern Hemisphere breeding areas.

We were able to identify two other pairs of cows and calves in both 1978 and 1979, one pair 11 months 1 day apart, the other 10 months 6 days later. In 1979 the first pair returned to the exact location it was photographed in 1978. The second pair was approximately 10.5 km south of its original location.

We identified whale #1015 three times in 1977, once in March with a pod of eight adults, and twice in April, alone. In 1978, we sighted this same whale again on three occasions in March, each time as an escort; the identity of the mother/calf pair is known on only one of these occasions. In 1979, we photographed it once in March, this time alone.

We have also been able to identify eight pairs of cows and calves and two escorts either on two or three occasions within the same year. Appendix Table 2 shows the times between successive sightings which ranged from 1 to 27 days.

Growth Rates

From our photographs we were able to measure 12 mothers and calves to determine the length of the calf as a percentage of its mother's length. We used only photographs in which the cow and calf were parallel to each other and were judged to be perpendicular to the line of sight from the camera to the whale.

We measured each whale from the tip of its snout to the notch in its flukes, repeating our measurements on each photograph three times. We were able to measure four cow and calf pairs from two different photographs each. We calculated the average percent error in measuring two photographs of the same whale to be 2.7%. Following the measuring technique of Whitehead and Payne (1981), we calculated the ratio of the length of the calf to the length of the mother, L_c/L_m. Figure 5 shows a plot of these ratios versus time and the computed regression line. We then found the mean L_c/L_m for each month. We plotted the mean ratio of lengths versus time and calculated a regression line for this graph as shown in Figure 6. From the slope of this regression line, we determined the initial growth rate of calves to be 47.5 cm/month. This figure corresponds closely to Tomilin's (1967) suggested fetal growth rate of 40-45 cm/month.

Nishiwaki (1959) measured the length of 70 mature females caught in the waters around the Ryukyu Islands and found their average length to be 13.2 m. Using 13.2 m as an average length for mothers, we determined the average length of calves we measured for each month. We next plotted the mean length on Nishiwaki's (1959) growth curve. According to the curve, our average calf length in January corresponded to a two-month-old, in February to a three-month-old, and in March to nearly a four-month-old. From fetal length data of humpbacks taken in Aleutian waters, Nishiwaki calculated that the height of parturition occurs in November. Birth in November would put our January calves at two months old. By comparing the distribution of near-term fetuses with the presence of newborn calves off of western Australia, Chittleborough (1953, 1958) found that the height of parturition occurs at the end of July or early August in the Southern Hemisphere. It is possible that an exact six-month lag does not occur between the two hemispheres. Tomilin (1967), stating that mating occurs year round, found two peaks in the time of conception in humpback whales in the North Pacific, one in February-March-April and one in September-October. More data are needed to confirm the birthing peak in Hawaii.

Both Chittleborough (1958) and Nishiwaki (1959) found that the mean length of calves at birth is 4.0 m - 4.3 m. Assuming that birth occurs in November, we extrapolated our curve and found the mean length at birth to be 4.4 m.

On 22 February 1981, a humpback whale calf was found alive on Punaluu Beach, Oahu, Hawaii. The calf was transported to Sea Life Park, an oceanarium at Makapuu Point, Oahu, and placed in a saltwater tank. On 23 February 1981, one of us (D.G.) measured the length of the calf from

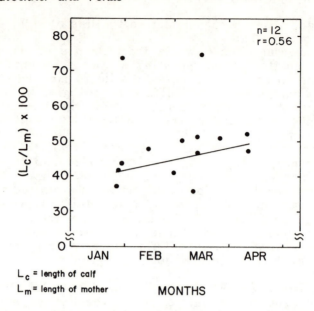

Figure 5. Growth curve for calves. Ratio of length of calf, L_c, to length of mother, L_m, versus time.

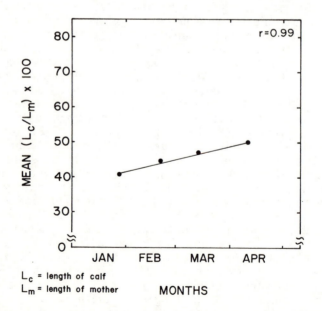

Figure 6. Mean growth curve for calves. Mean ratio by month of L_c/L_m versus time.

the tip of its snout to the notch in its flukes and found it to be 4.65 m. The calf died on 2 March 1981. Illness in the calf could have caused it to have below normal growth. Appendix Table 3 lists the body measurements taken on 23 February.

We were able to measure whale #1006C8 both as a calf and as a yearling relative to the length of its mother. In exactly one year it had grown an additional 30% of its mother's length. Using 11.9 m to 13.7 m as the range in lengths of mature females (Nishiwaki 1959; Chittleborough 1965), we obtained a range of 8.7 m to 10.1 m for the possible length of this yearling. Chittleborough determined the average length of five yearlings to be 9.1 m.

Behaviors of Cows and Calves

Edel and Winn (1978) found that humpback whale calves in the West Indies most often swam above their mothers. We observed calves swimming above their mothers in 59 out of 64 cases, or 92.2% of the time we photographed. In 7.8% of our sightings, or 5 out of 64 cases, the calf swam below the mother's tail. In all five of the latter cases, the cow and calf were travelling rapidly. We observed seven different calves swimming upside down. We never observed a cow-calf pair travelling with another cow-calf pair.

The majority of cows and calves usually travelled slowly, parallel to the coastline, temporarily stopping at various points. The calf would spout three or four times and then sound, remaining in the same general area, repeating the pattern approximately every two to six minutes. When a calf sounded, it usually did not lift its flukes vertically above the water. Occasionally, a calf would sound by lifting a portion of its flukes sideways above the water. The cow would spout three or four times and sound approximately every ten to fifteen minutes. After the cow spouted, the cow and calf would usually travel 50 to 100 m further along the coastline.

We observed both cows and calves breaching, flipper slapping, and tail lobbing. Scammon (1874) first described these behaviors as typical of humpback whales. We also observed calves touching their mothers with their flippers and cows touching their calves.

At times, we observed both cows and calves resting at the surface of the water. Both the cow and calf would be stationary, lying side by side perpendicular to shore, with their eyes closed. On many occasions, whales were observed swimming with their eyes closed.

Calves were observed nursing on three occasions. The cow was in a stationary horizontal position at a depth of approximately thirty to forty feet. The calf was positioned vertically, head up, below the cow's tail. Its mouth was pointed toward the mammary grooves, its flippers pointed forward. Lillie (1915) notes that a compressor muscle along the mammary gland forces milk through the nipple into the mouth of the calf.

Presence and Behavior of Yearlings

We identified three yearlings accompanying their mothers. We had photographed these yearlings as calves the previous year. We identified two additional yearling-size whales accompanying adult whales we had not previously identified. The behavior of yearlings resembled that of the calves. Like the calves, the yearlings remained close to their mothers. The yearlings would usually spout three or four times every five to ten minutes. Chittleborough (1953) reported the presence of partly digested milk in the stomachs of two yearlings, proving they were still nursing.

Presence and Behavior of Escorts

The presence of an adult whale, the "escort", accompanying a cow and calf has been observed throughout the breeding and feeding areas (Chittleborough 1953; Nemoto 1964; Herman and Antinoja 1977). An escort whale accompanied a cow and calf in 40 out of 58 (69.0%) of our sightings. An escort accompanied a cow and its yearling in 2 out of 3 (66.7%) of our sightings. Of the 40 escorts seen accompanying a cow and calf, 90.9% of the escorts swam behind and below the cow and calf, while 9.1% swam ahead and below the cow and calf.

On four occasions, one or more whales joined a cow and calf after we were already in the vicinity. On seven occasions, two or more whales were present with the cow and calf, the number ranging up to ten accompanying adults. On each occasion in which more than one whale accompanied a cow and calf, the escort, while travelling, was seen head-lunging[1] behind the cow and calf, with the remaining adults following. Samaras (1974) relates a similar behavior of gray whales (*Eschrichtius robustus*) during mating behavior. In three out of four underwater observations, the escort was blowing out bubbles underwater, as first reported by Darling, Gibson, and Silber (1983). In these three cases, only one other accompanying adult was present in addition to the cow and escort.

In 1978, we observed two different escorts singing the song of the humpback whale, as described by Payne and

---

[1] Head above water at 45°, much forward motion, whitewater; often mouth open, throat grooves expanded, and water streaming from mouth.

McVay (1971). Also in 1978, we observed an escort on two separate occasions with different cows and calves.

Humpback whales have been observed to exhibit epimeletic (care-giving) behavior on numerous occasions (Caldwell and Caldwell 1966). Chittleborough (1953) and Herman and Antinoja (1977) have suggested that the escort whale may have a protective function. Chittleborough (1953) reported one occasion of an escort defending a mother and calf against a pod of killer whales (*Orcinus orca*).

On 19 April 1979, one of us (D.G.) observed a cow, calf, and escort from below the surface of the water. The calf was alone at the surface. The mother was beneath the calf at a depth of approximately 15 fathoms. The escort swam directly to the calf. The calf and escort then swam down towards the mother, the calf directly above the escort. The mother rose, and the calf changed positions to above the mother's back. The mother and calf swam into the distance with the escort following.

Association of Cows, Calves, and Escorts with
Dolphins and Fish

Wolman and Jurasz (1977) reported an interaction between several dolphins and a pair of humpback whales. On five occasions we photographed bottlenose dolphins (*Tursiops truncatus*) swimming with a cow, calf, and escort.

Angot (1951) reported the presence of remoras attached to humpack whale calves. We noted two remoras, one attached to the ventral surface and one to the dorsal surface of one calf's flippers.

In 32.8% of our underwater sightings, a school of leather-back runners (*Scombroides sancti-petri*) accompanied the cows, calves, and escorts. It is possible that a symbiotic relationship exists, since these fish travel along with the whales and may act as "cleaner" fish grazing along the flanks of the whales. Gooding and Magnuson (1967) reported the attraction of pelagic fish to drifting objects in the ocean, stating that the object affords the fish a food supply and protection from predators.

Summary

1) We identified 49 cows, 49 calves, and 23 "escort" whales through the pigmentation pattern on their flippers, flanks, and the undersurfaces of their flukes, through the distinctive spacing and number of lip grooves, through the shape of the dorsal fin, and through the presence of scars and other markings.
2) We resighted 26 individuals and identified 3 yearlings accompanying their known mothers.

3) We found the change from a predominantly southward to a predominantly northward movement of the cows and calves to occur in March.
4) We measured the lengths of the calves as a percentage of their mother's length and determined an initial growth rate of 47.5 cm/month for the calves.
5) We described various behaviors of the cows, calves, and escorts. We noted the presence of bottlenose dolphins (*Tursiops truncatus*), remoras, and leather-back runners (*Scombroides sancti-petri*) with the whales.

The techniques we used to study humpback whales in their natural environment can easily be adapted for use in studying other endangered species of baleen whales. We feel very optimistic that whale biology can become a science of the living whale.

Acknowledgements

We would like to gratefully acknowledge the tremendous contribution and advice of Roger and Katharine Payne of the New York Zoological Society, and the Roger Payne Laboratory, Lincoln, Massachusetts in producing this manuscript. We are especially grateful to James Bird, Eleanor Dorsey, and Victoria Rowntree for their many suggestions and contributions. We would like to thank James Bird, Mark Ferrari, and Victoria Rowntree for typing versions of this manuscript. We thank James Bird for his library research and for compiling and typing the references. We thank Victoria Rowntree, James Bird, and Elizabeth Mathews for printing and drawing of graphs and figures. We thank Eleanor Dorsey for her statistical consultations. We thank Gregory Silber and Victoria Rowntree for developing and processing of prints, and Mark Ferrari for other photographic work. We are indebted to Roger and Katharine Payne for the use of their laboratory facilities.

We would like to thank Maui Whalewatchers for their tremendous contributions to our research. We would like to thank James Darling and Gregory Silber for contributing their photographs and for taking us out on their Boston Whaler. We would like to thank Reginald Gooding of the National Marine Fisheries Service and Phil Labell for their contributions in particular with fish references. We are sincerely grateful to Wilfred and Elrita Glockner, David Glockner, Ronald Glockner, Alma Seward, Chuck and Connie Sutherland, and Nelson and Leslie Hiraga for their contributions, advice, and assistance, especially in purchasing and constructing equipment.

Appendix Table 1. Classification of morphological features of cows, calves, and escorts photographed in 1977-79. L=left, R=right, M=mother, C=calf, Y=yearling, E=escort. Note: Escorts present but not photographed are not designated in table.

Whale	Role	Date	Flippers Dorsal	Flippers Ventral	Flukes Ventral	Flanks Ventral	Numerical Pattern Lip Grooves
1201	M	Feb 6	R4 L5	RL2	4	4	R3L3
1201C7	C	Feb 6	RL4	--	--	4	--
1401	E	Feb 6	RL5	--	2	4	--
1202	M	Feb 9	R3 L4	RL2	5	4	--
1202C7	C	Feb 9	RL2	R2	--	4	--
1203	M	Feb 18	L5	R3	5	4	L2
1203C7	C	Feb 18	L5	R3	5	4	L4
1205	M	Mar 13	R3	L2	--	4	--
1205C7	C	Mar 13	R5	L2	--	4	R3
1405	E	Mar 13	RL4	--	4	4	--
1406	E	Mar 15	RL5	R2	--	4	--

Appendix Table 1 (continued).

| Whale | Role | Date | Flippers | | Flukes Ventral | Flanks Ventral | Numerical Pattern Lip Grooves |
			Dorsal	Ventral			
1207	M	Apr 10	L2	--	--	4	--
1207C7	C	Apr 10	L3	--	--	4	--
1208	M	Apr 10	RL5	RL2	3	4	L3
"		Apr 20	RL5	RL2	--	4	L3
1208C7	C	Apr 10	RL4	RL2	5	4	L3
"		Apr 20	RL4	RL2	--	4	--
1408	E	Apr 10	RL5	RL2	2	4	R3 L4
1209	M	Apr 11	RL2	R2	--	4	L2
1209C7	C	Apr 11	L2	--	--	4	--
1210	M	Apr 14	RL5	L2	2	4	L4
1210C7	C	Apr 14	RL4	RL2	1	3-4	L1
1410	E	Apr 14	RL5	R2	--	4	--
1211	M	Apr 15	RL5	L2	--	4	--
1211C7	C	Apr 15	R2	L2	2	4	--
1411	E	Apr 15	RL4	--	--	4	--
1212	M	Apr 26	R5	RL2	--	4	--
1212C7	C	Apr 26	R5	L2	--	4	--

Appendix Table 1 (continued).

Whale	Role	Date	Flippers Dorsal	Flippers Ventral	Flukes Ventral	Flanks Ventral	Numerical Pattern Lip Grooves	
1213	M	Apr 26	R2	L2	3	4	--	
1213C7	C	Apr 26	R5	L2	5	4	R3	
1413	E	Apr 26	R5	--	--	4	--	
1214	M	May 3	RL5	R2	2	3	--	
1214C7	C	May 3	RL2	L2	2	4	--	
1215	M	May 16	R5	--	4	4	--	
1215C7	C	May 16	R4	L2	--	4	R3	

Morphological Features of Cows, Calves, and Escorts Photographed in 1978

Whale	Role	Date	Flippers Dorsal	Flippers Ventral	Flukes Ventral	Flanks Ventral	Numerical Pattern Lip Grooves	
2201	M	Jan 27	L5	R2	5	4	L2	
2201C8	C	Jan 27	L4	R2	3	4	--	
2401	E	Jan 27	L4	R2	3	4	--	
1006	M	Jan 29	R2	--	--	4	--	
1006C8	C	Jan 29	R2	--	--	4	--	
2203	M	Jan 30	RL5	L2	4	4	R4	L3
"		Feb 26	RL5	--	4	4	R4	L3
2203C8	C	Jan 30	R3 L4	L2	2	4	L3	
"		Feb 26	R4 L4	L2	--	4	L3	

Appendix Table 1 (continued).

Whale	Role	Date	Flippers		Flukes Ventral	Flanks Ventral	Numerical Pattern Lip Grooves
			Dorsal	Ventral			
2204	M	Feb 13	L2	--	--	4	--
2204C8	C	Feb 13	L2	R2	--	4	--
2404	E	Feb 13	L4	--	--	--	--
2205	M	Feb 13	L5	R2	3	4	L3
2205C	C	Feb 13	RL2	--	--	4	--
2405	E	Feb 13	L4	--	--	4	--
2206	M	Feb 27	RL5	RL3	4	4	R3 L2
2206C8	C	Feb 27	RL4	RL2	5	4	R3
2207	M	Feb 27	L5	--	--	4	L4
2207C8	C	Feb 27	L5	--	--	4	L2
2208	M	Mar 10	RL2	L2	2	4	--
2208C8	C	Mar 10	R2	RL2	2	4	--
2209	M	Mar 15	RL4	R2	4	4	--
2209C8	C	Mar 15	RL4	R2	2	4	--
2409	E	Mar 15	RL5	--	5	4	--
2210	M	Mar 17	RL4	--	--	4	--
2210C8	C	Mar 17	L2	--	--	4	--
2410	E	Mar 17	RL5	--	--	--	--

Appendix Table 1 (continued).

| Whale | Role | Date | Flippers | | Flukes Ventral | Flanks Ventral | Numerical Pattern Lip Grooves |
			Dorsal	Ventral			
2211	M	Mar 17	RL5	R2	5	4	--
2211C8	C	Mar 17	L4	RL2	3	4	--
2212	M	Mar 26	RL5	RL2	4	4	R2
2212C8	C	Mar 26	RL4	RL2	3	4	--
1015	E	Mar 26	RL5	L3	5	4	L3
2213	M	Mar 27	RL2	RL2	--	4	L4
2213C8	C	Mar 27	RL2	RL2	--	4	--
2413	E	Mar 27	RL2	--	--	--	--
2214	M	Mar 29	R2	L2	4	4	--
2214C8	C	Mar 29	R4	L2	4	4	--
2215	M	Apr 2	R5	L2	5	4	R3
"		Apr 11	RL5	RL2	--	4	R3
"		Apr 12	L5	RL2	--	4	--
2215C8	C	Apr 2	R5	RL2	5	4	R3
"		Apr 11	RL5	R2	5	4	L2
"		Apr 12	L5	RL2	5	4	L2
2415	E	Apr 12	RL5	--	4	4	--

Appendix Table 1 (continued).

Whale	Role	Date	Flippers Dorsal	Flippers Ventral	Flukes Ventral	Flanks Ventral	Numerical Pattern Lip Grooves
2216	M	Apr 9	RL5	RL2	5	4	R3 L4
2216C8	C	Apr 9	RL5	RL2	5	4	L3
2416	E	Apr 9	R2	R2	--	4	L5
2217	M	Apr 17	RL5	L2	--	4	L5
2217C8	C	Apr 17	RL5	RL2	4	4	R2
2413	E	Apr 17	RL2	--	--	4	R3
2218	M	Apr 27	R4	L2	4	4	R4
2218C8	C	Apr 27	R5	RL2	4	4	R3

Morphological Features of Cows, Calves, and Escorts Photographed in 1979

Whale	Role	Date	Flippers Dorsal	Flippers Ventral	Flukes Ventral	Flanks Ventral	Numerical Pattern Lip Grooves
3201	M	Jan 26	R5	L2	--	4	R3
3201C9	C	Jan 26	R4	--	3	4	R3
1016	M	Jan 29	R2	--	--	4	--
1016C8	Y	Jan 29	R2	--	--	4	--
2208	M	Feb 11	RL2	--	2	4	--
2208C8	Y	Feb 11	RL2	--	2	4	--
3202	M	Feb 18	R5	L2	--	4	R2
3202C9	C	Feb 18	L4	--	--	4	--

Appendix Table 1 (continued).

Whale	Role	Date	Flippers Dorsal	Flippers Ventral	Flukes Ventral	Flanks Ventral	Numerical Pattern Lip Grooves
3203	M	Feb 27	RL5	RL2	4	4	R3 L2
3203C9	C	Feb 27	RL4	RL2	4	4	R4
3204	M	Mar 3	L4	R2	4	4	R6 L3
"		Mar 13	RL4	--	--	4	L3
3204C9	C	Mar 3	RL2	R2	2	4	--
"		Mar 13	RL2	L2	2	4	--
3404	E	Mar 3	R5	--	3	4	--
3205	M	Mar 4	RL4	--	4	4	L2
"		Mar 24	RL4	RL2	4	4	L2
3205C9	C	Mar 4	RL4	RL2	3	4	R2
"		Mar 24	RL4	--	3	4	--
2218	M	Mar 5	R4 L5	RL2	4	4	L3
2218C8	Y	Mar 5	L5	R2	4	4	L3
3003	E	Mar 5	R4	--	--	4	R2
3206	M	Mar 5	L5	R2	--	4	L3
3206C9	C	Mar 5	L4	R2	4	4	L3
3207	M	Mar 11	RL5	L3	5	4	--
3207C9	C	Mar 11	R4	L2	--	4	--

Appendix Table 1 (continued).

Whale	Role	Date	Flippers		Flukes Ventral	Flanks Ventral	Numerical Pattern Lip Grooves
			Dorsal	Ventral			
3208	M	Mar 12	RL4	R2	1	4	L3
"		Mar 13	RL4	RL2	1	4	R4 L3
"		Mar 24	RL4	RL2	1	4	R4 L3
3208CXX	C	Mar 12	RL4	RL2	2	4	L2
"		Mar 13	RL4	RL2	2	4	R3 L2
"		Mar 24	RL4	RL2	2	4	R3 L2
3408	E	Mar 12	L5	R3	5	4	L4
"		Mar 13	RL5	R3	5	4	--
3209	M	Mar 25	RL5	--	5	4	--
3209C9	C	Mar 25	RL4	--	4	4	R3
3210	M	Mar 31	R5	--	2	4	--
3210C9	C	Mar 31	R4	L2	2	3-4	R2
3211	M	Mar 31	RL5	--	5	4	--
"		Apr 1	RL5	--	5	4	--
3211C9	C	Mar 31	RL4	RL2	--	4	L3
"		Apr 1	RL4	--	--	4	--
3212	M	Apr 2	RL5	L2	4	4	R5
3212C9	C	Apr 2	R4	R2	5	4	R3
3412	E	Apr 2	RL4	R2	3	4	L3

Appendix Table 1 (continued).

Whale	Role	Date	Flippers		Flukes Ventral	Flanks Ventral	Numerical Pattern Lip Grooves
			Dorsal	Ventral			
3213	M	Apr 11	RL5	RL2	5	4	L4
3213C9	C	Apr 11	RL2	RL2	5	4	R4 L2
3413	E	Apr 11	RL4	L2	3	4	R3
3214	M	Apr 15	R5	R2	4	4	--
"		Apr 22	L5	--	--	4	--
3214C9	C	Apr 15	RL4	R3	--	3	R4
"		Apr 22	L4	--	--	3	--
3215	M	Apr 19	RL2	--	--	4	--
3215C9	C	Apr 19	RL2	L2	1	4	R3
3415	E	Apr 19	RL4	--	--	--	--
3216	M	Apr 22	RL5	--	--	4	R3
3216C9	C	Apr 22	R5	RL2	5	4	R3
3416	E	Apr 22	RL3	RL3	--	4	--
3217	M	Apr 22	RL2	--	4	4	--
3217C9	C	Apr 22	L4	R2	5	4	--
3218	M	Apr 25	--	--	5	4	R3
3218C9	C	Apr 25	R4	RL2	4	4	R2

Appendix Table 2. Resightings of individual whales. A=adult, C=calf, E=escort, M=mother, S=singleton, Y=yearling.

Whale #	Date of Sighting		
	1977	1978	1979
1006	Feb 14 (8a)	Jan 29 (M,C)	Jan 29 (M,Y,E)
1006C8		Jan 29 (M,C)	Jan 29 (M,Y,E)
1015	Mar 17 (8A), Apr 1,8 (S)	Mar 8,25,26 (M,C,E)	Mar 6 (S)
1208	Apr 10,20 (M,C,E)	--	--
1208C7	Apr 10,20 (M,C,E)	--	--
1601	Feb 27 (S)	Mar 4 (8A)	--
2203	--	Jan 30 (M,C); Feb 26 (M,C)	--
2203C8	--	Jan 30 (M,C); Feb 26 (M,C)	--
2208	--	Mar 10 (M,C)	Feb 11 (M,Y)
2208C8	--	Mar 10 (M,C)	Feb 11 (M,Y)
2413	--	Mar 27 (M,C,E); Apr 17 (M,C,E)	--
2215	--	Apr 2 (M,C); Apr 11,12 (M,C,E)	--
2215C8	--	Apr 2 (M,C); Apr 11,12 (M,C,E)	--
2218	--	Apr 27 (M,C)	Mar 5 (M,Y)
2218C8	--	Apr 27 (M,C)	Mar 5 (M,Y)
3204	--	--	Mar 3,13 (M,C)

Appendix Table 2 (continued).

Whale #	Date of Sighting		
	1977	1978	1979
3204C9	--	--	Mar 3,13 (M,C)
3205	--	--	Mar 4,24 (M,C,E)
3205C9	--	--	Mar 4,24 (M,C,E)
3208	--	--	Mar 12,13,24 (M,C,E)
3208C9	--	--	Mar 12,13,24 (M,C,E)
3408	--	--	Mar 12,13, (M,C,E)
3211	--	--	Mar 31, Apr 1 (M,C,E)
3211C9	--	--	Mar 31, Apr 1 (M,C,E)
3214	--	--	Apr 15 (M,C); Apr 22 (M,C,E)
3214C9	--	--	Apr 15 (M,C); Apr 22 (M,C,E)

Appendix Table 3. Measurements of beached humpback whale calf taken on 23 February 1981, in centimeters, with percent of body length.

Total Length	465 cm	100.0%
Length – Snout to anterior insertion of dorsal fin	267	57.4
Length – Snout to center of blowhole	71	15.3
Length – Blowhole	15	3.2
Length – Dorsal fin base	41	8.8
Length – Flipper (anterior origin to tip)	137	29.5
Width – Flipper	38	8.2
Width – Flukes (tip to tip)	117	25.2
Fluke notch to nearest point on the leading edge	51	11.0

Literature Cited

Angot, M.
1951. Rapport scientifique sur les expeditions baleinieres autour Madagascar (saisons 1949 et 1950). (Scientific report on the whaling expeditions around Madagascar '1949 and 1950 seasons') Mem. Inst. Rech. Sci. Madagascar Ser. A Biol. Anim. 5: 439-486.

Balcomb, K.C. and G. Nichols.
1978. Western North Atlantic humpback whales. Int. Whaling Comm. Rep. Comm. 28: 159-164.

Caldwell, M.C. and D.K. Caldwell.
1966. Epimeletic behavior in Cetacea. *In* K.S. Norris (ed.), Whales, dolphins, and porpoises, p. 775-789. Univ. Calif. Press, Berkeley.

Chittleborough, R.G.
1953. Aerial observations on the humpback whale, *Megaptera nodosa* (Bonnaterre), with notes on other species. Aust. J. Mar. Freshw. Res. 4: 219-226.

1958. The breeding cycle of the female humpback whale, *Megaptera nodosa* (Bonnaterre). Aust. J. Mar. Freshw. Res. 9: 1-18.

1965. Dynamics of two populations of the humpback whale, *Megaptera novaeangliae* (Borowski). Aust. J. Mar. Freshw. Res. 16: 33-128.

Darling, J.D., K.M. Gibson, and G.K. Silber.
1983. Observations on the abundance and behavior of humpack whales (*Megaptera novaeangliae*) off West Maui, Hawaii, 1977-79. *In* R. Payne (ed.), Communication and behavior of whales. AAAS Selected Symposia Series, p. 201-222. Westview Press, Boulder, Colo.

Dawbin, W.H.
1956. The migration of humpback whales which pass the New Zealand coast. Trans. R. Soc. N.Z. 84: 147-196.

1960. An analysis of the New Zealand catches of humpback whales from 1947 to 1958. Norsk Hvalfangsttid. 49: 61-75.

1966. The seasonal migratory cycle of humpback whales. *In* K.S. Norris (ed.), Whales, dolphins, and porpoises, p. 145-170. Univ. Calif. Press, Berkeley.

Edel, R.K. and H.E. Winn.
1978. Observations on underwater locomotion and flipper movement of the humpback whale *Megaptera novaeangliae*. Mar. Biol. (Berl.) 48: 279-287.

Gooding, R.M. and J.J. Magnuson.
1967. Ecological significance of a drifting object to pelagic fishes. Pac. Sci. 21: 486-493.

Herman, L.M. and R.C. Antinoja.
1977. Humpback whales in the Hawaiian breeding waters: Population and pod characteristics. Sci. Rep. Whales Res. Inst. Tokyo 29: 59-85.

Ivashin, M.V. and Yu. P. Golubovsky.
1978. On the cause of appearance of white scars on the body of whales. Int. Whaling Comm. Rep. Comm. 28: 199.

Kraus, S. and S. Katona (editors).
1977. Humpback whales in the Western North Atlantic - A catalogue of identified individuals. College of the Atlantic, Bar Harbor, Maine, 26p.

Lillie, D.G.
1915. Cetacea. British Antarctic ("Terra Nova") Expedition, 1910. Nat. Hist. Rep. Zool. (Lond.) 1: 85-124.

Matthews, L.H.
1937. The humpback whale, *Megaptera nodosa*. Discovery Rep. 17: 7-92.

Nemoto, T.
1964. School of baleen whales in the feeding areas. Sci. Rep. Whales Res. Inst. Tokyo 18: 89-110.

Nishiwaki, M.
1959. Humpback whales in Ryukyuan waters. Sci. Rep. Whales Res. Inst. Tokyo 14: 49-87.

1962. Ryukyuan whaling in 1962. Sci. Rep. Whales Res. Inst. Tokyo 16: 19-28.

Omura, H.
1953. Biological study on humpback whales in the Antarctic whaling Areas IV and V. Sci. Rep. Whales Res. Inst. Tokyo 8: 81-102.

1955. Whales in the northern part of the North Pacific. Norsk Hvalfangsttid. 44(6): 323-405. Continued in 44(7): 395-405.

Payne, R.S. and S. McVay.
1971. Songs of humpback whales. Science (Wash. D.C.) 173: 585-597.

Pike, G.C.
1951. Lamprey marks on whales. J. Fish. Res. Board Can. 8: 275-280.

1953. Colour pattern of humpback whales from the coast of British Columbia. J. Fish. Res. Board Can. 10: 320-325.

Rawitz, B.
1897. Ueber norwegische bartenwale. (On the Norwegian humpback whale) Sitz Berg. Nat. Fr. (Berl.): 146-150.

Rice, D.W.
1974. Whales and whale research in the Eastern North Pacific. *In* W.E. Schevill (ed.), The whale problem - a status report, p. 170-195. Harvard University Press, Cambridge, Mass.

Samaras, W.F.
1974. Reproductive behavior of the gray whale *Eschrichtius robustus*, in Baja California. Bull. South. Calif. Acad. Sci. 73: 57-64.

Scammon, C.M.
1874. The marine mammals of the northwestern coast of North America. J. Carmany & Co., N.Y., 317p.

Schevill, W.E. and R.H. Backus.
1960. Daily patrol of a *Megaptera*. J. Mammal. 41: 279-281.

Shevchenko, V.I.
1970. A riddle of white scars on the body of whales. Priroda (Mosc.) 6: 72-73.

Tomilin, A.G.
1967. Mammals of the U.S.S.R. and adjacent countries, vol. 9: Cetacea. (translated by O. Ronen from the 1957 Russian edition) Israel Program for Scientific Translations, Jerusalem, 717p.

True, F.W.
1904. The whalebone whales of the Western North Atlantic. Smithson. Contrib. Knowl. 33: 332p.

Whitehead, H. and R. Payne.
1981. New techniques for measuring whales from the air. U.S. Dept. Commer. NTIS PB81-161143, 36p.

Wolman, A.A. and C.M. Jurasz.
1977. Humpback whales Hawaii: Vessel census, 1976. U.S. Natl. Mar. Fish. Serv. Mar. Fish. Rev. 39(7): 1-5.

Kenneth S. Norris, Bernardo Villa-Ramirez,
George Nichols, Bernd Würsig, Karen Miller

8. Lagoon Entrance and Other Aggregations of Gray Whales (*Eschrichtius robustus*)

Abstract

Localization of life activities such as shore migration and lagoon breeding is assessed as possibly relevant to the resiliency of gray whales to exploitation. Lagoon entrance aggregations are composed of males, non-parturient females, and juveniles. Near slack tide these groups aggregate and courtship and mating take place. Near flood tide animals line up along current margins, dive regularly and appear to be feeding. Food organisms present in high densities are euphausiids and galatheid swimming crabs.

No population of gray whales was found at Bahia Reforma. An aggregation at Cabo Falso, Baja California, consisted of non-parturient animals. Tidally related behavior and breaching are assessed.

Introduction

The gray whale seems uncommonly resilient in the face of whaling. Twice in the last 123 years it has been heavily fished. In 1875, American whalers had reduced its numbers to an estimated 4400 animals (Ohsumi 1976). It recovered to a degree, to have its advance stopped in the 1920's when modern factory ships were used. After its protection in 1946 by the International Convention for the Regulation of Whaling, the species recovered further, nearly to primordial levels.

To what degree can we assign this resiliency to features of the animal's life history, and to what extent is it merely a result of there having been an adequate breeding reservoir after exploitation? This question is, of course, complex. Too many pieces of the puzzle are missing for a valid

259

judgment to be made now. But by framing the question we may sharpen the things we ask about gray whale life history.

For comparison one might first look at the Northern Pacific black right whale (*Eubalaena glacialis*). Rice (1974) notes:

> "The nineteenth-century American whalers almost succeeded in completely exterminating the right whale in the eastern North Pacific. How close they came is apparent from the fact that from 1905, when modern whaling methods were introduced on the west coast, to 1937, when right whales were given legal protection, only 24 were killed by the whaling stations in Alaska and British Columbia."

He concludes: "....this stock numbers only a few individuals and has not noticeably increased in the past 35 years." Only sporadic sightings of the species have recently been made in the eastern North Pacific (Rice and Fiscus 1968), and somewhat greater numbers are thought to persist in the western Aleutians, Kuriles, and Hokkaido (Reeves 1979).

In the northeastern Pacific the right whale is not known to nurture its young in nearshore shallows as is the case in some other places in the world (Rice 1974). Perhaps this fact is relevant to its present precarious state, but more obvious is the fact that the right whale was a richer prize than the gray whale, and safer to take, and hence a better target for relentless pursuit.

But to understand resilience we must dig deeper. We must come to understand the reproductive arrangements of the species, and we must understand its energy sources. A view through geologic time needs to be taken, since the species geography of today is unlikely to have been invariant in the past. Since the purpose of this paper is to look at only one feature of this broader set of questions, we will mention only briefly these other issues before focussing on events at lagoon entrances.

Present day gray whale breeding lagoons in Baja California are almost certainly transient features. They lie upon young alluvium and inland of each are regular strand lines of higher water stages, indicating recent emergence of the land (Beal 1948). The whole region is tectonically active, since all of Baja California and much of Southern California have been profoundly altered by plate evolution and deformation from the interaction of the North American and Pacific plates (Atwater 1970). Clearly, gray whales must possess sufficient flexibility to accommodate to the disappearance and development of lagoons adequate for breeding.

It is also becoming clear that the gray whale must be flexible with regard to food and feeding grounds. Major food supplies at present come from the great continental shelf extending across the northern Bering and Chukchi seas (Rice and Wolman 1971). The species is commonly considered stenophagous (Pike 1962), relying almost wholly upon benthic amphipods grubbed from the bottom of this huge shallow area. Yet in the recent past this area was repeatedly unavailable to gray whales, as sea ice cyclically persisted in the area, and as continental glaciation extended to the margins of the continental shelf in southeastern Alaska (Nelson, Hopkins, and Scholl 1974; Hall 1980). Work on small populations of gray whales that summer far south of these northern grounds and which seem to consume a variety of different foods therefore assumes considerable importance (Hatler and Darling 1974; Darling 1977). They may represent what the bulk of gray whales did when ice excluded them from the northern regions.

We will present evidence which suggests that part of the gray whale population may feed near the southern terminus of its range, and that spatial separation between calving and breeding portions of lagoon populations exists, and could be an important part of species resilience.

The search for such correlations is heightened in urgency when the threats to the six calving lagoons are considered (Reeves 1977). Oil and gas explorations are taking place on the Magdalena Plain-Vizcaino Desert area, including drilling in the vicinity of Laguna Ojo de Liebre (Scammon's Lagoon), and such efforts could involve three major lagoons: Guerrero Negro, Ojo de Liebre, and San Ignacio. Tourism oriented toward whales has increased rapidly in recent years (Reeves 1977), and in spite of closures and stringent regulations, concerns remain. The southern lagoons of Estero Soledad, Santo Domingo, Las Animas and Magdalena Bay are, at this writing, largely without regulation. Phosphate mining and the cutting of canals to accommodate barge traffic is a new concern in the Magdalena Bay-Estero Soledad region. Of special concern is the potential use of these lagoons for sport boat traffic, involving local marinas. This threat, mentioned at the Mexican-American marine mammal meetings in La Paz, Baja California, in February 1977, could reduce the usability of the lagoons to the whales. Commercial boat traffic typically tends to ignore whales, and this is reciprocated for the most part by the whales, but uncontrolled small boats can become a severe source of harassment.

How important are these lagoons to the whales? To answer this question we must know how the whales use the lagoons. A good deal is now known about mother-calf use of the lagoons, thanks to the work of Swartz and Jones (1980a and 1980b). They and their coworkers showed that there are two major overlapping events taking place in or near the calving lagoons. Pregnant females arrive first, mostly in early January, enter lagoons, give birth within them, and nurture their young in the calm waters. Somewhat later in January, males, juveniles, and non-parturient females arrive and concentrate largely around lagoon entrances. This same order of migration has been described off California on the southward migration (Rice and Wolman 1971). Entrance aggregations persist until March, when they begin to dissipate (Swartz and Jones 1980a) and the whales presumably migrate north. At the same time that lagoon entrance groups are dissipating, congregations of mother-calf pairs shift from the deepest recesses of the lagoons, and move, in part, into the vacated entrances (Swartz and Jones 1980a). The mother-calf aggregation exists as a collection of fairly discrete mother-calf couplets that tend to remain mostly separate from one another, and in certain places to flood back and forth with the tides (Norris, Goodman, Villa-Ramirez, and Hobbs 1977). This tidal effect is very obvious in lagoons that are simple channels, such as Estero Soledad, while it is less evident in broader lagoons with eddies and bars. We present new observations on this effect later in the paper.

When entrance groups and aggregations of mother-calf pairs meet, as they sometimes do near lagoon entrances, strong reactions such as chases, courtship, and high-speed swimming sometimes result (Norris et al. 1977).

Nonetheless, the behavior of whales at or near lagoon entrances, which is presumed to include courtship, is poorly known, and is the main subject of this report. Gilmore (1961) first suggested the occurrence of entrance courting-mating groups, though Scammon, (1874) described aggregations at entrances. Samaras (1974) described what he called staging areas, at lagoon entrances, and provided a photograph and description of mating at Laguna Ojo de Liebre.

Courting and mating seem to occur throughout the migratory path. Sauer (1963) reports watching courtship and mating from St. Lawrence Island in the Bering Sea. Hatler and Darling (1974) describe a possible mating including the erect penis of the participating male at Vancouver Island, British Columbia, Canada and Samaras (1974) describes a mating seen off southern California. Rice and Wolman (1971) comment on such behavior during migration. Because they detected ovulation in gray whales taken off central California on the southward migration (but detected no *conceptus*), and because of the prelinear growth rate of gray whale fetuses, they project that effective mating takes place off California, and that gray whales therefore typically have a 13-month gestation period. If correct, this means that lagoon entrance groups are not contributory to fertilization, and that they may be functionless, reproductively speaking. It means that all gray whales except mothers and calves swim from California to the breeding lagoons and back without utility to the species. Such a 13-month gestation period seems not to occur elsewhere in the Cetacea; other baleen whales are thought to have a 12-month period (Laws 1959, 1961) related at least in some, including the gray whale, to an annual photoperiod (Dawbin 1966; Norris 1967). In our opinion it is difficult to assign a function to most such apparent courtship during migration or at lagoon entrances, except in the cases where intromissions or erections have been noted. Juveniles are prominent members of these groups and their behavior can best be considered as exploratory of later relationships, but it is often difficult to be sure if adults are involved in a given activity, and what their sexes might be.

The reason so little is known of entrance aggregations is simple. Lagoon entrances are typically dangerous places to be, with shallow bars, strong tidal currents, and lines of breakers. Most are open to prevailing winds without significant shelter.

But two entrances in our experience promised a chance for observations, because one or more of these conditions does not exist. The southernmost calving lagoons at Almejas and Santa Marina Bays reach the sea through the sheltered Canal Rehusa. This channel is bounded on the north by the steep east-west trending shoreline of Santa Margarita Island and is thus sheltered from prevailing northwest winds. Nonetheless, it remains a dangerous place for navigation with very strong tidal currents, and waves breaking over submerged bars. An observer can, however, work the main channel in calm weather.

At the northern tip of Santa Margarita Island is the main entrance to Magdalena Bay. It is a deep water passage, averaging about 30 m depth, where the bay sill crosses the 5.4 km entrance between Punta Redonda on Santa Margarita Island and Punta Entrada at the southern tip of Magdalena Island. Throughout this paper we will refer to this entrance as La Entrada. Vessels of very large size regularly enter this channel in any weather. The rocky cliffs bounding La Entrada are neither so high nor so close to areas frequented by whales as those at the Canal Rehusa and most observations need to be done from shipboard. We found it convenient to anchor amongst the whales.

Materials and Methods

Two research cruises to study entrance aggregations are the basis of this paper. First, in the winter of 1978, Norris and Villa-R. attempted to establish a shore camp at

Canal Rehusa, but because of bad weather switched observations to the northern tip of Isla Magdalena, at Boca Soledad where 8 days of observations were carried out in the entrance area. Counts of whales were made through the Estero Soledad, reactions of animals in the entrance were noted, and tidal behavior was observed. Then, in 1979, a cruise of the research vessel *Regina Maris* was taken under the captaincy of Nichols, to study entrance aggregations. The first leg (2 January to 2 February 1979) visited Bahia Reforma on the Sonora-Sinaloa Coast, Cabo San Lucas - Cabo Falso, Canal Rehusa, Bahia Magdalena and Bahia Almejas, under the scientific leadership of Norris. The second leg (3 to 14 February 1979) traversed Devil's Bend to the lower reaches of Estero Soledad and made additional population counts at Bahia Magdalena, and at Cabo San Lucas - Cabo Falso, under the scientific leadership of Würsig.

Shore watching stations were established at Punta Entrada, Bahia Magdalena, and at Colina Coyote in Estero Soledad, using a portable Schmidt reflecting telescope. In 1978 the 8 m twin engine fiberglass vessel *Nai'a* was used to make counts in Estero Soledad, and to study bottom topography and tidal relations.

The cruise of the *Regina Maris* provided several unique opportunities. Her track is shown in Figure 1. She is a 44 m barkentine, with three masts 25 m high that can be climbed for observation. She was able to anchor in the midst of entrance aggregations and to take long counting traverses in or near lagoon mouths, and at Cabo San Lucas - Cabo Falso.

Counts of whales were made visually. In Estero Soledad two observers on the *Nai'a* recorded, one to port and the other to starboard. From the *Regina Maris* counts were made by a single observer aloft on the foremast. While the ship ran a predetermined course, the observer counted animals that passed imaginary lines 45° to both starboard and port, taking special care to allow consort animals to surface and be counted. While there is undoubtedly error due to animals that failed to surface during our passage, we expect this error to be modest, and feel that this kind of count is far more accurate than aerial counts in which many submerged animals may be missed in the brief passage. Due care was taken to avoid counting animals twice where track legs paralleled each other, by recording only those half the distance or less to the other track. Tide rips helped in this assessment. Distances were estimated by eye, and by reference to the chart and ship's position. Counts are probably conservative.

Figure 1. Track chart, R/V *Regina Maris*, Cruise No. 12, 1 January - 15 February 1979.

$\longrightarrow$

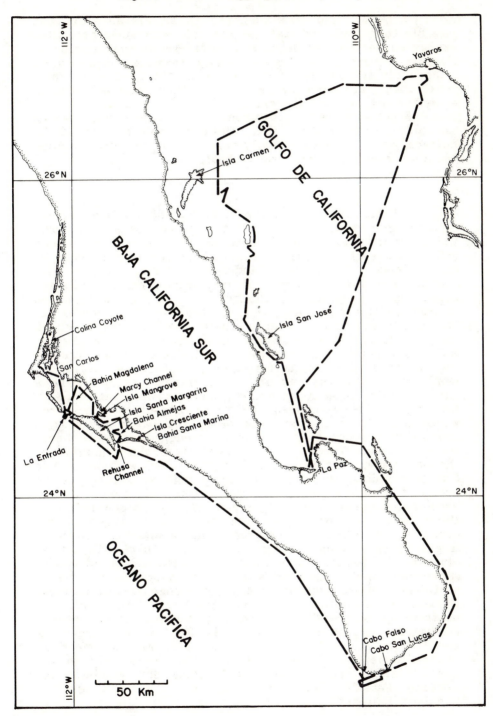

When the ship was anchored in an entrance, it was possible to specify the state of the tide with some precision, except on very windy days, and to determine the position of tide rips. Surface tidal speed was determined regularly by using chips thrown into the water near the bow. The chips were then timed by stopwatch as they passed along a measured segment of the rail.

Breach data were obtained by continuous mast watch assisted by observers on deck. Data on whale headings in La Entrada were obtained visually relative to the ship's heading and then plotted on the track chart.

Results

In 1967, Gilmore, Brownell, Mills, and Harrison (1967) reported gray whales in Bahia Reforma near Yavaros, Sonora, Mexico, and in adjacent offshore waters. Rice and Wolman (1971) had earlier reported the species far to the south at Punta Mita, Bahia de Banderas, Jalisco, Mexico. No additional published records of gray whales on the mainland coast have appeared, though in management-related discussions of the species one often hears of the "Sonoran" or "Mainland" population of gray whales. We sought to check the the status of gray whales near Yavaros. The *Regina Maris* traveled across the Gulf of California and anchored in the entrance of Bahia Reforma, on 5 to 6 January 1979. A skiff transect of the bay and its entrance was made and a number of inhabitants including fishermen were interviewed. No gray whales were seen and no reports of their presence were obtained. A beach walk along the outer bay margin produced no bones of gray whales, though odontocete bones were common. The bay is now dominated by a cannery, and the bay waters are very polluted and heavy with algae and fish fragments. Our visit was early, and we probably came before the arrival of the whales. One informant (Robert Brownell) suggests they may arrive as late as February, and there have been unpublished summer sightings of what were probably gray whales in the area. Informants were aware of larger whales offshore, where we too found them. They were blue and fin whales and were concentrated at the edge of a submarine shelf that parallels the shore and extends northward toward Guaymas. A larger concentration of these whales than we noted was recorded earlier off Guaymas, also on the edge of the shelf, and also among schools of fish and flocks of birds, on the cruise of the *Regina Maris* preceding ours (Cruise #11). It may be that these animals are members of the same group that is frequently encountered in the midriff island area of the Gulf (Islas San Esteban, San Lorenzo, Partida, Raza, and Angel de la Guarda, and the Canal de Ballenas).

Cabo San Lucas - Cabo Falso

An aggregation of 82 juveniles and adults was noted in the Cabo Falso - Cabo San Lucas area on 9 February 1979 by Norris. Behavior taken to be courtship was observed and only one uncertain record of a young of the year was noted. Since there is no calving lagoon near this location, the composition of this group posed questions about its origins and possible function.

Because of this earlier sighting, the cruise of the *Regina Maris* was designed to include two transects of the area to determine the distribution, composition, and numbers of whales found there (Figure 2). The group was found to congregate over the relativly shallow bank extending southward from the tip of Baja California. The presence of courting groups, juveniles, and single adult whales was confirmed, but no mother-calf pairs were seen. The easternmost whales were located just to the east of San Lucas Bay at Cabeza Ballena. The number of whales present varied from a low of 18 animals seen on the transect on 13 January 1979, to 66 animals seen on 11 February 1979. Though the heaviest concentration was found in the transect area, many other whales were seen in transit to and from the area and as far as 32 km to the north. A total of 174 whales seen in the entire area was composed only of juveniles and adults, without mother-young pairs.

Two days after our 1979 count at Cabo Falso - Cabo San Lucas 16 animals were counted at Punta Tosca, all adults and juveniles with no calves. Then on 18 January 1979 a transect was run across the entire entrance at La Entrada, and 183 adults and juveniles, many in courting groups, were counted. No mothers and young were seen. The La Entrada population stayed more or less stable after this count while the Cabo San Lucas group grew markedly.

Thus the Cabo San Lucas - Cabo Falso group is clearly composed of juveniles and courting adults, the same classification seen at lagoon entrances elsewhere. It seems to build up later than entrance aggregations immediately to the north. Two readily identifiable individuals were seen at La Entrada during our entire stay there, suggesting, but of course not proving, stability. More study is needed to uncover the dynamics of these entrance groups of whales.

Punta Tosca - Canal Rehusa - Bahia Santa Marina

The southernmost calving lagoon in Baja California is southern Bahia Almejas - Bahia Santa Marina, which opens to the sea through the Canal Rehusa at Punta Tosca, at the southern tip of Isla Santa Margarita (Figure 3). Though the channel is dangerous to most navigation, it is protected from

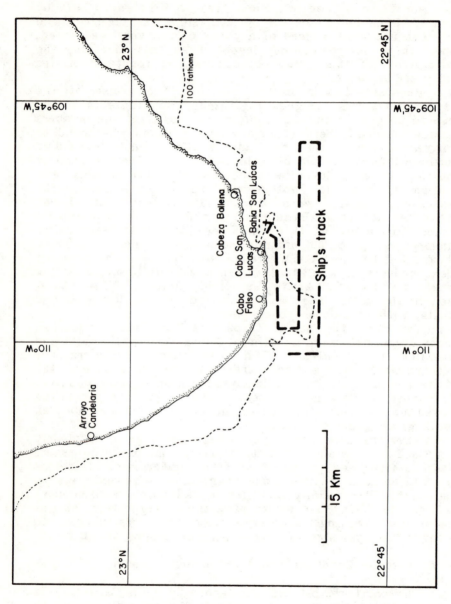

Figure 2. Transect Map, R/V *Regina Maris*, Cabo Falso-Cabo San Lucas, Baja California Sur, Mexico, January – February 1979.

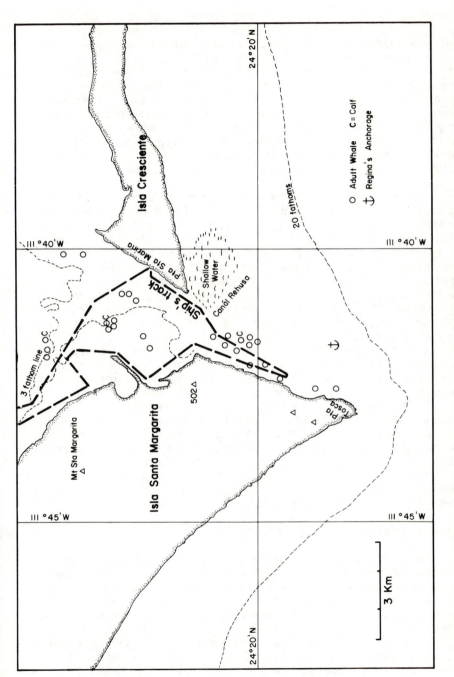

Figure 3. Transect Map, Canal Rehusa, Baja California Sur, Mexico, January 1979, showing the route taken by a small boat and the positions of whales sighted. The anchorage of R/V *Regina Maris* is shown.

prevailing northwest winds by the steep rocky ridge of Punta Tosca. Two connected sand bars are located along the southern margin of the Canal Rehusa and produce a large area of swells and breaking waves at low tide, even in calm seas. Whales are common both in the channel and over the bars. Canal Rehusa curves northeastward along the western tip of Isla Cresciente, broadens to a considerable embayment inside, and then ramifies into a series of narrow but rather deep channels that connect to both Bahia Santa Marina and Bahia Almejas. This complex is occupied both by an entrance aggregation of whales, and mother-calf pairs, the latter concentrating in the Canal Rehusa and inside Isla Cresciente.

While the *Regina Maris* was at anchor, whales moved by us, moving into Canal Rehusa with the flood tide, especially across our stern and along the outer edge of the main sand bar. Whales also travelled in the channel itself, some within an estimated 100 m of the Punta Tosca bluffs. A later reconnaissance by inflatable boat on 23 January 1979 of the inner and outer Canal Rehusa showed approximately the same distribution of animals, with mother-calf pairs scattered throughout the inner embayment and inner Canal Rehusa. The outermost pair overlapped the entrance group, and was seen at the outer bar (Figure 3).

Whales were commonly seen wallowing in the shallow turbulent water over the bars, and pitchpoling (whales extending their heads vertically out of the water) was frequent, an indication that water depth was less than the length of the whale (Walker 1975; Norris et al. 1977). Whales were also noted to pitchpole with water pouring from the corners of their mouths, sometimes clear and sometimes brown with bottom sediments (Norris et al. 1977). Courtship groups were commonly seen in this entrance.

Though we had planned extensive observations at Punta Tosca, bad weather forced us out, and into Bahia Magdalena where the remainder of our work was done.

Bahia Magdalena and La Entrada

La Entrada is a 5.4 km wide entrance defined by Punta Entrada on the WNW and Punta Redonda on the SE. Both points are tipped with submerged shoals that narrow the entrance an additional 1.3 km. Underwater, these shoals cross La Entrada as a sill averaging 28 m deep; the effective entrance width at 18 m is about 3.3 km. Inside, the sill depths vary from 28 to 36 m. This restricted opening serves the very large water area of Bahia Magdalena and much of Bahia Almejas. The result is a pattern of persistent and clearly visible tide rips at both ebb and flood tide. Two major rips, approximately off the tips of the submerged extensions of the two points, are closely associated with whale distribution in La Entrada. The ebb rip at Punta Entrada joins with a longshore current to curve southward across La Entrada, while a similar rip off Punta Redonda tends to extend farther to sea, probably pushing the longshore current with it in a long arc bending southward. Inside the bay as many as three rips were seen, the major northern one extending toward Punta Belchers, another more minor one more or less bisecting the bay, and the other major southern one curving off toward Marcy Channel and Mangrove Island.

Flood rips begin at La Entrada and flood into the bay, arcing in from the Punta Redonda side as tidal currents outside the bay change direction. Both major rips have wavery cusped margins at moderate velocities, tending to straighten as speed increases. Both birds and whales congregate along the rips at ebb and flood tides.

An eddy of considerable size was found to exist between Punta Entrada and the northern tide rip. This proved important in attempting to understand whale activity in the area. Much of the water of the eddy ran in the opposite direction from the main flow in La Entrada, and the whales there responded to its flow. Before we understood this, much data had been gathered of total movement patterns in La Entrada, and hence the relationships shown in Figure 4 are less clear-cut than was actually the case. Two relationships were involved in the eddy: reversed flow and reduced current speed. Both affected whale behavior. Most whales, however, occupied the rip areas.

The eddy was found to extend in along the Punta Entrada shore to Punta Belchers, where it curved out into the bay and met the northern tidal rip. Punta Belchers, in fact, gives the appearance of having been formed by this eddy. Eddy waters moved much more slowly than those of the rips, a fact borne out by bottom samples -- fine-grained muds from the eddy area and coarse clean sand from the rips.

A large entrance group of gray whales was present in La Entrada (Figure 5), and it proved possible to anchor the *Regina Maris* almost anywhere in the area. Her high masts served as fine observation platforms from which to study these whales. Four surveys on a standard track were run to determine the abundance and distribution of whales in La Entrada. Once the entrance group was defined, another track was run on 21 January 1979, to determine the separation or overlap of mother-calf aggregations that normally stay inside lagoons, and the entrance group, which included no mother-calf pairs. Then a series of tasks was run from the vessel anchored in the Punta Entrada rip. Included were tidally related movements of whales, behavioral periodicities including breaching, possible feeding by both whales and birds, and the behavioral details of

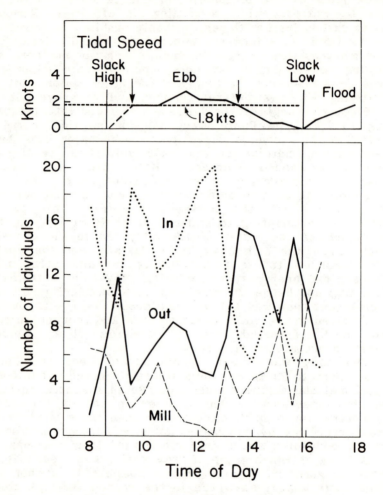

Figure 4. La Entrada, on the ebb tide of 27 January 1979. The separation between animals facing in and out was much sharper at the rip than in the eddy. (The figure is based on pooled data.)

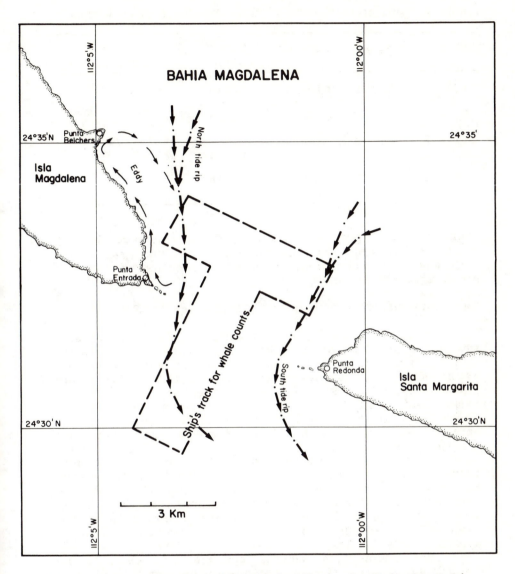

Figure 5. Transect Map, R/V *Regina Maris*, La Entrada, Bahia
Magdalena, Baja California Sur, Mexico; January -
February 1979; showing the approximate positions
of north and south tide rips, the north eddy and
its relations to Belchers Point, and the whale
counting transect.

Table 1. Population Counts of Gray Whales at La Entrada.

Date	Observer	Weather	Tidal State	No. of Whales
18:I:79	Norris	Clear, moderate breeze	Rising, near slack water	183
30:I:79	Norris	Clear, calm	Ebbing, 2 hrs to slack water	104
9:II:79	Würsig	Clear, calm	Rising	131

courting-mating groups. Taken together, our understanding of these patterns lets us comment on the gray whale migration as a whole, and the possible function of entrance group aggregations in it.

The Distribution and Numbers of Whales at La Entrada

During the time the *Regina Maris* worked in La Entrada, whales exhibited a regular distribution that varied significantly with the tide. At high tidal velocities most, but by no means all, whales were concentrated in the Punta Entrada rip, with lesser numbers in the Punta Redonda rip. Some could usually be seen in the eddy behind Punta Entrada. Those in the Punta Entrada rip extended from just inside the entrance sill far into the bay. These animals were not precisely lined up along the rip, but were found across the rip to a few hundred meters to either side of it. In the Punta Entrada eddy whales tended to defy the tidally related regularities shown by whales nearer the rip. For instance, courting and mating behavior was commonly seen there in the slowly revolving water throughout a tidal cycle, while it was essentially absent in the rip area above current velocities of approximately 1.9 knots (1.0 m/sec.).

Three population surveys spanned our visit to La Entrada (25 days). They do not clearly demonstrate changes in numbers of whales during this period (Table 1). The lowest count was made on ebb tide after a very extreme tide the day before and many animals were far outside the bay in extensions of the tide rips there. Two observers were involved in the surveys (Norris and Würsig). A hint that the stays of some animals are rather long at a given lagoon mouth comes from repeated observations of two easily identified whales, one with fluke tips shorn off on both sides, and the other with a dinner plate-sized white patch just beneath the first dorsal knuckle on the right side. These animals were sighted on the day of arrival and on the last count, and were seen many times in between.

Tidally Related Behavior

Whether whales at La Entrada swam into or out of the bay, or milled, proved to be related to tidal velocity. Most whales responded to a tidal current by swimming into it above about 1.9 knots surface velocity (1.0 m/sec.) (Figure 4). This is unlike the reactions of mother-calf pairs deep in lagoons who responded directly to the change of tide, rather than its velocity (Norris et al. 1977).

The effect of entrance behavior was to change the behavior of whales from milling at low tidal velocities to aligned diving of animals oriented into the current at higher velocities. Since the animals that remained in the large eddy occupied slow moving water throughout the tidal cycle, they often continued milling while most of the population had begun to orient to tidal rips. The orientation to rips seemed independent of a fixed position on the bottom, since at ebb tide considerable portions of the population could be found out to sea from La Entrada, orienting to the current, while on the incoming tide nearly all were inside.

The pace of orienting whales matched or was slightly greater than the current velocity, causing them to dive and swim almost in place, or to proceed slowly along against the current.

As tidal velocity slacked, the animals began to wander away from rip areas even though true slack water was many minutes away. Sometimes, however, when the prevailing northwest wind reinforced tidal flow, the picture became less clear, with many animals continuing to orient into the surface current throughout the tidal cycle. Typically, animals gathered at the rips and did not occupy the entire mass of moving water at the entrance. This produced two major lines of animals, one off Punta Redonda, which tended to have fewer animals, and a larger assemblage at the rip off Punta Entrada.

A more detailed picture of these events is as follows. As tidal flow increases, the tide rip straightens and strengthens, and whales that have been engaged in courtship patterns begin to orient into the current, usually in two's and three's, and nearly always in echelon formation. If three animals are involved, there is usually a close echelon of two animals nearly touching, and a somewhat more distant consort animal also in echelon. These animals swim cleanly forward into the current. They do not tip onto their sides, or throw pectoral fins free of the water as they do during courtship. After a few breaths, and while swimming forward at or slightly in excess of the current speed, they dive one by one, each throwing its flukes deliberately upward in a moderately steep dive. They dive for 1.5 to 6.7 minutes in our records, often emerging close to the submergence point but sometimes travelling forward 100 m or more. They then respire several times before diving again; often about two times for each minute submerged. One could see animals gathering into the rips, swimming toward the turbulent water from calmer water to the sides, and then beginning the resolute diving pattern just described. The appearance is of animals seeking out and diving in this swift water.

Such orientation to currents by gray whales has been observed by others (Darling 1977) and by us elsewhere (Norris et al. 1977). Gray whales often enter the surf along their migration path and navigate easily through turbulent entrances such as that at Boca Soledad. From our 1978 camp at this entrance we watched the passage of whales from the lagoon to the sea. Most stayed in the two channels, though some were seen in small eddies or actually in turbulent breaking water. One animal was followed from the calm water inside until it passed the final line of breakers. It floated passively at times, and sometimes swam slowly toward the entrance. When floating passively, it often bobbed head up in the water as it drifted along. At the outermost bar it bobbed its way through partly breaking water, moved wholly, it seemed, by the force of the current. It should be emphasized that circumstances at Boca Soledad and La Entrada are wholly different. Boca Soledad is a mass of breaking water tumbling over shallow bars intersected by two rather narrow and very swift tidal channels, while La Entrada is rather deep open water without breakers. If current-oriented diving occurs at Boca Soledad, and we could not observe it from our low sandy beach vantage point, we would expect it outside the entrance. The spouts of whales could be seen outside Boca Soledad but the behavior of these whales could not be seen. Flights over the outside area of this Estero and Estero Santa Domingo to the north revealed what seemed to be diving and perhaps feeding whales, leaving long accurate muddy paths in their wakes (Norris et al. 1977).

Swartz and Jones (1980a) found some indications of tidal orientation and movement near the entrance of San Ignacio Lagoon, but in the area furthest from the entrance, they found a very strong correlation with tidal movement (Swartz and Jones 1980b). Therefore, we decided to repeat our previous tidal observation at Estero Soledad (Norris et al. 1977) and to do so from a vantage point in the inner lagoon where it is a long, simple, and rather narrow tidal channel for several km in either direction. This vantage point is called the Second Shell Mound, and it is located a short distance to the south of the fishing town of Lopez

Mateos. The first day's observations extended from near high water to slack low tide. During this time a continual procession of animals passed our vantage point in the 0.8 km wide channel moving toward the entrance. Forty-eight mother-young pairs and 23 animals judged to be single passed our view, all moving with the tide. Most passed swimming slowly with the current, some drifted passively, and a few faced upstream as they passed. Mother-calf pairs tended to remain spaced many meters apart, as though a very low order of repulsion, or tendency to remain separate, existed between them. Most such pairs were separated from their nearest neighbors by 100 m or more, though some drifted within a few meters of one another, though usually only transitorily. The exception to this regularity was a group of whales that entered an eddy adjacent to the west bank and did not move further with the tide. They milled and pitchpoled repeatedly in an area estimated at 200 m in diameter.

The observation was repeated the following day, spanning a tidal change. The whales acted as before, but passed in both directions with the tide, first out toward the entrance of the Estero, then followed by a period with few passing whales around slack tide, and then in a stream moving into the inner Estero. The milling animals that had occupied the eddy the day before were absent.

Probably all animals in the central lagoon were involved in this movement as counts of animals there closely matched the numbers of moving whales counted. Tidal events at Colina Coyote in the innermost lagoon reported on earlier (Norris et al. 1977) were reobserved by Würsig. The figures are shown in Table 2.

These figures show that as the tide went out many whales swam actively with it, although many others drifted in, with no consistent orientation with the current ("Milling" category, Table 2). During low tide and the beginning stages of rising tide there were fewer whales in the area than during high tide. And finally, during rising tide the whales did not actively swim into the area with the tide, but instead drifted in with the moving water.

Possible Feeding Behavior

Gray whales are generally supposed to feed very little while south of polar feeding grounds (Pike 1962). This view was reinforced by Rice and Wolman (1971) who failed to find a significant difference in blubber thickness between 26 southbound and 26 northbound whales and attributed a considerable weight difference to utilization of body fat. They felt that girth changes and oil yield of north and south moving whales were likely to be more reliable indicators of nutritive state than blubber thickness. We concur, feeling

Table 2. Tidally Related Movements of Whales at Colina Coyote, 4-7 February 1979.

	# Mother-calf pairs within 1 km of boat								
	Moving North	%	No.	Moving South	%	No.	Milling	%	No.
Ebb Tide (water moves north)	3/hr	38	48	1.1/hr	14	18	3.8/hr	48	61
Flood Tide (water moves south)	1.1/hr	20	13	1.0/hr	19	12	3.3/hr	61	40
Slack Water	1.3/hr	23	9	1.0/hr	18	7	3.3/hr	59	23

that in addition to first utilization of body fat during inappetence the blubber itself will not necessarily collapse in direct proportion to oil removal for metabolic needs.

Rice and Wolman (1971) reviewed the literature on gray whale feeding. Like Scammon (1874), Andrews (1914), and Pike (1962), they confirm that whales taken by whalers during migration almost invariably have empty stomachs. They found exceptions though. Two northbound whales (of 136) contained food. An anoestrous female taken on 20 March 1964 contained about:

> "20 liters of the zoea stage of the littoral crab *Pachycheles rudis* (Anomura, Porcellanidae) and a few brachyuran zoeae, probably of the genus *Fabia* (Brachyura, Pinnotheridae). The other animal was an immature female taken on 11 April 1968. Its stomach contained about 50 liters of the zoeae stage larvae of a pinnotherid crab, probably of the same species found in the preceding specimen, and a few scattered porcellanid zoea, which were in too poor condition to identify further."

These samples indicate the ability of the gray whale to filter quite small food. They further report that migrating whales sometimes had other miscellaneous items in their stomachs, ranging from gravel mixed with ascidian tunics, fragments of hydroid stems and polychaete worm tubes, a few gastropod opercula, one pelycypod shell and miscellaneous bits of wood. They go on to note, in a fashion contradictory to their thesis of inappetence, that "Gravel and sand are probably ingested accidentally while the whale is feeding."

Even though the great majority of the Rice and Wolman sample showed empty stomachs, the tendency to feed south of the polar feeding grounds is indicated by their data. Baleen whales are batch feeders, adapted to periods of inappetence between bouts of feeding. Collection sites, such as the San Francisco location of their sample, may be in areas without significant food resources or feeding circumstances for moving animals, while other areas or locations may be more suited to feeding. Certainly, the summering populations of gray whales south of the Arctic, some of whose members return year after year to the same location, feed. Darling (1977) has produced such evidence for gray whales summering along the west coast of Vancouver Island. These animals spend much time in apparent feeding, presumably on benthic invertebrates. The whales move in shallow waters over sand bars, and in areas of strong tidal or wave-generated currents, and grub in the bottom producing boils and trails of muddy water pouring from their mouths. Other such summering populations are known from Langara, Queen Charlotte Islands (Pike and MacAskie 1969) and various loci along the Washington, Oregon, and northern California coasts, as far south as the Farallon Islands (Rice and Wolman 1971; Ainley, Huber, Henderson, Lewis, and Morrell 1977).

The assumption has been made that whales during the migration move so resolutely at speeds of 7-9 km/hr that little or no time is left for feeding (Rice 1962; Rice and Wolman 1971). Such velocities have been recorded for whales passing observation stations where the moving stream of animals becomes concentrated close to shore by the proximity of deep water offshore. Movement there may not be typical of the migration path in its entirety. Assuming a 4 mph (6.4 km/hr) migration speed, and a 5,000 mile (8,050 km) path, such velocities, if constant, would cover this migration distance in 41 to 52 days. The southward migration, however, appears to take a minimum of 60 days, judging from times of leaving the Arctic (late October) to arrival at calving lagoons (early January). The northward migration is bimodal; the first northward migrants are single males, pregnant females, and juveniles, and these are followed thirty to sixty days later by females with calves (Poole in press; Herzing in press). These figures support somewhat lower overall rates of migration than have been assumed, and are themselves supported by the frequent observation of whales along the migration route circling and diving, sometimes for hours, behind points of land where the migration course changes, or wallowing in the surf. Migration is not always a resolute and continuous motion along the path. Especially on the northward movement there is considerable time when whales might be feeding (note that Rice and Wolman's two whales with significant stomach contents were from anoestrous or immature females on the northward migration). An immature animal observed on the northward migration was seen mouthing kelp and was presumed to be feeding on mysids (Wellington and Anderson 1978). Other observations on the migratory path implicate schooling fish (Gilmore 1961; Sund and O'Connor 1974) and euphausiids (Howell and Huey 1930).

None of this embraces the problem of potential feeding at calving lagoons. Various reports have shown that some level of feeding occurs there. Whales have been seen with muddy water spilling from their baleen, and muddy boils have been seen to follow lagoon whales (Walker 1975), especially those near entrances (Norris et al. 1977; Sprague, Miller, and Sumich 1978). The secondhand report of Matthews (1932) states that Norwegian whalers (operating from a whaling station at Punta Belchers) found gray whales feeding on the "red crab" (*Pleuroncodes planipes*; Anomura, Galatheidae) at Bahia Magdalena, Baja California Sur, in 1926. Walker (1975) reported a mother-calf pair "feeding on bait fish." But baby whales at this age are nursing, and no evidence is given to substantiate the claim that the mother was feeding. Swartz and Cummings (1977) and Swartz and Jones (1981) reported behavior from San Ignacio Lagoon strongly suggesting feeding.

Even now we cannot assert with complete security that feeding does occur in the vicinity of calving lagoons. Observations on Cruise No. 12 of the *Regina Maris*, however, strongly suggest daily feeding by entrance aggregations of gray whales.

As tidal currents at La Entrada rise in velocity, courtship-mating gives way to what we believe are probably feeding dives. This is the activity we found concentrated along tidal rips, both inside and out of Bahia Magdalena. The resolute and regular dives of these whales without major social interactions between animals seem best explained as feeding dives. As the whales rose, sea birds usually hovered over them; in our observations juvenile and adult Heermann's Gulls (*Larus heermani*), California Gulls (*Larus californicus*), and Ringbill Gulls (*Larus delawarensis*). These birds, which were sometimes extremely numerous, followed the diving whales, and often settled in the rear edges of boils caused by the whales as they rose. Several times in this circumstance we saw gulls picking small red crabs (*Pleuroncodes*) from the water, but most frequently they pecked at the surface for prey we could not see. This behavior is closely similar to that observed by Harrison (1979) in the Bering Sea where most common marine birds feed in the boils created by feeding gray whales. He speculates that this feeding method is of great importance in providing food for sustaining the large number of birds involved. He also notes that Bedard (1969) recovered gammarid amphipods from St. Lawrence Island auklets, and raised the question of how these birds obtained bottom-dwelling crustaceans.

From time to time we also saw schools of fish in the vicinity of the surfacing whales. The fish were small (2-4 cm) and sometimes cascaded into the air or dimpled the surface near the whales. Neuston tows behind rising whales produced a quantity of planktonic forms, especially euphausiids of the species *Nyctiphanes simplex* (3-6 mm), fish eggs, crab larvae, copepods and phytoplankton. It is possible the whales were ingesting such planktonic forms.

Dr. Edward Brinton of Scripps Institution of Oceanography (pers. comm.) who identified the planktonic forms, notes that *Nyctiphanes simplex* is almost always dense in the Vizcaino Bay-Magdalena Bay areas (the water can be soupy with younger stages), and that it can be locally as great or greater in density than the food of whales in northern waters. He also notes that our samples contained only juveniles and larvae of *Nyctiphanes*, but that adults, which reach 10-13 mm in length, can avoid plankton nets in the daytime (as can *Pleuroncodes*). The recorded density of *Nyctiphanes* in our sample of 182 organisms per m^3 is regarded as dense (Table 3).

On 25 January 1979, the beach at Belcher's Point was windrowed with dead *Pleuroncodes* at high tide line, whitening in the sun, and thus many days old. The thousands of gulls resting there during slack water, when chased into the air, left scats across the beach that proved to be composed wholly of fragments of *Pleuroncodes*. Judging from the size of the chelipeds in these scats, as compared to adult crabs in the windrows, the birds seemed to be feeding on approximately half grown crabs, the same as we saw them take behind whales in the rips of the Entrada. These same birds were found along the rips at the height of flood and ebb tide.

Our supposition is that these entrance group whales in the mouth of Bahia Magdelena congregate in or near rips for feeding, and that they dive and move only slowly against the

Table 3.   Plankton samples taken near diving whales at the entrance of Bahia Magdalena, Baja California Sur, Mexico, 29:I:79.

Net	Species	No./m^3
Neuston (taken directly behind surfacing whale)	Crab larvae; incl. zoeas and megalops of Grapsids, pasiphaeids, alpheids, callianassids and cancrids	150
	copepods	34
	mysids	6
	Nyctiphanes simplex	182
Borego (taken near tidal rip during diving by whales)	crab larvae	3
	copepods	22
	chaetognaths	39
	Nyctiphanes simplex	13

current, allowing the moving water to pass food organisms past them. The entrances and their attendant rips may serve to concentrate food both from inside the bay and from the ocean outside. We saw no evidence that these whales were grubbing in the bottom, though it could have occurred. Neither *Nyctiphanes* nor *Pleuroncodes* are bottom dwellers.

The short coarse baleen of the gray whale seems an inefficient feeding device for very small prey. Very small prey items are, however, sometimes taken, and these include planktonic forms unlikely to be grubbed from the bottom, as for example the zoea reported by Rice and Wolman (1971). Nonetheless, the shortness of gray whale baleen is probably related to the grubbing habit. Kasuya and Rice (1970) point out that the baleen on the right side of gray whales typically shows more wear than the left, hinting that these animals swim on their right sides when feeding. Curved mud trails or elongate muddy boils produced by gray whales grubbing in shallow water suggest that the animal moves along, sometimes on its side poking or sucking at the benthic surface. The observations of Ray and Schevill (1974) suggest that effluent water is ejected out of the corner of the animal's mouth by an unexplained pumping mechanism.

The view that begins to emerge is that at least non-parturient, non-suckling gray whales may feed, perhaps at important levels, on the migratory path at localities where food may be concentrated, as for example in the current rips at lagoon entrances. A variety of prey, some quite small, and not limited to organisms grubbed from the bottom, may be included. The evidence that pregnant or nursing gray whales feed south of polar feeding grounds is less compelling but suggestive of some level of intake.

Breaching

The cause of breaching remains unclear. In this behavior the whale bursts from the surface at about a 45-60° angle in normal orientation, rotates about a quarter turn, arches and plunges back into the water partly on its side, landing in a great welter of white water. An attempt was made to see if breaches correlated with tidal state or behavioral state.

Watches were kept for breaching both in La Entrada and the lower reaches of Estero Soledad. Five days of Entrada beach data and four days of Estero Soledad data are summarized with respect to time of day in Figure 6. Breaches occur in series, with 2.5 (N=301) breaches per animal in La Entrada, and 1.9 (N=236) breaches per animal in Estero Soledad. The cumulative record shows a peak in breach frequency around midday. In La Entrada more breaches occurred during rising and lowering tides (11.2

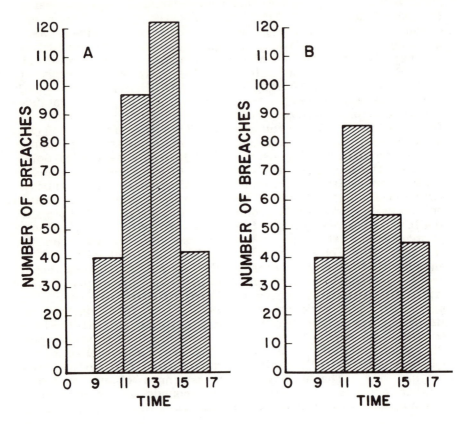

Figure 6. (A) Numbers of breaches in relation to time of day
        (M.S.T.) from five days (20, 25, 26, 27, and 29
        January 1979) in La Entrada.
        (B) Numbers of breaches in relation to time of
        day (M.S.T.) from four days (4, 5, 6, and 7
        February 1979) in the calving lagoons of Colina
        Coyote (Estero Soledad).
        Numbers of breaches during the beginning and end
        of the day (0900-1100 and 1500-1700) are
        significantly lower than those during midday
        (1100-1500) in (A) and (B). (t-test of
        percentages with arcsin transformation, p<0.001
        for (A) and p<0.01 for (B)).

breaches per hour; SD=7.32, N=21 hrs) than during periods of slack low and high water (3.6 breaches per hour; SD=3.15, N=7 hrs) (p<0.02, t test). In Estero Soledad this trend appeared reversed with 3.8 breaches per hour (SD=6.40, N=32 hrs) during tidal flow and 8.1 breaches per hour (SD=6.40, N=4 hrs) during slack tide (p<0.01, t test). But the few hours of slack tide at Estero Soledad coincided with the midday peak in breaching activity, and it is therefore difficult to assess whether tide or time of day was the dominant factor influencing breaching.

The highest breaching incidence in La Entrada occurs during highest tidal velocities, after a change in behavioral state. It tends to occur less frequently during courtship periods at lower water velocities. This supports Walker's thesis that breaching and feeding are linked since he saw water streaming from the baleen of leaping whales on many occasions (see Walker 1975, p. 37). We too have several times seen water streaming from the baleen of leaping whales, but it does not always occur and is seldom noted in breaching whales deep in lagoons. The occurrence could well be adventitious; that is, if whales are feeding at high water velocities, their mouths could be full of water when they breach and this might not relate to the reason for the breach. It is worth noting that while less frequent at slack water, breaching does occur then at considerable levels.

The pattern is reminiscent of the aerial behavior of odontocetes, which has been implicated in the production of underwater sound (Norris and Dohl 1980; Würsig and Würsig 1980). The gray whale pattern is almost identical in detail to that of *Steno*, with a long curving leap, and a turning onto the side as the animal falls, thus cupping water under the falling body and producing a welter of spray as the animal enters the sea, and presumably producing underwater sound as well.

We cannot sort out these potential causes with our preliminary data, but suggest that its use in underwater signalling is not unlikely.

Courtship and Mating Groups

We use the terms "courtship" and "mating" tentatively with regard to the behavior of entrance groups of gray whales. While we identified both males and females in such groups, and we saw sexual behavior, including extended penises, and much caressing of one animal by another, juveniles were members of many such groups. It may be, as in odontocetes, that sexual patterns begin very early in life and that they assume in part a social signalling function, quite apart from the needs of reproduction. For this reason, in the field we used the unfreighted term **wuzzle** to describe such activity.

With the coming of slack water, either in eddies, or as the rip areas began to slow and wander, whale behavior changed markedly. This was clearly indicated by the headings of animals (Figure 5). From most animals being oriented into the current and swimming and diving slowly and regularly into it, animals began to wander in the Entrada, and to mill. Shortly, the regular swimming was supplanted by animals in courting-mating patterns. Swimming oriented to current direction was lost. In place of regular dives, animals began to roll in the water, throwing pectoral fins, or parts of flukes into the air. Dives became irregular. Bodies were turned and flukes tipped as animals submerged, curving their paths underwater. Such dives were often shallow excursions bringing the diver up under an adjacent whale. Whales probed the genital areas of other whales. Often a whale swimming at the surface was approached from behind by another animal, often followed by probing the genital area of the lead animal. Often the first animal then arched its tail over the snout of the probing animal, and its flukes either rose above the water or skimmed the surface. Animals swam over others, in contact. Swift lateral swipes of tail and flukes were common, directed at but generally not contacting another whale. We had previously come to know this as an aggressive movement. Animals often travelled in close echelon pairs, and were often joined in echelon by a third animal, farther away. As slack water ensued, these duos or trios tended to coalesce from all points of the compass, until as many as 7 or 8 groups might come to occupy a disc of sea 100 m in diameter, the individual groups separated by a dozen meters or less. At times in such aggregations we noted groups swimming head on, sometimes touching snout tips and sometimes diving beneath others.

Though we noted extended penises in such groups, only one case of intromission was noted by us. Others have noted it, presumably in such groups (Gilmore 1968; Samaras 1974).

Friendly Whales

On two occasions (27 January 1979) near slack water, whales approached the *Regina Maris* while she lay at anchor in the Entrada. The first encounter started with a single juvenile of unknown sex, approximately 10.2 to 10.8 m in length, approaching the ship astern, investigating her hull, stern post rubbing lightly against her hull, and circling around her for a period of slightly more than 1 hr. During this time the animal repeatedly rose to blow within a meter or two of the vessel. Three observers had the animal literally blow in their faces, and reported that the blow was salt water, and that no foul or ketone odor was involved. From time to time one to three additional animals, also juveniles, came in with the first individual, though they generally stayed outboard of the first animal. On the first encounter the primary animal twice was seen to swish its flukes toward the *Regina Maris*, but in neither case made contact.

Our impression of these encounters was, first, that the object of the encounter was the ship's hull, especially below the waterline, and especially aft. The attentions, we felt, could either have been sexual in nature, or the approaches of a juvenile to a mother whale. The fluke blows seemed petulant, and the pace of swimming of the animals, especially the three that did not come close, seemed more rapid than other animals in the area, perhaps indicative of heightened excitement. We saw no reason to think that the presence or absence of people on the ship was involved in the event.

Discussion

It is instructive to think of the gray whale migration as composed of two more or less separate parts: There is a group which leaves the Arctic first, pregnant females who move resolutely down the coast, and enter calving lagoons to give birth and nurture their young. They seem to avoid entrance whales, and even tend to remain separated from each other. They float quietly, sometimes remaining still for long periods, nurse, and move back and forth with the tidal cycle. Occasionally they breach or pitchpole, but mostly behavior is that between mother and calf. The second group is composed of resting females, males, and juveniles of both sexes. They stay somewhat longer in the Arctic, move southward in a column that to some degree overlaps the front running pregnant animals. All along the migratory path they court and mate. Once at lagoons they tend to aggregate at entrances, only a few penetrating into the lagoons themselves. At the entrance of Bahia Magdalena, courting and mating are daily tidally-related occurrences, interspersed with long bouts of what appears to be feeding, when water velocity into or out of the bays is strongest. These whales seem to distribute themselves at lagoon entrances probably starting at the northern ones first, and late arrivals move further and further south, until the latest comers aggregate over banks at the very southern tip of Baja California at Cabo Falso - Cabo San Lucas, which are the last shallow waters available unless the whales round the cape.

The northward migration for both groups is leisurely compared to the southward movement, and in the case of entrance groups at least, may involve some feeding along the way. Some whales do not reach the Arctic at all, but distribute themselves out to small subsidiary feeding grounds along the way (Darling 1977). We may regard them as alternate feeding sites to the main Arctic grounds, and thus a sort of population overflow mechanism that may have been especially important during times of ice coverage of Arctic feeding grounds.

While we could not actually observe feeding by whales either at Canal Rehusa or in the entrance of Bahia Magdalena, the circumstances suggested strongly to us that when the current was strong whales congregated, without major social interaction (except that they came and dove in echelon-oriented pairs or trios), and dove into a place where food, including small schooling fish, lobster krill, and smaller planktonic life such as euphausiids were common. Their dives did not take them long distances underwater but instead seemed to station the animals, swimming at about tidal flow speed, for a period of minutes. We expect that these rips concentrate food in the entrances. At ebb tide, much of the volume of the bay flows through the entrance, and organisms distributed within the watermass of the bay should find themselves swirling and concentrated in spiralling penants in these rips. An animal might collect significant food simply by stationing in such a current. Gulls were noted feeding extensively behind the surfacing whales on small lobster krill. At flood tide the animals once again became oriented. The flood tide, too, could be a concentrating mechanism for food, only this time for the krill or other organisms swept in from the sea outside. The windows of *Pleuroncodes* on Belchers Point suggest such a mechanism.

A mother-calf pair was seen only once in La Entrada, though they do mix occasionally with entrance groups at other, more restricted lagoons such as Estero Soledad or Estero San Ignacio. Courting-mating aggregations are almost certainly disruptive places for mother-calf pairs. The attentions of males, as we know from occasional observation of encounters inside lagoons, can separate mother-calf pairs temporarily or permanently (Norris et al. 1977) and the attentions of juveniles as exemplified by our "friendly whale" encounters would be conducive to the peace normally experienced by mother-calf pairs deep in lagoons. It seems reasonable that a modest degree of repulsion exists between entrance whales and mother-calf pairs, as was shown in the almost whale-free buffer zone at Bahia Magdalena between these two kinds of whale aggregations. The circumstances of departure on the northward migration suggest the same thing. Swartz and Jones (1980a) found that once entrance animals leave, mother-calf pairs move lower in the lagoon (San Ignacio) and occupy the vacated space.

We saw no evidence of complex social structure amongst entrance whales. There was no hint of stationing or territory by specific whales, and group processes seemed to flow simply between fluid courtship-mating groups, including echelon pairs or trios, and suspected feeding patterns. The tide and currents seemed to be the major determinants of social events. If feeding does occur in these entrance animals, it could reflect on the otherwise inexplicable movement southward of animals whose effective mating is supposed to occur off California on the southward migration (Rice and Wolman 1971), and who therefore seem to undergo an inexplicable use of energy. A feeding ground, even if less than optimum, is valuable once Arctic grounds are covered with ice and of low productivity in the dark polar winter. But we suspect that the coastal procession and the daily courtship and mating at entrances are more than that and that they probably contribute to the high resiliency of gray whales. We regard also with some skepticism the idea that some level of effective mating does not occur at these loci.

From the standpoint of conservation, it seems quite clear that the entire structure of whale populations at the calving lagoons in Mexico is crucial to the well-being of gray whales. Mother-calf pairs seek solitude and are disrupted without it, thus the deep bays require protection. The proximity of whales at entrances may promote complete mating and probably contributes importantly to energy budgets as well. The planktonic food that seems to be involved is wholly dependent upon adequate water quality. Thus, when policy is being made for protection of the gray whale, not only must mothers and calves be protected, but also the integrity of bay ecosystems as sites for mating and probably feeding, and for the productivity of the food involved.

Acknowledgements

We wish to thank the crew of our 1978 cruise for their assistance: William Schevill, Barbara Lawrence, William Rogers, Mr. and Mrs. James Ruch, and Captain James Christman of the *Nai'a*. We are also grateful to the crew of the *Regina Maris* and the many passengers who helped us. We are grateful to the Port Captain Fernando Armas for helping with arrangements and to the Mexican Government for permission to work in their waters. We wish to thank Dr. Edward Brinton of Scripps Institution of Oceanography, La Jolla, for his thorough analysis of our plankton samples.

Literature Cited

Ainley, D.G., H.R. Huber, R.P. Henderson, T.J. Lewis, and S.H. Morrell.
   1977. Studies of marine mammals at the Farallon Islands, California, 1975-1976. U.S. Dep. Commer., NTIS PB-266 249, 32p.

Andrews, R.C.
   1914. Monographs of the Pacific Cetacea. The California
   gray whale (*Rhachianectes glaucus* Cope). Mem. Am. Mus.
   Nat. Hist. 1: 227-287.

Atwater, T.
   1970. Implications of plate tectonics for the Cenozoic
   Tectonic evolution of western North America. Geol. Soc.
   Am. Bull. 81: 3513-3536.

Beal, C.H.
   1948. Reconnaissance of the geology and oil possibilities of
   Baja California, Mexico. Geol. Soc. Am. Mem. 31: 138p.

Bedard, J.
   1969. Feeding of the least, crested and parakeet auklets
   around St. Lawrence Island, Alaska. Can. J. Zool. 47:
   1025-1050.

Darling, J.D.
   1977. Aspects of the behavior and ecology of Vancouver
   Island gray whales, *Eschrichtius robustus* Cope. M.S.
   Thesis, Univ. Victoria, Victoria, Can., 200p.

Dawbin, W.H.
   1966. The seasonal migratory cycle of humpback whales. *In*
   K.S. Norris (editor), Whales, dolphins and porpoises, p.
   145-170. Univ. Calf. Press, Berkeley.

Gilmore, R.M.
   1961. The story of the gray whale. Privately printed, San
   Diego, 17p.

   1961. The story of the gray whale, 2nd Edition. Privately
   published, San Diego, 17p.

   1968. The gray whale. Oceans Mag. 1(1): 9-20.

Gilmore, R.M., R.L. Brownell, Jr., J.G. Mills, and A. Harrison.
   1967. Gray whales near Yavaros, southern Sonora, Golfo de
   California, Mexico. Trans. San Diego Soc. Nat. Hist. 14:
   198-203.

Hall, J.D.
   1980. Aspects of the natural history of cetaceans of Prince
   William Sound, Alaska. Ph.D. Thesis, Univ. Calif., Santa
   Cruz, 148p.

Harrison, C.S.
   1979. The association of marine birds and feeding gray
   whales. Condor 81: 93-95.

Hatler, D.F. and J.D. Darling.
1974. Recent observations of the gray whale in British Columbia. Can. Field-Nat. 88: 449-459.

Herzing, D. and B.R. Mate.
In press. Gray whale (*Eschrichtius robustus*) migration along the Oregon Coast, 1978-81. *In* M.L. Jones, J.S. Leatherwood, and S.L. Swartz (editors), The Gray Whale. Academic Press, N.Y.

Howell, A.D. and L.M. Huey.
1930. Food of the gray and other whales. J. Mammal. 11: 321-322.

Kasuya, T. and D.W. Rice.
1970. Notes on baleen plates and on arrangement of parasitic barnacles of gray whales. Sci. Rep. Whales Res. Inst. Tokyo 22: 39-43.

Laws, R.M.
1959. The foetal growth rates of whales with special reference to the fin whale, *Balaenoptera physalus* Linn. Discovery Rep. 29: 291-308.

1961. Reproduction, growth and age of southern fin whales. Discovery Rep. 31: 327-486.

Matthews, L.H.
1932. Lobster krill, anomuran crustacea, that are the food of whales. Discovery Rep. 5: 467-484.

Nelson, C.H., D.M. Hopkins, and D.W. Scholl.
1974. Cenozoic sedimentary and tectonic history of the Bering Sea. *In* D.W. Wood and C.J. Kelly (editors), Oceanography of the Bering Sea, p. 485-516. Univ. Alaska Inst. Mar. Sci., Fairbanks.

Norris, K.S.
1967. Animal orientation and navigation. *In* R. Storm (ed.), Oregon State 27th Ann. Biol. Colloq., p. 101-125.

Norris, K.S., R.M. Goodman, B. Villa-Ramirez, and L. Hobbs.
1977. Behavior of California gray whales, *Eschrichtius robustus*, in southern Baja California, Mexico. Fish. Bull., U.S. 75: 159-172.

Norris, K.S. and T.P. Dohl.
1980. Behavior of the Hawaiian spinner dolphin, *Stenella longirostris*. Fish. Bull., U.S. 77: 821-849.

Ohsumi, S.
   1976. Population assessment of the California gray whale.
   Int. Whaling Comm. Rep. Comm. 26: 350-359.

Pike, G.C.
   1962. Migration and feeding of the gray whale
   (*Eschrichtius gibbosus*). J. Fish. Res. Board Can. 19:
   815-838.

Pike, G.C. and I.B. MacAskie.
   1969. Marine mammals of British Columbia. Fish. Res.
   Board Can. Bull. 171: 1-54.

Poole, M.M.
   In press. Preliminary assessment of annual calf production
   of the California gray whale *Eschrichtius robustus*, from
   Pt. Piedras Blancas, California. *In* Proceedings of
   symposium on cetacean reproduction, La Jolla, CA, 28
   Nov.-7 Dec. 1981.

Ray, G.C. and W.E. Schevill.
   1974. Feeding of a captive gray whale. U.S. Natl. Mar.
   Fish. Serv. Mar. Fish Rev. 36(4): 31-38.

Reeves, R.R.
   1977. The problem of gray whale (*Eschrichtius robustus*)
   harassment: At the breeding lagoons and during
   migration. U.S. Dep. Commer., NTIS PB-272 506, 60p.

   1979. Right whale: Protected but still in trouble. Natl.
   Parks Conserv. Mag. 53(2): 10-15.

Rice, D.W.
   1974. Whales and whale research in the eastern North
   Pacific. *In* W.E. Schevill (editor), The whale problem, a
   status report, p. 170-195. Harvard Univ. Press,
   Cambridge, Mass.

Rice, D.W. and C. Fiscus.
   1968. Right whales in the southeastern North Pacific. Nor.
   Hvalfangstid. 57: 105-107.

Rice, D.W. and A.A. Wolman.
   1971. The life history and ecology of the gray whale
   (*Eschrichtius robustus*). Am. Soc. Mammal. Spec. Publ. 3,
   142p.

Samaras, W.F.
   1974. Reproductive behavior of the gray whale,
   Eschrichtius robustus in Baja California. Bull. South.
   Calif. Acad. Sci. 73: 57-63.

Sauer, E.G.F.
1963. Courtship and copulation of the gray whale in the Bering Sea at St. Lawrence Island, Alaska. Psychol. Forsch. 27: 154-174.

Scammon, C.M.
1874. The marine mammals of the northwestern coast of North America. J. Carmany & Co., N.Y., 317p.

Sprague, J.G., N.B. Miller and J.L. Sumich.
1978. Observations of gray whales in Laguna de San Quentin, northwestern Baja California. J. Mammal. 59: 425-427.

Sund, P.M. and J.L. O'Connor.
1974. Aerial observations of gray whales during 1973. U.S. Natl. Mar. Fish Serv. Mar. Fish Rev. 36(4): 51-52.

Swartz, S.L. and W.C. Cummings.
1977. Gray whales, *Eschrichtius robustus*, in Laguna San Ignacio, Baja California, Mexico. U.S. Dep. Commer., NTIS PB-276 319, 38p.

Swartz, S.L. and M.L. Jones.
1980a. Gray whales, *Eschrichtius robustus*, during the 1977-1978 and 1978-1979 winter seasons in Laguna San Ignacio, Baja California Sur, Mexico. U.S. Dep. Commer., NTIS PB 80-202 989, 35p.

Swartz, S.L. and M.L. Jones.
1980b. Gray whales, *Eschrichtius robustus*, in Laguna San Ignacio and its nearshore waters during the 1979-1980 winter season. Final technical report to World Wildlife Fund, U.S. 37p.

Walker, T.J.
1975. Whale primer. Cabrillo Historical Society, 65p.

Wellington, G.M. and S. Anderson.
1978. Surface feeding by a juvenile gray whale *Eschrichtius robustus*. Fish. Bull., U.S. 76: 290-293.

Würsig, B.G. and M.A. Würsig.
1980. Behavior and ecology of the dusky dolphin *Lagenorhynchus obscurus*, in the South Atlantic. Fish. Bull., U.S. 77: 871-890.

Roger Payne, Eleanor M. Dorsey

9. Sexual Dimorphism and Aggressive Use of Callosities in Right Whales *(Eubalaena australis)*

Abstract

From a long-term study of a population of southern right whales (*Eubalaena australis*) wintering in the waters of Península Valdés, Argentina, we have evidence from 20,000 photographs that individuals are distinguishable by differences in the pattern of their callosities. Determining the sex of these animals has been a problem. Whereas it is easy to identify a mature female (the only consistent companion of a calf), we only rarely get a ventral view of a male to confirm his sex. Nevertheless, we have been able to study sexual dimorphism in this species. We have compared our known females with a group of whales that contains the few known males plus other animals that are more likely males, a judgment based on behavioral characteristics of males. By examining photographs of their heads, we find that these probable males have statistically more and larger callosities than the females. Another external feature that differs significantly between the sexes is the occurrence of temporary scrape marks on the body. Males have more scrape marks than females. We suggest that the scrape marks are caused by the callosities of conspecifics and that callosities function as weapons for intraspecific aggression.

Introduction

Callosities, the raised, thickened patches of skin on the heads of right whales, are relatively smooth in fetuses and calves but become rougher with age (Lönnberg 1906). In adults the true color of most callosity tissue is gray, but from a distance callosities usually appear white due to a partial covering of cyamids, ectoparasitic amphipod

295

crustaceans (see Payne 1976, p. 334 for a color photograph). The variability of callosity patterns is sufficient to make recognition of individual whales possible (Payne, Brazier, Dorsey, Perkins, Rowntree, and Titus 1983).

The function of callosities has been a matter of speculation for years. One hypothesis is that callosities act as a substrate to promote large populations of cyamids and that the cyamids benefit the whales by roaming over their bodies and removing larval stages of sessile organisms which, if they matured, might cause more biological damage to the whales than do the cyamids themselves (Lönnberg 1906). However, numerous underwater observations of free-ranging right whales by ourselves and others indicate that in daylight, at least, cyamids rarely leave the callosities. Furthermore, during a long-term study of southern right whales along the coast of Argentina (Payne, in press), we gained the impression that callosities were larger in males than in females. We describe here 1) our method of testing this suspicion in free-ranging animals, 2) the results, confirming our suspicion of sexual dimorphism, and 3) further results that led to a new hypothesis for the function of callosities -- that callosities function as weapons in intraspecific fights and that these fights occur primarily between males (probably for access to females).

Methods and Results

Data Base

Our principal data base, described by Payne et al. (1983), is about 20,000 aerial photographs of right whales taken over the nearshore waters of Peninsula Valdés, Argentina. Observations were also made from shore, from small boats, by diving underwater, and by examining beached corpses. From 1971 through 1977, 557 whales were identified, 411 of which were seen 2 or more (up to 18) times.

Studying Sexual Dimorphism When Only One Age Class of One Sex Can Be Determined

In order to test our impression of larger callosities in male right whales, we needed to know the sex of as many whales as possible, and we wished to do this in a benign way (their endangered status, furthermore, makes any take illegal). Mature females were easy to recognize by their close association with calves, but the situation with males was much more difficult. Because we have seen occasional reversals of expected copulatory roles both by males and by females, the only certain indicator of a male's sex is anatomical evidence obtained from a ventral view (see below for a description). Adult males rarely roll over except in active mating groups, where it is very difficult to connect with certainty a clearly sexed ventral view of a whale with a dorsal view of the same whale, showing its callosity pattern.

When we started our comparison of callosity patterns of females and males, we knew for sure the identity (i.e., the callosity pattern) of only nine males, not a large enough sample to compare with our many known females. We got around this problem by selecting <u>probable</u> males ("males") to add to the known males. Probable males were whales with no evidence of being female and with behavioral indications of maleness pieced together from our long-term study of this population of right whales (presented below). We reasoned that a statistical comparison of 50 definite females with 50 "males" that included only a few definitely sexed males would uncover any obvious sexual differences. The error introduced by the accidental inclusion of females on the "males" list would always be in the same direction, causing the impression that the differences between the sexes are less than they actually are. The presence of females on the "males" list could never support the conclusion that the sexes were more different than they actually are or, for example, that males had structures that are missing in females.

We feel that this approach could be used to study sexual dimorphism in field studies on other species in which it is difficult to determine one of the sexes. In fact, if it were possible to know members of only one sex, with none of the opposite sex being recognizable, one could compare a selection of animals of the single known sex with a list of animals of unknown sex and discover any prominent sexual dimorphism if the sex ratio was not strongly skewed away from the unknowable sex. It would be important in such a situation to try to compare animals within the same age class, as much as possible, so as not to confuse any developmental changes with sexual differences.

In our population of southern right whales, however, we did not know the sex ratio, so we attempted to weight our list of probable males as heavily as possible toward actual males.

Sexing Free-Ranging Right Whales

Anatomical Field Marks. While right whale females are slightly larger than males, there is broad overlap, so size alone is an ambiguous determinant of sex (see Omura 1958 for discussion). A male can be immediately recognized by a penis, but it is rare to see one. Ventral apertures are visible more often, and their configurations are unambiguously

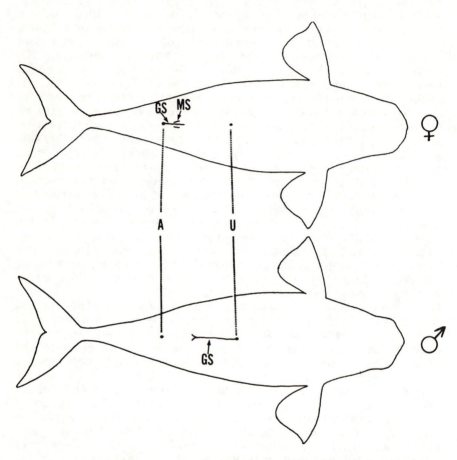

Figure 1. Comparison of anogenital configuration of female
and male right whales. A=anus, GS=genital slit,
MS=mammary slit, U=umbilicus.

different in males and females (Figure 1; Matthews 1938; Omura 1958). The male's ventral slit is more than twice as long as the female's and is located further forward, originating at or near the umbilicus. The male's ventral slit also bifurcates into a Y at the posterior end, whereas the female's does not (Figure 2). Relative lengths of the genital slits and their positions on the body are often difficult to determine, however, under field conditions. In our experience, the most useful indicator of male sex, outside of a penis, is a visible, separate anus, because the anus, in males, is separated from the genital slit by a distance about equal to the length of the genital slit (Figure 2). In females, on the other hand, the genital slit is more caudal, and the anus is not distinct from the genital slit, but appears to be part of the same continuous opening.

Also, in males, the anus is at the summit of a slightly raised pedestal, reminiscent of the crater of a low volcano (Figure 2f). When whales are in active social groups, one sometimes gets a glimpse just of the anus without seeing a genital slit. Such sightings are definite males. The converse (a glimpse of a genital slit with no obvious anus nearby) is not reliable evidence that the animal is a female unless the observer can be certain that a separate anus, if present, would definitely have been seen.

Mammary slits flank the genital slits of females and are easy to see. They are a good indicator of sex even though they are reported to be present in some males (Slijper 1962, p. 380), because they are, in our experience, invariably inconspicuous in males under field conditions. The nipples themselves are seen only in reproductively mature females and may protrude prominently even after a calf is weaned. One nipple may be noticeably larger than the other (Figure 2e), perhaps indicating that calves nurse more on one teat than the other.

Many whales have a white ventral blaze into which the forward end of the ventral slit extends. Because ventral slits are harder to see against white skin (Figure 2b), inexperienced observers may fail to notice the full length of a genital slit, and under such circumstances, score a male as a female. When a blaze is present, the genital slit can still be used to determine sex by looking for its Y-shaped posterior end. Finally, in males, the margins of the genital slit are smoother and more uniform than in reproductive females, which often have margins that are uneven and somewhat scalloped in appearance.

Unfortunately, we have only rarely been able to identify the callosity pattern of a whale whose ventral side we have seen and sexed. Using anatomical criteria, we have been able to sex only 23 males definitely from all data through 1977.

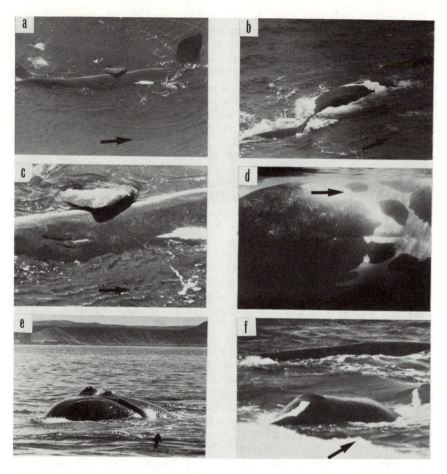

Figure 2. Anatomical features useful for sexing right whales
in the field (see also Figure 3). (In all cases the
arrow points anteriorly.)
a) Female, showing genital slit, mammary slits,
and no separate anus;
b) Male, showing more anterior genital slit and
separate anus;
c) Close-up of female genital slit, showing nipple
slits flanking it;
d) Close-up of male ventral slit, showing origin of
genital slit at umbilical scar and bifurcation at
posterior end;
e) Partially everted nipples on a female; the one
on the left (the female's right) is larger. A
male is pressed against her side (his erect
penis can be seen lying on her abdomen;
f) Raised anus of a male.

Behavioral Cues. When we did not have a good ventral view of an animal to indicate sex, we resorted to behavioral indicators of sex. Reproductive females are easy to know by their association with calves. First-year calves are invariably accompanied by an adult, and the adult-calf pairs stay close to shore, often travelling very little, with the adult frequently turning ventral side up either in play or to avoid nursing. On every occasion when we could see the configuration of the ventral slits of such adult companions of calves (more than 100 occasions, but usually with unidentified individuals), we found the adults to be females, often with obviously enlarged nipples.

Furthermore, calves appear to stay with the same adult. Whenever we could positively identify both adult and calf on more than one occasion in the same year, we found that both members of the pair were always the same (29 pairs with up to 7 sightings per pair). We assume that the accompanying female is the calf's mother. Out of 557 known whales, we had evidence that at least 124 of them were mature females from this kind of association with calves.

The situation with males was less straightforward. Our first behavioral indicator of maleness involves the number of consecutive years a whale has been seen as an adult without a calf. When an apparently full-grown right whale has been seen for three or more consecutive years without a calf, we consider it a likely male. Several lines of evidence support this judgment. First, after a female gives birth, she is rarely filmed on a survey flight for the rest of the year without her calf, and she is often seen repeatedly in that year (because she is so often in shallow water). So the chances are very slim that a whale filmed without a calf does in fact have a calf in that year. Secondly, mature females tend to bear calves every three years (Payne, in press). Finally, mature females tend to be seen in the years that they bear calves and not in other years (Payne, in press). In our data, only 11 out of 94 known mature females have been seen for three consecutive years, while 6 out of 17 known mature males have been seen for three consecutive years, a statistically significant difference (chi-square = 6.197, df = 2, $p<0.05$). A full-grown whale seen for three or more consecutive years without a calf, therefore, is more likely to be a male than a female.

The second behavioral indicator that we use to select likely males is participation in mating groups. A mating group is an active group of three or more non-juveniles separated by less than an adult whale length. Most of the action is at the surface and there is much white water. It is caused by rolling, rapid turning and diving, rubbing of bodies and stroking of flippers. Within these groups, there is frequently a whale that lies belly up, at the surface, for long periods. If the sex of this individual can be determined, it is seen to be a female. She may often have another adult whale beneath her, also belly up and holding its breath for long periods. When she rolls into a position to breathe, with her back uppermost, other adults approach the female from behind, swimming on an even keel. As they pull along side and parallel to her, they may stroke along her back with a flipper, perform a roll, and sink under her to try to mate. Occasionally an unsheathed penis can be seen in the midst of all the action. In these groups, as the bellies are turned up, they show male after male, so we have concluded that these groups are composed of more males than females. Because of the confusion in these groups, however, we have been able to identify by callosity pattern only a very few of the males. A whale seen in a mating group, therefore, is more likely to be a male than a female.

This contention was supported by an examination of the individuals found in the mating groups. In 39 mating groups, we identified 98 individual whales, some of which were seen in two or three such groups. For 29 of these 98 whales, we could assign a sex, either certain or likely. (For this analysis, a certain male was known anatomically, and a likely male was a whale that had been seen for three or more consecutive years as an adult without a calf). The 29 sexed participants in mating groups consisted of 16 certain or likely males and 13 certain or likely females, or 55.2% males. Because of the difficulty of identifying males, however, only 21.2% of all sexed whales at the time were males (33 certain or likely males in 156 sexed whales). The percentage of males in mating groups was significantly greater than the percentage of males in the overall population (arcsin-transformed t-test for comparing percentages: $p=3.549$, $df=183$, $p<0.0001$). Thus the identified male:female ratio in the mating groups was 4.6 times that ratio in the total identified population.

The third behavioral indicator of maleness is close companionship to a known male in a non-mating situation. There is reason to believe that males tend to associate with other males. When females are spaced along the coastline, each with a calf, we have observed small groups of whales move from female to female; when a female does not repulse them, one or more members of the group may try to mate with her before the group moves on to the next female. These groups are apparently composed of males travelling together. We have often observed such groups from the shore, where we can determine the sex of participants but cannot see their head patterns well enough to identify them. In such cases, most members of these groups have proven to be males. In addition to this, we know that females with calves usually stay relatively isolated from other whales, while mature females in non-calf years are usually not seen in the areas. Therefore, we felt that any full-grown whale seen close to a known male in a non-mating context was more likely to be a male.

An analysis of the whales seen with the known males outside of mating situations supported this belief. We identified 29 companions of known males, adult whales photographed within their body length of a male where there was no evidence of mating. 13 of these could be sexed: 7 (53.8%, as above) were certain or likely males (as above). Comparing this figure to the percent of males in all sexed whales at the time (21.2% of 156 sexed whales), we found that the incidence of males among the non-mating companions of known males was significantly higher (arcsin-transformed t-test for comparing percentages: $t=2.390$, $df=167$, $p<0.02$).

We list in Table 1 the categories we have used in determining the sex of animals, in decreasing order of certainty. Each category in this list is graded from A to C, with A being at or close to certainty and C being a good guess.

Applying the criteria in Table 1 to our data from 1971 through 1977, we have 106 females of grade A, 29 of grade B and 3 of grade C; we also have 23 males of grade A, 27 of grade B and 17 of grade C.

Sexual Dimorphism in Callosities

In order to compare the callosities of males and females, we took 50 grade A females and selected for comparison the most definite males that we had at the time – 9 grade A males, 12 grade B males and the 29 best grade C males. These whales are listed by identification number and sex grade in Figure 3. The callosity pattern of each whale was scored for 61 features by close examination of our best photographs of it. These features refer to number, approximate size, and shape or configuration of the major callosities -- bonnet, rostral islands, lip patches, mandibular islands, coaming, nostril islands, and post-blowhole islands. For the locations and typical shapes of these callosities, see Figure 4; for a more complete description, see Payne et al. (1983). Appendix 1 defines in detail the callosity features included in Figure 3.

A solid circle in Figure 3 indicates definite presence of a feature and an open circle indicates possible, but not certain, presence. Because we were conservative in ascribing certainty, we suspect that most features indicated by open circles were actually present. As an example of how we scored the callosity patterns, lip patches (the callosities occuring along the upper margin of the lower lip) could be

Table 1. Sex grades for female and male right whales. The categories were derived from long-term observations and are in decreasing order of certainty.

FEMALE SEX GRADES

A1: Photograph shows short genital slit without a separate anus or with nipple slits or protruding nipples.

A2: Field notes indicate definite female from ventral slit configuration.

A3: An adult-sized whale seen on three different days in the same year, each time alone with a calf.

A4: An adult-sized whale seen on two different days in the same year, each time alone with a calf.

A5: An adult-sized whale seen on two days in the same year with the same identified calf.

A6: An adult-sized whale seen once accompanied by a calf in water so shallow (as indicated by a shadow or the observer's notes) that it can definitely be stated that no other adult is out of sight underwater within the immediate vicinity.

B1: An adult-sized whale seen on two different days in the same year with a calf, one sighting including another adult.

B2: An adult-sized whale seen once with a calf late in the season (after November 10).

B3: A whale seen alone with a calf in two different non-consecutive years, during both of which it was at adult size.

B4: An adult-sized whale seen once alone with a calf.

C1: An adult-sized whale seen once with a calf, either not alone or with behavior that makes the relationship uncertain.

C2: A whale, sub-adult size or larger, obviously avoiding mating attempts, e.g. lying belly-up at the surface with another adult-sized whale belly up beneath it or chasing it.

Table 1 (continued).

MALE SEX GRADES

A1: Photograph shows a penis, or a separate anus, or an anteriorly placed ventral slit with a forked posterior end.

A2: Field notes indicate definite male from anogenital configuration.

B1: A whale seen for four or more consecutive years at adult size without a calf.

B2: A whale seen for three consecutive years at adult size without a calf and seen on three or more occasions when there were social indicators of maleness: (a) or (b) below.

B3: A whale seen for three consecutive years at adult size without a calf and seen on one or two occasions when there were social indicators of maleness: (a) or (b) below.

B4: A whale seen for three consecutive years at adult size without a calf.

C1: A whale seen for two consecutive years at adult size without a calf and seen on one or more occasions when there were social indicators of maleness: (a) or (b) below.

SOCIAL INDICATORS OF MALENESS

(a) Seen in a mating group

(b) Seen within one adult body length of an adult-sized male of sex grade A when there is no evidence of mating.

WHALE IDENT	SEX GRADE	BONNET																						ROSTRUM						LIP PATCHES											MND IS		COAMING									NSTRL IS			POST BLOW-HOLE IS						
		XCL PEN		NOTCHES ANT				NOTCHES LAT			POS MAR		ORIG OF PENIN					FORM OF PENIN					PEN IS		RST IS		ROSTRAL IS CHN																																	
		1 SYM.	2 ASYM.	3 NONE	4 SLIGHT	5 MED.	6 EXTEN.	7 L. ONLY	8 R. ONLY	9 BOTH	10 # BUMPS	11 # PENINS.	12 L.	13 LM.	14 M.	15 MR.	16 R.	17 ELONG.	18 CURVES L	19 CURVES R	20 STRAIGHT	21 BIFUR.	22 # L.	23 # R.	24 # L.	25 # R.	26 CONJ. PR.	27 L. LEADS	28 R. LEADS	29 EVEN	30 NONE	31 L. ONLY	32 R. ONLY	33 SHORT	34 MED.	35 LONG	36 FAT	37 L. BROKEN	38 R. BROKEN	39 L. LONGER	40 R. LONGER	41 # L.	42 # R.	43 NONE	44 SMALL	45 MED.	46 LARGE	47 ◆ (posterior)	48	49 ●	50 ▲	51 ▼	52 L. ONLY	53 R. ONLY	54 BOTH	55 DARK	56	57	58	59	60 THIN	61 COMPLEX
4	A		○	●																					2	2	●			○					○				●		●	2?	2?			●		○									○	○				
13	A	●		○								1		●						●					2	2	○	●								●						5	4				○	○	○								●					
33	A		●			●						1			●			○			○				2	2	●		●						●		●					4	1		●					●										●	○	
43	A		●	●					○		3														3	2		●								●	○					3?	3?			●		●							●				●			
44	A	○		○								1		●							●			1	1	1		●			●											3	4			○		○				○						○	○			
52	A	○		●				○				1			●			●			●		1	2	2	1		○						●							●	4	4		●					●							●					
54	A	○		○								1			●						●		1	1	2	2	●	○							●							2	2			●		○	○										○			
65	A	○		○	○						3														3	3			●						●			○				0	1?				●	●	●						●				○	○		
67	A		○	●						○		1			●						●		1	1	1	2			●						●				○			U	U			●						●				○	●					
68	A		●				○					1			●						●		1		3	3	○	●							●						●	2	2			●		○	○											●		
70	A		●	○	○			○			2												1		2	3		●								●						2	2			●		○	○										●			
83	A		●				●					1			●							●			1	1				○				●								2	2			●				○		○			○				○	○		
87	A		●				●			●	2													1	1	1			●		●											2	2		●					○								●			○	
91	A		●				●					1		●							●				3	3	●			●			○	●								4	3				○					●								○		
92	A		○	○								1			●						●		1		3	4			●		●											3?	4?			○		●										○	○			
94	A	○					●					1			●						●			1	2	1		●			●											2	2			●		○	○								●					
99	A		●			●					2													1	4	3			●						●			●			●	4	4			●						●										●
105	A		●		●					○	2														5	6		●								●						4	4			●		○	○						○				●			
106	A		●	○						○		1		●					●					1	3	2		●			●											4	2?		●			●									●					
108	A		●				●				2?														2	2		●							●		●	●	●			1?	1			●		○									●					
111	A		●			●						1				●			●						2	2	●			●			○	●								5	4		●			○				○					●					
119	A		●	●				○				1					●		○						1	2			●					●					○		●	2	3			●				○		○					●					
123	A		●	●						○	2												1		2	2	●			○					●							2	2			●						●						○				
125	A		●	●					○			2	●		●			●			●				3	2	●		●						●							3	1		●					●									○	○		
131	A		●			●					3														4	4	●		●		○											3	2		○							●					●					
139	A		○	○							1	1		●						●				1	1	2			●							●					●	0?	0?		●					○							○	○				
145	A		○	○							3														2	2			●					●							●	4	4			○		●							○			○				
153	A		●	○						●		2	●		●			●			●		1	1	3	3		●							●						○	4?	4			●			○			○							○	○		
154	A		●			●						1			●					●					2	2		●				○		●								2?	4			●						●	○				○	○				
155	A		○	○							1												1		1	2	●		●					●							●	0?	0?		●			○		○									●		○	
159	A		○			●			●			1			●						●		1		2	3	●		○						○		●				○	4?	5			●			○			○			●				○	○		
161	A	●				●						2	●				●		●	●					3	4			●		●											5?	4?		●			○									●					
176	A		●	●				●				1			●				●						7?	3?	●	●						○								U	4		●					●					●		○	○				
179	A		●				●				2														2	3	●	○						●							●	5?	6?		●						●									●		
184	A	○		●				●			2														2	2		●								●	●					4?	5?			●						●								○	○	
191	A	○				●			●		0?														3	2	●	○						●					●		●	4?	3?			●						●							●			
195	A		○				○		○		1												1		2	2	●			●	●											3	2			●				○							●					
196	A	○					●					1			●						●			1	2	4			●						●							3?	4?			●		○				○			●					●		
199	A	○		●								1			●			○	●					2	1	1			●						○		○	○				1?	1		●				●										○	○		
202	A		○	○								1					●	●			●			1	1	1			●					●							●	3?	3?			●		●									●					
205	A		○			●						1			●				●				1	1	1?	2?			●		○											1?	2?		●				○		○						●					
216	A	●				●						1			●						○		1	1	3	3		●			●											2?	2?			●				○					●		●					
229	A		●	○							3													1	3	2		●								●		●	●			2	1		●					○		○			●				○	○		
237	A		●	○								1			●						●		1	1	3	2		●							●			●	○			3?	U				●		●								●					●
249	A		○	●								1			●						●		1	1	1	0								●				●			○	0?	0?				●					●								●		
251	A		●				●		●			2	●				●			●	○				2	2	●		○		●										○	3?	1?		○					○									●			
252	A		○	●								1			●						●		1	1	1	2			●						●						○	1?	4		●				○			○					●					
253	A	○			○					○		1	●									●	1		2	3			●						●							4?	U			●						○						○	○			
256	A		●		●							1		○								●	1		2	1		●						●					●		●	4	U			●						○					○	○				
264	A		○	○								1?			○				○			.			3	3			●							●		●	○		●	3	3			●						●							●			

F E M A L E S

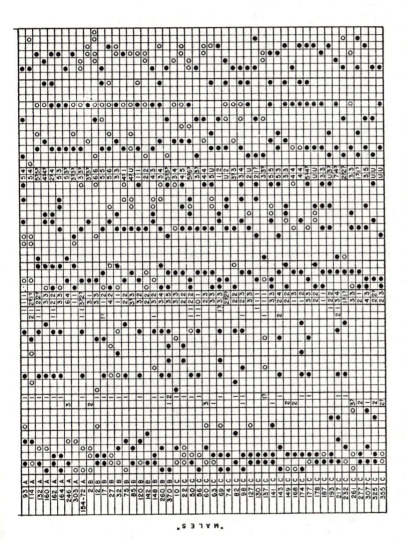

"MALES."

Figure 3. Callosity features in 50 female and 50 "male" right whales. See Table 1 for an explanation of the sex grade. See Appendix for an explanation of the callosity features and how they were scored. See Figure 4 for a sketch of typical callosities and their names. ● = feature definitely present, O = feature probably present, no symbol = feature not present, ? = number undetermined because obscured by water. See Tables 2 and 3 for statistical analysis.

absent entirely, present on only one side, short or medium or long, longer on one side than the other, unusually wide or not, and discontinuous (broken) on one or neither side or both sides.

Since completing the analysis of the data in Figure 3 (presented below), we have acquired more information on the sex of seven males graded B or C. For six of them, we have definite anatomical confirmation of sex and all six are indeed males. The sex of the seventh (a grade C male in Figure 3) is still uncertain, but the whale is now classified as a grade B4 female. Thus we appear to have been successful in skewing our selection of probable males toward true males on the basis of behavioral evidence.

To test for sexual dimorphism in callosities, then, we analysed callosity features one at a time (e.g. size of coaming or frequency of one coaming shape), using either the chi-square test of independence or, where categories were ordered, the Wilcoxon rank sum test. In this latter test, corrections were made for ties, since there were always many, and the levels of significance used were all two-tailed, because we were looking for any difference between the sexes. We treated the uncertain entries in Figure 3 ("o" and "?") in two different ways. In one case, we considered them to be positive entries ("o" = "●") and included them in the calculations. In the other case, we considered them to be undeterminable entries ("o" = "u") and omitted them from the calculations.

The results of these tests are presented in Table 2. We found differences between males and females, at significance levels ranging from $p < 0.10$ to $p < 0.005$, in 10 of the 25 callosity features analysed and in a total of 17 of the 19 tests possible on those ten features. Of the 17 results suggesting sexual dimorphism, 14 were at a significance level of $p < 0.05$, and 7 of those were highly significant at a level of $p < 0.01$. We have chosen to consider and report all results where $p < 0.10$, rather than the more usual $p < 0.05$, because of the probability that our sample of males is diluted with some females.

Of the callosity features showing evidence for sexual dimorphism, the first four in Table 2 concern the bonnet, the largest and most anteriorly placed callosity. There is, first, a tendency for males' bonnets to be more symmetrical than the females'. Secondly, the males have fewer anterior notches in their bonnets than the females do, this at a high level of significance. The males also have bonnets with significantly smoother posterior margins than the females do. And lastly, the males have significantly fewer peninsula islands, while having the same frequency of peninsulas as the females. The form and origin of peninsulas do not differ between the sexes. Most of the observed differences in the bonnet suggest that males tend to have a bonnet that is more filled in than the females, resulting in a smoother bonnet outline.

The number and conformation of rostral islands and the number of nostril islands show no differences between the sexes, but the mandibular islands do. The males have significantly more of these islands than the females.

Another dimorphic feature is the size of the lip patches. The males have significantly longer lip patches than the females and show a tendency to have wider ones as well.

The last callosity feature showing sexual dimorphism concerns the coaming, or splash barrier, just in front of the blowholes. Although there is no significant difference between the sexes in the size of this callosity, one of the shapes, wide in back and narrow in front, occurs more often in males than in females, at a high level of significance. Concomitantly, two other coaming shapes, diamond-shaped and thin, occur less often in males than in females.

The four areas of callosity dimorphism on the whales are thus the bonnet, the lip patches, the mandibular islands and the coaming. But rather than just listing the features that show sexual dimorphism, we found it was more instructive to look at the kinds of differences we were finding between the sexes. We found that the callosity features considered fell naturally into 3 categories: degree of coverage, shape (or configuration), and complexity of bonnet outline (Table 3). Only two out of seven features pertaining to shape differed between males and females, whereas five out of nine measures of degree of coverage proved to be different, always with males showing greater coverage. In two out of five measures of bonnet outline complexity, males had bonnets whose outline was less complex -- i.e. more filled in, suggesting again that the male bonnets may be larger.

Since many of our results indicated larger and more callosities in males, we decided to test that possibility by another approach. We measured the differences in the percent of the head covered by callosities in males and females, using high quality aerial photographs of their heads. In order to avoid the distortions of foreshortening, we selected the photographs so as to insure a consistent angle between the photographer's line of sight and the top of the whale's head. In order to achieve this we took advantage of the fact that the eyebrow callosities of surfacing right whales show up clearly in most aerial photographs even though they are near the bottom of the head and thus usually seen through more than a meter of water. In aerial photographs taken from a point within that vertical plane which includes the whale's long axis, both eyebrow callosities can be seen. If a line is drawn between these callosities, the point (in a photograph) at which it intersects with the

Table 2. Statistical analysis of sexual dimorphism in southern right whale callosities, from data presented in Figure 3.

CALLOSITY FEATURE		predominant in which sex	test	result (o = ●)	level of significance (o = ●)	result (o = u)	level of significance (o = u)
BONNET	symmetry	♂♀	χ²	–	.25<p<.50	+	.05<p<.10
	anterior notch	♀	r s	**	p=.0096	**	p=.0014
	side notches	–	r s	–	.1362	–	.3844
	smooth posterior margin	♂	χ²	*	.025<p<.05	**	.005<p<.01
	frequency of peninsulas	–	χ²	–	.25<p<.50	–	.25<p<.50
	frequency of bumps	–	χ²	–	.75<p<.90	–	.25<p<.50
	origin of peninsulas	–	χ²	–	.10<p<.25	–	.10<p<.25
	form of peninsulas	–	χ²	–	.75<p<.90	–	.75<p<.90
	# of peninsula islands	♀	r s	*	p=.0348	*	p=.0220
ROSTRAL ISLANDS	number	–	r s	–	p=.6384	–	p=.4472
	freq. of conj. pairs	–	χ²	–	.10<p<.25	–	.25<p<.50
	evenness of chains	–	χ²	–	.50<p<.75	–	.10<p<.25
LIP PATCHES	length	♂♀	r s	*	p=.0178	*	p=.0300
	width	♂	r s	+	p=.0574	+	p=.0734
MAND. IS.	number	♂	r s	**	p=.0034	*	p=.0114

				p=.1676		p=.1286
COAMING	size	–	rs	–	p=.1676	–	p=.1286
	frequency of ◆ shape	♀	χ²	*	.01<p<.025	–	.25<p<.50
	frequency of ●● shape	♀	χ²	**	.001<p<.005	/	.10<p<.25
	frequency of ◄ shape	–	χ²	–	.75<p<.90	–	.10<p<.25
	frequency of ► shape	♂	χ²	**	.001<p<.005	**	.005<p<.01
NSTR. IS.	number	–	χ²	–	.25<p<.50	–	.10<p<.25
POST-NOSTRIL ISLANDS	frequency of ● shape	–	χ²	–	.50<p<.75	–	.75<p<.90
	frequency of ●●● shape	–	χ²	–	.50<p<.75	/	.75<p<.90
	frequency of ● shape	–	χ²	–	.75<p<.90	–	.75<p<.90
	frequency of ● shape	–	χ²	–	.75<p<.90	–	.50<p<.75

**: highly significant difference between the sexes (p<.01)
 * : significant difference between the sexes (.01≤p<.05)
 + : tendency without usually accepted statistical significance (.05≤p<.10)
 - : no difference between the sexes (p≥.10)
 / : sample size too small to do test
rs: Wilcoxon rank sum test, corrected for ties, at two-tailed level of significance
χ^2: chi-square test of independence
callosity features: abbreviations and explanations as in Figure 3.
o = ●: uncertain entries are treated as certain and included in totals
o = u: uncertain entries are treated as undeterminable and excluded from totals
p : the probability of the observed distribution occurring by chance assuming the null
    hypothesis of no difference between males and females

Table 3. Sexual dimorphism results from Table 2 with callosity features grouped into 3 categories. Note that some features fall into two different categories. ●●● = highly significant (p<0.01), ●● = significant (0.01≤p<0.05), ● = tending toward significance (0.05≤p<0.10), - = not significant (p≥0.10)

DEGREE OF COVERAGE

Anterior Bonnet Notch	●●●
Side Bonnet Notch	-
# Peninsula Islands	●●
# Rostral Islands	-
# Mandibular Islands	●●●
# Nostril Islands	-
Length of Lip Patches	●●
Width of Lip Patches	●
Size of Coaming	-

SHAPE

Form of Peninsula	-
Origin of Peninsula	-
Rostral Islands: Frequency of Overlapping Pairs	-
Rostral Islands: Evenness	-
Shape of Coaming	●●●
Shape of Post-Blowhole Islands	-
Symmetry of Bonnet	●

COMPLEXITY OF BONNET OUTLINE

Anterior Bonnet Notch	●●●
Side Bonnet Notch	-
Frequency of Smooth Posterior Bonnet Margin	●●●
Frequency of Peninsulas	-
Frequency of Bumps	-

midline of the head depends on how the whale was holding its head. Unless the whale was on an even keel with its body held straight, the head may have been nodded (moved about its pitch axis) or twisted (moved about its roll axis). We could ensure that all pictures showed the head in the same orientation relative to the camera by requiring that all photographs met two criteria: 1) the eyebrow-to-eyebrow line passed through the nostrils (corrects for pitch of head), and 2) a line passing from the tip of the snout to the center of the coaming bisected the eyebrow-to-eyebrow line, within about 10 percent (corrects for roll of head).

We chose 15 known females and 15 known or suspected males for which we had high quality photographs meeting these criteria. As before, the list of males may have included some females, but 11 of the 15 "males" were sex grade B or better, the rest were grade C. Each acceptable photograph was greatly enlarged in area (2000X-3000X) and traced onto paper. We defined a consistent measure of total head area as the area enclosed by four lines connecting the centers of each of the eyebrows and each of the chin callosities. The area of each callosity and the total area of the head were then measured using a surveyor's compensating polar planimeter (Keuffel and Esser Co., model 620010). The planimeter contributed a negligible error to the data, measurements being reproducible to about 0.05%. Larger errors were introduced by the enlarging procedure (reproducibility about 2%), and by the angle from which the photograph had been taken (12% or slightly more). These were considered to be random errors.

From the planimeter measurements, the relative coverage of several kinds of callosity was calculated by dividing the area of each callosity by the total head area. The callosities studied were the bonnet, lip patches, coaming, post-blowhole islands, and rostral islands (see Figure 4). These callosities were chosen because they are at the top of the whale's head and are not distorted by water in aerial photographs, as lower callosities may be. The results are presented in Table 4. Males had a significantly larger area of their heads covered by callosities, confirming the results from Figure 3. The level of significance when all callosities are considered together ($p < 0.001$) is particularly noteworthy, given that some of the "males" might be females. The percent coverage in males was greater in each of the callosities measured separately, although not all of the differences were significant.

Although, as mentioned earlier, female right whales are bigger than males, the difference in length is only about 1%. The difference we have found here in total callosity coverage is about 45%, the square root of which, to get a linear measure for comparison, is almost 7%. The males

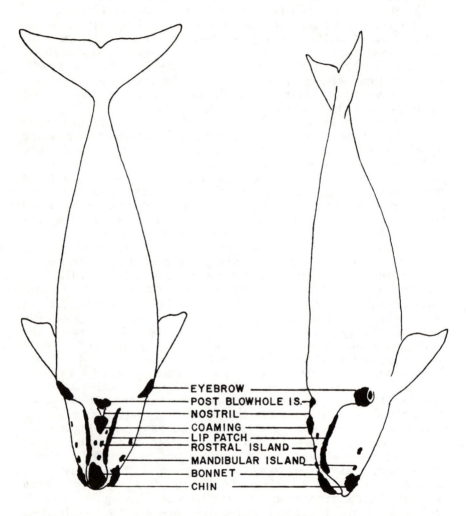

EYEBROW

POST BLOWHOLE IS.

NOSTRIL

COAMING

LIP PATCH

ROSTRAL ISLAND

MANDIBULAR ISLAND

BONNET

CHIN

Figure 4. Diagrammatic view of a right whale showing the name, position, and form of typical callosities.

Table 4. Percent of head area covered by callosities in male and female right whales.

	MEAN PERCENT COVERAGE		t STATISTIC	p
	FEMALES n=15	MALES n=15		
ALL CALLOSITIES	11.24	16.39	3.43	.001
BONNET	5.30	5.93	1.53	.068
ROSTRAL ISLANDS	1.25	1.32	0.28	.389
LIP PATCHES	2.27	5.51	2.88	.004
COAMING	1.16	2.02	3.13	.002
POST-BLOWHOLE ISLANDS	1.26	1.63	2.12	.022

clearly have, on average, a larger absolute area of their heads covered with callosities.

Both of the analyses described above yielded results confirming our original suspicion that male right whales have more and larger callosities than females. Some of the results, like significantly different coaming shapes, were unexpected. We want to stress that the dimorphisms we have discovered are all statistical in nature rather than absolute, with considerable overlap between the sexes, like height in humans. We have not found any callosity feature that indicates sex unfailingly by itself.

Scrape Marks

In many animals, secondary sexual characters which are more developed in the males are related to fighting -- antlers in moose, horns in sheep, the mane of a sea lion, the enlarged claw of fiddler crab males, for examples. We thought that the callosities might likewise function sometimes as weapons. In this light, it may be significant that the height of callosity tissue in southern right whales varies. When viewed from the side, they can appear low and relatively smooth (Figure 5a) or jagged (Figure 5b). We do not have enough side views of animals of known sex to relate height of callosities to sex, but suspect that the unevenness increases with age, for the callosities of calves are always smooth, and very rough contours occur only in large adults. Callosities are not very hard tissue to the human touch, more like the skin of a cantaloupe than a human fingernail, for example, but the normal skin of mysticetes lacks the protection of a cornified epithelium (Ling 1974). We therefore suspect that at least the roughened callosities might be capable of inflicting injury to the skin of another whale.

Right whales at times make extensive physical contact with one another -- rubbing bodies together, stroking with flippers, and rubbing with heads. Usually they appear to take great care not to make contact with their callosities: they will slide the underside of their chin along another's back touching only with the smooth area that contains no callosities. But sometimes in active mating groups, we will see a whale turn its head over and run the dorsal side of the head against another whale, making clear contact with its callosities. The animal on the receiving end immediately twists or writhes, so as to move its body away, while the whale doing the scraping adjusts so as to keep its callosities in contact with the recipient. All indications are that the behavior is deliberate and that the result is painful. We have not seen any marks actually result from such an incident, but it would require very exceptional viewing conditions to do so. One would have to be sure that the portion of the recipient's back being scraped was originally free of any marks and that

Figure 5. Side views of two southern right whale heads showing the variation in the height of the callosities. a) Low, relatively smooth callosities; b) Tall, rough callosities.

Table 5. Incidence of scrape marks by sex in adult right whales. A chi-square test of independence gave the following result: χ^2 = 29.014, df=2, p<<0.001.

	males	females
scrapes present	16	14
scrapes absent	3	59

Figure 6. Scrape marks, probably caused by the callosities of another whale, on the back of a southern right whale.

what was visible immediately following contact was not streaks of foam or some other artifact.

However, we have found that many right whales carry suggestive scrape marks. Figure 6 shows a close-up of the back of a right whale. In addition to amorphous skin mottling, a few sets of gray parallel lines can be seen. These gray parallel lines are temporary marks and are distinctly different from the scars we have seen on right whales from the teeth of killer whales (Payne, in press) or sharks. The tooth scars are evenly spaced and strictly parallel lines that are deeper and much whiter and more nearly permanent than the marks in Figure 6. The gray lines in Figure 6 are irregularly spaced and not strictly equidistant within each set, and they sometimes fade in and out as though they were made by projections of irregular height bearing unevenly against the surface. We believe that these scrape marks were made by callosities of another whale -- by irregular callosity projections such as those shown in Figure 5b. Sets of scrape marks occur in widths comparable to the length of a bonnet or a coaming and never wider. We have also noted that the spacing between two or more narrower sets of scrape marks whose lines are mutually parallel is similar to the spacing between mandibular island callosities.

Another possible explanation for these scrape marks is contact with hard surfaces like rocks. We have three reasons to believe, however, that the origin of the scrape marks is not inanimate. First, the areas where whales spend the vast majority of their time at Peninsula Valdés have soft bottoms (mud or tiny, smooth pebbles). When they do get close to large rocks or cliffs, they appear to move carefully and slowly so as to avoid touching any surface. Second, when whales are most active, they move into deeper water than they stay in when they are quiet (Payne, in press). And third, the incidence of these scrape marks is highly correlated with social factors, as we will now demonstrate.

To analyze the incidence of scrape marks in Argentine right whales, we considered only photographs showing at least half of the whale's dorsal surface above water with sufficiently clear detail, proper lighting and contrast to show scrape marks if any were present. Pictures of 92 adults were scored for presence or absence of three or more parallel lines -- 73 females (sex grade B or better) and 19 "males" (2 of grade A, 16 of grade B, and 1 of grade C). Table 5 shows a comparison of scrape marks found on males and females. Most of the males had scrape marks, most of the females did not, and a chi-square test of independence shows a negligibly small possibility ($p \ll 0.001$) of these results occurring by chance if there is no sex difference.

Adult females alternate between two different social conditions. In the years that they have calves, they keep

Table 6. Incidence of scrape marks by reproductive condition in adult female right whales. A chi-square test of independence gave the fllowing results: $X^2=9.204$, df=2, p=0.002.

	females with calves	females without calves
scrapes present	7	7
scrapes absent	5 I	8

Table 7. Incidence of scrape marks in adult right whales: males vs females in non-calf years. A chi-square test of independence gave the following result: $X^2=5.399$, df=2, p=0.020.

	males	females without calves
scrapes present	I 6	7
scrapes absent	3	8

mostly separate from other whales and are seldom involved in mating groups. One would thus expect a difference between calf years and non-calf years for adult females in the incidence of these scrape marks, if they are acquired principally in the context of mating groups. When we analyzed the females by reproductive condition (Table 6), we found a difference in the expected direction: females without calves had significantly more scrape marks than females with calves (p=0.002).

If scrape marks are mostly the result of males fighting over females, one would expect males to direct most of their callosity attacks against other males, with the result that males would be more scraped up than the females that are probably in those groups, which are females without calves. Once again, we found a significant difference (p=0.020) in the predicted direction (Table 7).

Discussion

We have no trouble calling the interactions that produce these scrapes aggression, since they are interactions principally between males, involving physical contact which leaves marks on the skin, in a context where competition for females is to be expected. In right whales, males appear to make a much smaller investment in the care of the young than the females, because females do all of the rearing as far as we can tell (Payne, in press). From studies on other animals in which this is the case, we might therefore have expected both the greater development of a sexually dimorphic character in males and the male-to-male competition (Wilson 1975).

In odontocetes, sexual dimorphism related to male-to-male aggression is well known. A few examples are: the large size of sperm whales (*Physeter catadon*) with battles observed (Shaler 1873); the tusk of the male narwhal (*Monodon monoceros*) and the scars it leaves on other males (Silverman and Dunbar 1980; Best 1981); and the special teeth found only in males throughout the family Ziphiidae with scars that match the arrangement of those teeth (McCann 1974; Mead, Walker and Houck 1982).

In contrast, among baleen whales, we believe that our findings on right whale callosities are the first example of a sexually dimorphic structure used for intraspecific aggression. Nemoto (1962) has observed a sexual dimorphism in the shape of the upper jaw of fin whales (*Balaenoptera physalus*) and, to a lesser extent, Bryde's whales (*B. edeni*), but there is no evidence that the observed difference is related to intraspecific aggression. Humpback whales (*Megaptera novaeangliae*) are the only other baleen whales for which we know of reports of clear aggression (Tyack 1981; Darling, Gibson, and Silber 1983). We suggest that it might be interesting to look for evidence of sexual dimorphism or aggressive function for the peculiar growths found on the anterior ventral surface of the lower jaw of humpback whales, or for the knobs on their flippers or heads, or perhaps even for the size and number of barnacles on them (as well, perhaps, as on other whale species). In this light, it is not entirely unreasonable to suppose that, if cyamids eat callosity tissue, as we and others have proposed (Payne et al. 1983; Leung 1976), they may actually be of use to male right whales in keeping their weapons "sharp" or roughened.

We do not intend to imply that callosities are useful to right whales only for fighting. Because they vary so much from whale to whale, they could be used by the whales for individual recognition. Ventral blazes are also individually specific (Payne et al. 1983) and, if used by right whales in combination with the callosities, might make it possible for them to identify each other visually from every direction but the rear.

Summary

1) We have described a benign technique for studying sexual dimorphism in a free-ranging whale species when the sex of only some members of one sex can easily be determined.

2) We present evidence that the extent of callosity coverage is a sexually dimorphic character in southern right whales, with males having more and larger callosities than females.

3) We present evidence that the callosities function at least in part as weapons for intraspecific aggression, with that aggression occurring mostly between males.

Acknowledgements

We wish to thank Victoria Rowntree for her help in the analysis of data, as well as for her input in many useful discussions at all stages of this work. She also did the graphics work for all the figures. We are also grateful to Judith Perkins and Nancy Cison for data analysis and for helping to collect the information presented in Table 2; and to Kevin Chu for measuring the percent coverage of callosities on males and females; Katy Payne for the drawing of Figure 1, Figure 4 and the sketches in Figure 2; Chris and Jane Clark for helping to sex whales from their field observations; Jim Bird for help in many ways, particularly with references and with manuscript preparation; Linda Guinee for typing the manuscript.

Funds for this research were provided by grants from the National Geographic Society, the New York Zoological Society and the People's Trust for Endangered Species.

Appendix 1

The callosity features presented in Figure 4 and the abbreviations used to describe them require explanation. For definition of terms used to describe callosities see Payne et al. (1983). For a diagram of typical callosities see Figure 5.

WHALE IDENT.
(Whale identity) Each whale is given a unique identifying number for life. Calves are given their mother's number followed by the year of their birth, e.g., 154-72 is a calf born to whale #154 in 1972.

SEX GRADE
Animals of whose sex we are certain are given an A grade, while those for which we can only guess the sex from behavioral data are given a B or C rating. (See Table 1 for explanation).

BONNET
The bonnet is the most anterior as well as the largest of the callosities and is an important clue to identity because it has a highly varied outline from individual to individual.

XCL. PEN. stands for "excluding peninsula". Although the bonnet is often rather oval in shape, there are frequently one or more projecting arms of callosity tissue called peninsulas, originating from its rear edge.
1. SYM. (symmetrical)
2. ASYM. (asymmetrical).
In these categories, we are noting whether the main body of the bonnet excluding the penin-sula(s) is symmetrical or asym-metrical about the line of intersection of the saggital plane with the top of the head (i.e., the midline of the head).

NOTCHES. Notches are indentations in the periphery of the bonnet.
ANT. (anterior). Refers to a notch in the anterior edge of the bonnet. It is placed in one of the following categories depending on its size:
3. NONE (no anterior notch in the bonnet)
4. SLIGHT
5. MED. (medium)
6. EXTEN. (extensive)

LAT. (lateral). refers to notches on the sides of the bonnet.

    7. L. ONLY. (left only). Present only on the left side.

    8. R. ONLY. (right only). Present only on the right side.

    9. BOTH. Present on both sides of the bonnet.

POS MAR. (posterior margin). Projections of more than one type can extend from the rear of the bonnet.

    10. # BUMPS (number of bumps). A bump is a projection that is wider than it is long. Here we are counting the number of bumps on the posterior margin of the bonnet.

    11. # PENINS. (number of peninsulas). A peninsula is a projection that is longer than it is wide.

ORIGIN OF PENIN. (origin of peninsulas). The base of a peninsula is connected with the bonnet at some point along the rear margin of the bonnet. Here we are noting whether it springs from the:

    12. L. (left)

    13. L.M. (left middle)

    14. M. (middle)

    15. M.R. (middle right)

    16. R. (right)

FORM OF PENIN. (form of peninsula). Peninsulas may be classified into several shapes as follows:

    17. ELONG. (elongated). The peninsula is longer than 3/4 of the length of the main body of the bonnet.

    18. CURVES L. (curves left). The peninsula projects towards the left side of the whale.

anterior

    19. CURVES R. (curves right). The peninsula projects towards the right side of the whale.

    20. STRAIGHT. The peninsula points straight, posteriorly.

    21. BIFUR. (bifurcated). The peninsula divides into a Y at its tip.

PEN. IS. (peninsula islands). Isolated callosities lying alongside the peninsula. When an island lies near the tip of the peninsula, it is considered a peninsula island (and not a rostral island) if and only if the midpoint of the island lies anterior to the most caudal point on the peninsula. In the example, the island on the right of the whale is a peninsula island, the one on the left is a rostral island.

ant. 22. # L. (number left). Number of peninsula islands to the left of the peninsula.

23. # R. (number right). Number of peninsula islands to the right of the peninsula.

ROSTRUM

The upper jaw from the blowholes forward.

RST. IS. (rostral islands). Small isolated callosities lying on the rostrum between the coaming and the rear of the bonnet. (see peninsula islands, above).

24. # L. (number left). Number of islands to the left of the midline of the head.

25. # R. (number right). Number of islands to the right of the midline of the head.

ROSTRAL IS. CHN. (rostral island chains). In many whales, the isolated callosities on the rostrum lie in lines like archipelagos or island chains stretching down each side of the head in graceful arcs. Where these chains come closest together near the midline of the head, two islands (i.e., the small isolated callosities) may touch or even partly fuse. The next category refers to a common juxtaposition of islands from the left and right chains.

26. CONJ. PR. (conjugated pair). Rostral islands which have a separation at their closest point not exceeding the width of the widest member of the pair and which are aligned such that their long axes overlap by 1/2 or more of their length. The pair on the left in the example is not conjugated, while the pair on the right is. The next three categories are concerned with how the chains are aligned along the length of the whale's head.

27. L. LEADS. (left leads). The anterior-most rostral island is on the left side of the whale's head.

28. R. LEADS. (right leads). The anterior-
    most rostral island is on the right
    side of the whale's head.
29. EVEN. The anterior-most rostral islands
    in both island chains lie side-by-
    side.

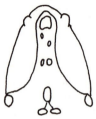      

L. LEADS.   R. LEADS.   EVEN.

LIP PATCHES

The lip callosities are very variable and often show
different degrees of development on the two sides of the
head. When asymmetries are present, it is virtually always
the right side which is more developed (see Payne et al.
1983). The following categories dissect out these
differences.

30. NONE
31. L. ONLY (left only)
32. R. ONLY (right only)
33. SHORT
34. MED. (medium)
35. LONG
36. FAT Some lip callosities are excep-
    tionally wide.
37. L. BROKEN (left broken). Some lip
    callosities are discontinuous.
38. R. BROKEN (right broken)
39. L. LONGER (left longer). The asymmetry
    alluded to above.
40. R. LONGER (right longer)

MND. IS. (mandibular islands)

These are small round callosities arrayed in rows along
the mandibles. They are often hard to see in aerial
photographs because they are usually underwater, but they
show considerable variability in number and placement.

41. # L. (number on left mandible).
42. # R. (number on right mandible).

COAMING
    The callosity lying just ahead of the blowhole.
            43. NONE
            44. SMALL
            45. MED. (medium)
            46. LARGE
anterior 47. ◆ (diamond-shaped)
            48. ▮ (cigar-shaped)
            49. ● (roughly circular)
            50. ▼ (triangular with the apex pointing back
                and the base forward.)
            51. ▲ (triangular with the apex pointing
                forward and the base caudal).

NSTRL. IS. (nostril islands)
    Small callosities lateral to the blowholes on either or
both sides of the head. They are caudal to the anterior
margin of the blowholes.
            52. L. ONLY (only on the left side of the
                head)
            53. R. ONLY (only on the right side of the
                head)
            54. BOTH (on both sides of the head)

POST BLOWHOLE IS. (post-blowhole islands)
    One or two (rarely more) small callosities directly
caudal to the nostrils. The diagrams show the shape of the
pair and how they fuse to make a single callosity.
            55. DARK. Some whales have few or no
                white cyamids covering their
                callosities. This is frequently
                the case with post-blowhole
                islands. When the cyamids are
                missing, the post-blowhole islands
                look dark and their form can be
                seen in only exceptionally good
                aerial photographs.
            56. ● ● The callosities are clearly separ-
                ated.
            57. ●● The callosities are just touching.
            58. ●━● The two callosities are fused by a
                narrow isthmus.
            59. ●━ Callosities appear as solid band
                with no sign of fusion.
            60. THIN. Post-blowhole callosities are
                sometimes very thin.
            61. COMPLEX. Rarely, the post-blowhole
                callosities have a bizarre and
                complex form

Literature Cited

Best, R.C.
    1981. The tusk of the narwhal (*Monodon monoceros* L.):
    interpretation of its function (Mammalia: Cetacea). Can.
    J. Zool. 59: 2386-2393.

Darling, J.D., K.M. Gibson, and G.K. Silber
    1983. Observations on the abundance and behavior of
    humpback whales (*Megaptera novaeangliae*) off West
    Maui, Hawaii, 1977-79. *In* R. Payne (editor),
    Communication and behavior of whales. AAAS Selected
    Symposia Series, p. 201-222. Westview Press, Boulder,
    Colo.

Leung, Y.M.
    1976. Life cycle of *Cyamus scammoni* (Amphipoda:
    Cyamidae), ectoparasite of gray whale, with a remark on
    the associated species. Sci. Rep. Whales Res. Inst.
    Tokyo 28: 253-260.

Ling, J.K.
    1974. The integument of marine mammals. *In* R.J. Harrison
    (editor), Functional anatomy of marine mammals, vol. 2,
    p. 1-44. Academic Press, London.

Lönnberg, E.
    1906. Contributions to the fauna of South Georgia: I.
    Taxonomic and biological notes on vertebrates. K. Sven.
    Vetenskapsakad. Handl. 40(5): 1-104.

McCann, C.
    1974. Body scarring on Cetacea-Odontocetes. Sci. Rep.
    Whales Res. Inst. Tokyo 26: 145-155.

Matthews, L.H.
    1938. Notes on the southern right whale, *Eubalaena
    australis*. Discovery Rep. 17: 169-182.

Mead, J.G., W.A. Walker, and W.J. Houck.
    1982. Biological observations on *Mesoplodon carlhubbsi*
    (Cetacea: Ziphiidae). Smithson. Contrib. Zool. 344, 25p.

Nemoto, T.
    1962. A secondary sexual character of fin whales. Sci.
    Rep. Whales Res. Inst. Tokyo 16: 29-34.

Omura, H.
    1958. North Pacific right whale. Sci. Rep. Whales Res.
    Inst. Tokyo 13: 1-52.

Payne, R.
1976. At home with right whales. Nat. Geogr. Mag. 149: 322-339.

In press. The behavior of southern right whales (*Eubalaena australis*). University of Chicago Press, Chicago, Ill.

Payne, R., O. Brazier, E.M. Dorsey, J.S. Perkins, V.J. Rowntree, and A. Titus
1983. External features in southern right whales (*Eubalaena australis*) and their use in identifying individuals. *In* R. Payne (editor), Communication and behavior of whales. AAAS Selected Symposia Series, p. 371-445. Westview Press, Boulder, Colo.

Scammon, C.M.
1874. The marine mammals of the northwestern coast of North America. J. Carmany & Co., N.Y., 317p.

Shaler, N.S.
1873. Notes on the right and sperm whales. Am. Nat. 7: 1-4.

Silverman, H.B. and M.J. Dunbar.
1980. Aggressive tusk use by the narwhal (*Monodon monoceros* L.). Nature (Lond.) 284: 57-58.

Slijper, E.J.
1962. Whales. Basic Books, N.Y., 475p.

Tyack, P.
1981. Interactions between singing Hawaiian humpback whales and conspecifics nearby. Behav. Ecol. Sociobiol. 8: 105-116.

Wilson, E.O.
1975. Sociobiology: The new synthesis. Belknap Press of Harvard University Press, 697p.

## Part 3

Migratory Destinations and Stock Identification

Roger Payne, Linda N. Guinee

10. Humpback Whale *(Megaptera novaeangliae)* Songs as an Indicator of "Stocks"

Abstract

Humpback whales (*Megaptera novaeangliae*) sing songs that change over time. They also sing different songs in different areas. In the eastern North Pacific, two widely separated wintering grounds for humpbacks are occupied by individuals singing the same song. We have analyzed over 70 humpback songs from Hawaii and the Revillagigedo Islands recorded in 1977 and 1979. Although these wintering grounds are 4700 km apart, the song in both places has remained the same throughout this period -- changes in one area paralleling changes in the other. This surprising result strongly suggests that the animals in these widely separated places are part of the same population.

The songs from the eastern North Pacific and those from the North Atlantic, on the other hand, are dissimilar, as would be expected in geographically isolated populations. In both oceans, however, the basic structure of the song is the same, indicating that songs in both oceans adhere to similar rules.

Introduction

Payne (1968), Payne (1979), and Payne and McVay (1971) demonstrated that the complex sounds produced by the humpback whale (*Megaptera novaeangliae*) near Bermuda in the breeding season are in the form of repeated patterns or songs. These results were confirmed with Caribbean humpbacks by Winn, Perkins, and Poulter (1971). Payne and Payne (in prep.) compared samples of humpback song recorded in April for 18 years near Bermuda. They found all whales singing the same song in each sample, but noted changes between the annual samples. Payne (1978) found changes in song in Hawaii, a finding which was confirmed by Winn, Thompson, Cummings, Hain, Hudnall, Hays, and Steiner (1981). Payne, Tyack, and Payne (1983) then demonstrated for Hawaiian humpbacks that changes are made continuously during the process of singing but that the song remains in a relatively stable form during the summer and fall feeding period, a time when singing is rarely heard, but not entirely absent (McSweeney, Dolphin, and Payne, in prep.). Guinee, Chu, and Dorsey (1983) have shown that individual singers change their songs to keep pace with the changes in the songs of the other singers sharing their breeding ground. Winn, Bischoff, and Taruski (1973); Tyack (1980); Darling, Gibson, and Silber (1983); and Glockner (1983) have demonstrated that male humpbacks sing, and Tyack (1981) has shown that the functions of humpback song are probably similar to the functions of bird song -- advertisement by breeding males to attract females as potential mates and to challenge males who are potential competitors.

Since songs change continuously and individual singers adopt new material sung by the other singers in an area, we can conclude that two whales singing the same song either are in direct acoustic contact or are linked by a chain of acoustic contact involving other whales. Thus we would expect variation to occur between populations that are geographically isolated, and the detection of dissimilarities between humpback whale songs in two different areas might be evidence for a lack of acoustic contact between the whales in those two areas.

We sought to exlore this question further by comparing songs sung by groups of humpback whales in the same and in different oceans. In 1977, Angelica Theriot and Mary Crowley recorded a sample of humpback whale songs in the Revillagigedo Islands which they made available to us for analysis. We also received a 1977 West Indies sample from the Ocean Research and Education Society and 1979 West Indies sample from David Matilla. In 1979, on r/v *Regina Maris*, we recorded, with the help of Katharine Payne, another sample of song from the Revillagigedos. This paper is a comparison of songs recorded in 1977 and 1979 in three areas: Hawaii and the Revillagigedos in the eastern North Pacific, and the West Indies in the North Atlantic.

Methods

The humpback whale songs discussed in this paper were recorded in three different areas (the Hawaiian Islands, the Revillagigedo Islands, and the West Indies) at two different times (February 1977 and March 1979) using the techniques and equipment described by Payne et al. (1983). The

Hawaiian sample was recorded in the protected waters surrounded by the islands of Maui, Molokai, Lanai, and Kahoolawe (centered at 20°54' N, 156°43' W). The Revillagigedo sample was recorded 4700 km away, around Isla Socorro (18°52' N, 110°50' W), off the west coast of Mexico. The West Indies sample was recorded in two different locations, separated by approximately 250 km. The 1977 sample was recorded on Silver Bank (20°42' N, 69°49' W) and the 1979 sample was recorded in Mona Passage (18°47' N, 68°00' W).

Spectrograms were made and song structure was classified in the manner described by Payne et al. (1983). The durations of phrases, the number of units in each phrase, the number of phrases in each theme, and the percent occurrence of each theme in each song were measured or counted directly from the spectrograms. Means for each of these measurements were calculated for each song session and were used as the basis of comparison in all statistical tests. The 1977 Hawaii sample includes some of the same song sessions analyzed by Payne et al. (1983).

Discriminant analysis was performed using an Apple II Plus computer and the Econometrics Linear Forecasting program.

Results

When we compared song sessions from the three areas, we found that the songs in the sessions from the two Pacific Ocean study areas resembled those from the West Indies only in the overall structure apparently common to all humpback songs. Regardless of where it is recorded, a humpback song is found to be composed of themes, which are in turn composed of repeated phrases, which are composed of subphrases, consisting of units and subunits (see Payne and McVay 1971; Payne et al. 1983). There are also structural features that are common to most humpback songs wherever sung -- such as the presence of transitional phrases and the frequency range of the units composing the song (Payne et al. 1983).

However, when we compared the mean duration of songs in 1979 in the three areas, we found that the two Pacific Ocean study areas were very similar (Hawaii 684.2 ± s.d. 138.2 sec., n=6; Revillagigedo 692.1 ± s.d. 175.3 sec., n=6), while the West Indies sessions (1086.2 ± s.d. 153.4 sec., n=3) were approximately 400 seconds longer. When we compared the mean number of themes in each song in 1979, the Pacific Ocean areas were very similar (Hawaii 4.96 ± s.d. .89; n=6; Revillagigedo 4.50 ± s.d. .59, n=6), and differed once again from the West Indies sample (7.00 ± s.d. 0.00, n=3). Spectrograms and tracings of representative songs from 1979 in all three study areas in both oceans are shown in Figures

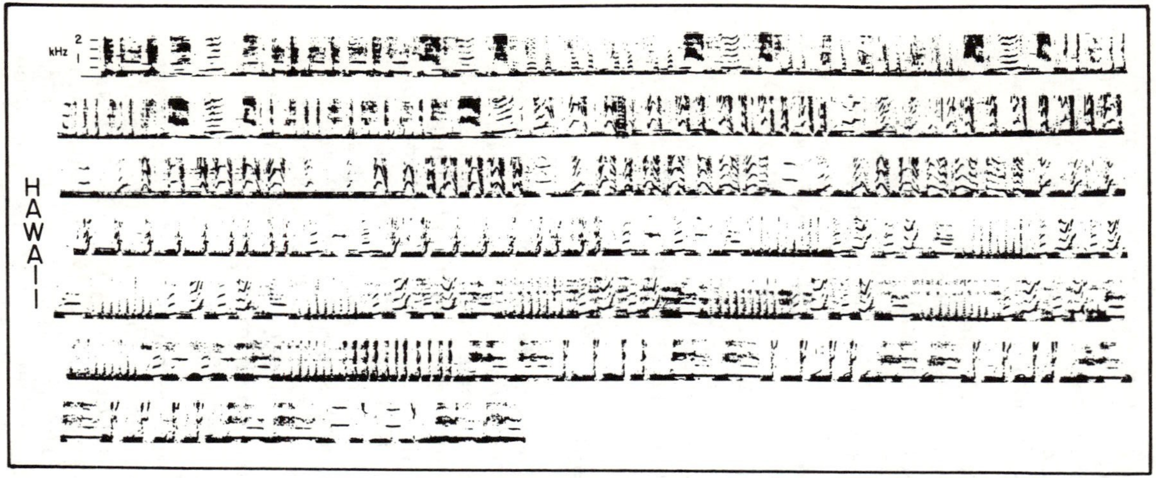
HAWAII
kHz
2
1

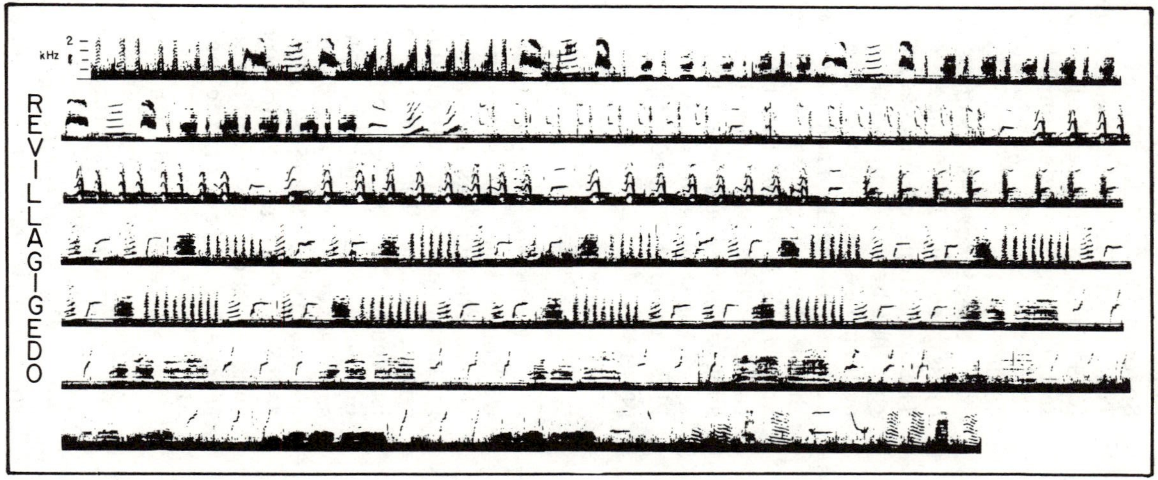
REVILLAGIGEDO
kHz
2

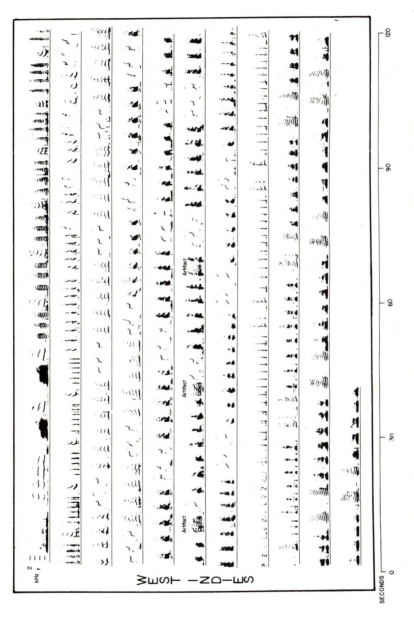

Figure 1. Continuous sound spectrograms of a single, typical 1979 song from each of the three areas studied.

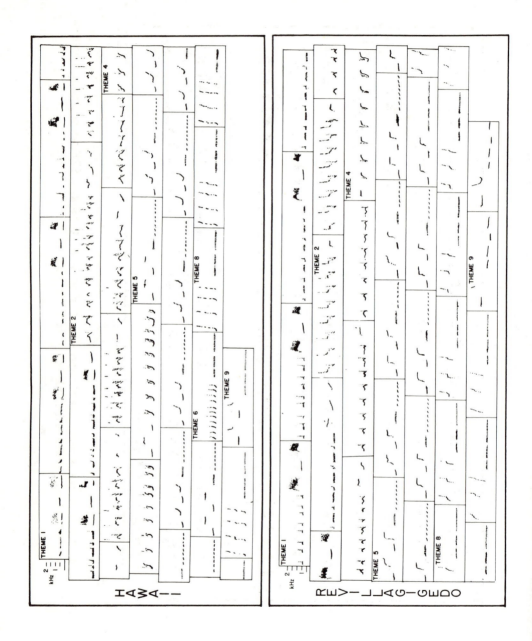

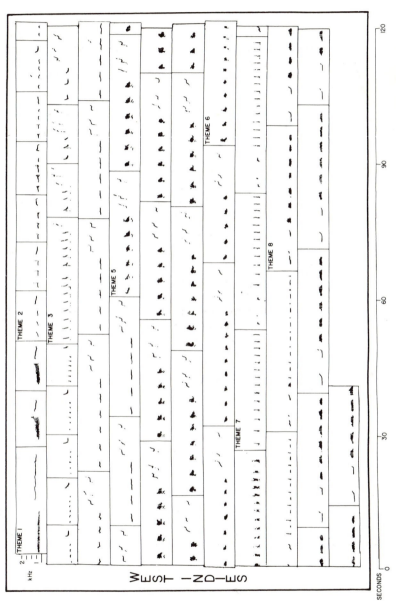

Figure 2. Tracings of the spectrograms shown in Figure 1. The tracings omit all extraneous sounds (e.g., ocean noise like ships, other whales, and underwater echoes) as well as harmonics.

Table 1. Our sample of analyzed songs from the Pacific
Ocean. Note that a plus (+) indicates an
incompletely recorded song in addition to the
number of complete songs recorded. 0+ indicates a
session during which we recorded less than a
complete song.

Location	Year	Date	Number of Songs Recorded in Session
Hawaii	1977	February 1	6+
"	"	February 9	4+
"	"	February 11	6+
"	"	February 12	3+
"	"	February 13	4+
"	"	February 14	4+
"	"	February 15	2+
"	"	February 18	1+
"	"	February 18	4+
"	"	February 19	4+
"	"	February 19	2+
"	"	February 25	6+
"	"	February 25	0+
"	1979	March 1	3+
"	"	March 5	1+
"	"	March 6	2+
"	"	March 7	4+
"	"	March 11	1+
"	"	March 24	2+
Revillagigedo	1977	February	1+
"	1979	March 3	2+
"	"	March 6	5+
"	"	March 9	1+
"	"	March 11	4+
"	"	March 24	1+

1 and 2. It is apparent from inspection that the Pacific
Ocean songs are very similar to each other and very
different from the West Indies songs; songs from the two
oceans thus represent strikingly different geographical
variations of humpback song.

We had originally expected to find differences between
the samples from the Hawaiian and Revillagigedo Islands as
well, since these two areas are so far apart. However, when
we listened to songs from these two regions, we realized at
once that they were remarkably similar. (See sample phrases
from each theme in each year, Figure 3.) In order to see
just how similar the songs were, we made quantitative
comparisons between songs from the two eastern North
Pacific study areas. (For sizes of samples and dates of
recordings from these two study areas, see Table 1.) In
every case, we made our analysis by song session so as to
give each singer equal weight in the analysis. If we had
instead analyzed the sample by song, our results would have
been skewed toward singers for whom we recorded very long
song sessions.

We have also listened to the songs we recorded in the
two Pacific Ocean study areas in the 1980 season and found
them once again to be similar. We have not yet analysed
these songs quantitatively.

The West Indies sample was not used in this
quantitative analysis because the differences in song between
the different oceans were too great for us to decide how to
align the songs for comparison.

Table 2 lists the 14 variables which we measured in our
sample of Pacific Ocean songs. We compared the mean
values of the variables from the 1979 Hawaiian song sessions
with those from the Revillagigedos in 1979. (The 1977

Table 2.   Variables measured for Pacific Ocean songs.

Theme 1: mean duration of phrase (seconds)
Theme 1: mean number of units per phrase
Theme 1: mean number of phrases per theme
Theme 2: mean duration of phrase (seconds)
Theme 2: mean number of phrases per theme
Theme 4: mean number of phrases per theme
Theme 5: mean duration of phrase (seconds)
Theme 5: mean number of units per phrase
Theme 5: mean number of phrases per theme
Theme 6: mean number of phrases per theme
Themes 8 and 9: mean duration of phrase (seconds)
Themes 8 and 9: mean number of phrases per theme
Themes 8 and 9: mean percent 8-type phrases (see Figure 3)
Themes 8 and 9: mean percent 9-type phrases (see Figure 3)

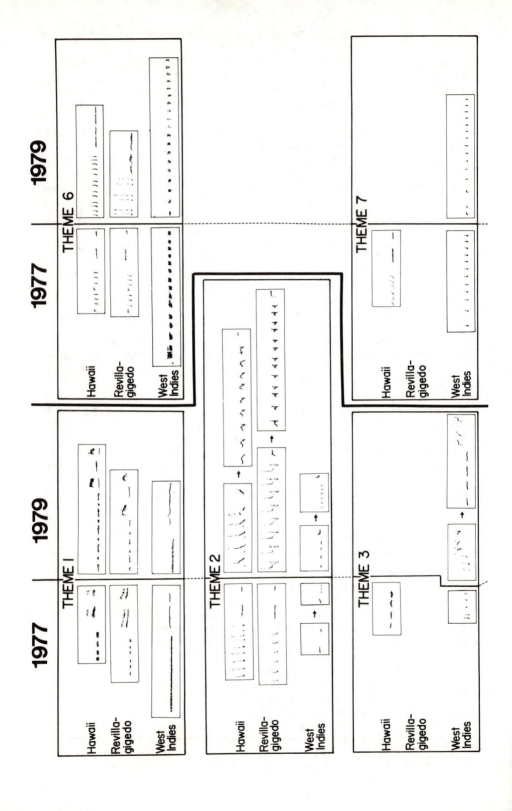

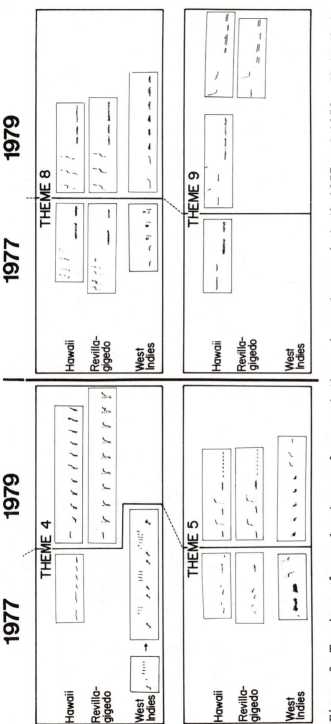

Figure 3. Tracings of sample phrases from each theme in each area in both 1977 and 1979. Note that there are strong similarities between Hawaii and the Revillagigedo Islands in each year, but that they are quite different from the West Indies song; and that changes between the 1977 and 1979 Hawaiian phrases are paralleled by changes in the Revillagigedo phrases. The designation of a theme as Theme 1 is arbitrary. Using what we call Theme 1 in the Hawaii song, we looked for the most similar theme in the West Indies song and designated it also as Theme 1.

Table 3. Results of t-tests on the 1979 Hawaii and Revillagigedo samples. The null hypothesis is that the mean values from the two areas are equal.

Theme	Variable	Null Hypothesis	
		Accepted	Rejected
1	Mean Phrase Duration		$0.05>p>0.02$
	Mean # Units/Phrase	$0.1>p>0.05$	
	Mean # Phrases/Theme	$0.2>p>0.1$	
2	Mean Phrase Duration	$0.4>p>0.2$	
	Mean # Phrases/Theme	$0.9>p>0.5$	
4	Mean # Phrases/Theme	$0.2>p>0.1$	
5	Mean Phrase Duration		$0.05>p>0.02$
	Mean # Units/Phrase		$0.01>p>.001$
	Mean # Phrases/Theme	$0.1>p>0.05$	
6	Mean # Phrases/Theme	$0.2>p>0.1$	
8	Mean Phrase Duration	$0.2>p>0.1$	
	Mean # Phrases/Theme	$0.1>p>0.05$	
	Mean % 8-type Phrases	$0.9>p>0.5$	
	Mean % 9-type Phrases	$0.9>p>0.5$	

samples were not used in this analysis because we have only one song session in our 1977 Revillagigedo sample.) Of the 14 variables measured, only 3 were significantly different in the two Pacific Ocean study areas (Table 3).

To investigate song change, we next compared the 1977 and 1979 Hawaii data. The null hypothesis was that the mean values of the measured song variables in the two years were equal. Our results are given in Table 4. They show a significant difference between the two years for some song variables. Of the 14 variables measured, 9 were significantly different between the 2 years, 3 gave inconclusive results, and only 2 (both measuring the mean number of phrases in a theme) were not significantly different in the two years. In more than half of the variables, then, there was change in the Hawaiian song between 1977 and 1979.

We next looked at the direction of change to see if the humpbacks were changing their songs in similar ways over time in the two study areas. With precise enough measurements, it was possible to classify every comparison of 1979 means to 1977 means as either an increase or a decrease. In Table 5, we signified an increase in the mean value of a variable with a "+" and a decrease with a "-". Only 3 of the 14 variables changed in different ways in the two areas from 1977 to 1979. We did not, however, attempt to find the statistical significance of these results since some of the variables are highly correlated with each other.

In addition to the 14 variables tested, there were at least four others in which a change was parallel in both study areas (see Figure 3). 1) Theme 6 has units which are shaped rather like the letters r and j on a spectrogram. Both r's and j's were present in Theme 6 in Hawaii and the Revillagigedo Islands in 1977 but only j's were present in 1979. 2) A second variable beyond the 14 given in Table 2 concerned miscellaneous and hybrid phrases in Theme 8 (not shown in Figure 3, but explained in more detail in Payne et al. 1983). Theme 8 was undergoing rapid change in these years and two common forms, known as miscellaneous and hybrid phrases, were sung in both Pacific study areas. In both areas the miscellaneous and hybrid phrases were present in 1977 but absent in 1979. 3) The "new" theme, shown in Figure 3 (where it is called Theme 9b) was absent in 1977 and present in both areas in 1979. This theme constitutes our only observation of a sound apparently introduced *de novo* to the song. As such, it is very interesting that it was heard in both areas in 1979. We do not know whether it was sung in the Revillagigedos in 1978; it was not part of the Hawaiian song in that year. 4) Theme 1 demonstrated a change in both areas involving unit length. In 1977 all units in subphrase one of Theme 1 were approximately the same duration in both areas. By 1979 this same subphrase

Table 4. Results of t-tests on the 1977 and 1979 Hawaii samples. The null hypothesis is that the mean values in the two years are equal.

Theme	Variable	Null Hypothesis	
		Accepted	Rejected
1	Mean Phrase Duration		$0.01 > p > .001$
	Mean # Units/Phrase		$p < .001$
	Mean # Phrases/Theme	$0.4 > p > 0.2$	
2	Mean Phrase Duration		$p < .001$
	Mean # Phrases/Theme	$0.2 > p > 0.1$	
4	Mean # Phrases/Theme	$0.1 > p > .05$	
5	Mean Phrase Duration		$0.01 > p > .001$
	Mean # Units/Phrase		$p < .001$
	Mean # Phrases/Theme	$0.1 > p > 0.05$	
6	Mean # Phrases/Theme		$0.01 > p > .001$
	Mean Phrase Duration	$0.1 > p > 0.05$	
8	Mean # Phrases/Theme		$0.05 > p > 0.02$
	Mean % 8-type Phrases		$p < .001$
	Mean % 9-type Phrases		$0.05 > p > 0.02$

Table 5. Direction of change from 1977 to 1979 in the two Pacific Ocean study areas. "+" is used to indicate an increase in the measured variable and "-" to indicate a decrease. Of the 14 variables, 11 changed in similar ways in the 2 areas.

Theme	Variable	Change in Hawaii	Change in Revillagigedos	Similar?
1	Mean Phrase Duration	+	+	yes
	Mean # Units/Phrase	+	+	yes
	Mean # Phrases/Theme	+	-	no
2	Mean Phrase Duration	+	+	yes
	Mean # Phrases/Theme	-	-	yes
4	Mean # Phrases/Theme	-	+	no
5	Mean Phrase Duration	+	+	yes
	Mean # Units/Phrase	+	+	yes
	Mean # Phrases/Theme	-	-	yes
6	Mean # Phrases/Theme	-	-	yes
8	Mean Phrase Duration	+	+	yes
	Mean # Phrases/Theme	-	-	yes
	Mean % 8-type Phrases	+	+	yes
	Mean % 9-type Phrases	-	+	no

Table 6. Results of discriminant analyses showing how well each variable in Table 2 discriminates between year and location for the two Pacific Ocean study areas. Because there is only one song session in 1977 in the Revillagigedos, we left it out of these calculations. In each category, Column A gives the number of song sessions which are correctly classified and Column B gives the number of song sessions which are incorrectly classified. Phr. Dur. is an abbreviation for Phrase Duration.

Variable	Hawaii 1977		Hawaii 1979		Revilla. 1979		%A	%B
	A	B	A	B	A	B		
# Units Theme 1	10	1	6	0	4	1	91	9
# Phrases Theme 1	2	9	3	3	3	2	36	64
# Phrases Theme 2	6	5	4	2	0	5	45	55
# Phrases Theme 4	6	5	2	4	5	0	59	41
# Phrases Theme 5	7	4	5	1	3	2	68	32
# Units Theme 5	11	0	6	0	4	1	95	5
# Phrases Theme 6	9	2	3	3	4	1	73	27
# Phrases Theme 8	4	7	5	1	1	4	45	55
Phr. Dur. Theme 1	10	1	5	1	4	1	86	14
Phr. Dur. Theme 2	11	0	6	0	3	2	91	9
Phr. Dur. Theme 5	7	4	5	1	3	2	68	32
Phr. Dur. Theme 8	10	1	4	2	4	1	82	18
% 8-type Phrases	9	2	3	3	3	2	68	32
% 9-type Phrases	7	4	2	4	3	2	55	45
							Mean: 68.7	31.3

contained units of alternately long and short duration in all songs in both Pacific Ocean study areas (the "long" units being more than twice as long as the "short" ones).

To verify our findings, we performed a series of discriminant analyses on the 14 variables listed in Table 2. (Discriminant analysis is a multivariable analysis which assigns an observation of unknown origin to the group most likely to contain that observation, based on the values of several variables.) In this case, we used discriminant analysis to classify individual song sessions by determining from which location, month, and year they were most likely to have come.

We performed discriminant analyses on each of the variables individually to see how well each variable discriminated between different locations and time periods (Table 6). Although there were a number of variables which did discriminate well between both years and locations, there were also many that resulted in misclassified song sessions.

Again considering each of the variables separately, we counted the number of song sessions which were classified into the correct year, regardless of location (Table 7). These results were more consistent with the true situation. In fact, with the exception of the variables measuring the number of phrases in each theme, song sessions could be correctly categorized into years based on any single variable. This seems to indicate that regardless of whether they were recorded in the Hawaiian or the Revillagigedo study area, sessions recorded within one season are more similar than sessions recorded in one study area in different years.

Discussion

Payne (1979) and Winn and Thompson (1979) note the similarity between songs recorded in different wintering grounds in the eastern North Pacific. What is striking about these results is that the song is the same over such a large area. Winn and Thompson (1979) and Winn et al. (1981) have found strong similarities between songs in two widely separated wintering grounds in the North Atlantic as well (the West Indies and Cape Verde). Our results indicate that not only are the Hawaii and Revillagigedo songs similar at any given moment but that they change together over the years. Since change is progressive (Payne et al. 1983) and individual whales keep pace with the changes (Guinee et al. 1983), we know that singers are constantly learning the newest version of the song. For a whale to keep track of and learn these changes, he must be in acoustic contact with other singers during the times when humpbacks are singing. Therefore, we know that humpbacks who are singing the same song in two separate areas must be in acoustic contact in at least part of the year.

Table 7. Results of discriminant analyses showing how well each variable in Table 2 discriminates between years, regardless of location. In each category, Column A gives the number of observations which are classified correctly and Column B gives the number of observations which are classified incorrectly. The 1977 Revillagigedo sample is included in this analysis. Phr. Dur. is an abbreviation for Phrase Duration.

Variable	1977 A	1977 B	1979 A	1979 B	% in A	% in B
# Units Theme 1	11	1	11	1	96	4
# Phrases Theme 1	3	9	9	3	50	50
# Phrases Theme 2	7	5	8	4	65	35
# Phrases Theme 4	5	7	10	1	65	35
# Phrases Theme 5	4	8	10	1	61	39
# Units Theme 5	12	0	10	1	96	4
# Phrases Theme 6	10	2	11	0	96	4
# Phrases Theme 8	4	8	9	2	57	43
Phr. Dur. Theme 1	10	2	11	0	91	9
Phr. Dur. Theme 2	12	0	10	1	96	4
Phr. Dur. Theme 5	7	5	11	0	78	22
Phr. Dur. Theme 8	10	2	11	0	91	9
% 8-type Phrases	9	3	11	0	87	13
% 9-type Phrases	7	5	10	1	74	26

Mean: 78.4   21.6

We do not know how far humpback song carries before it is lost in background noise. However, Levenson (1972) gives broadband maximum and source levels of 174 and 155 db re 1μPa for humpback sounds in the frequency range between 100 Hz and 8 kHz (with most energy concentrated below 4.0 kHz). Following the kind of arguments made by Payne and Webb (1971), let us assume that we have an omnidirectional source and receiver; and that, ignoring attenuation, transmission loss equals 20 log r (where r=range); and that the ear has adequate sensitivity and a detection threshold of 0 db signal-to-noise ratio; and that it can be tuned to a bandwidth of 1/3 octave; and that the background noise is from the usual shallow water traffic noise and sea state 3. With these assumptions in mind, we find that the loudest sounds in humpback song can probably be heard at less than 20 km and that under most circumstances, they probably go 1/8 to 1/4 this distance. The long range propagation for fin whale (*Balaenoptera physalus*) sounds calculated by Payne and Webb were dependent on a deep water sound path, whereas humpback singing sites seem generally to be in shallower water. If the assumptions of the range calculation for humpback song hold, then humpbacks must be within about 20 km of each other to achieve acoustic contact. There are at least three ways in which such physical proximity might occur between two distant wintering grounds.

After a winter in acoustic isolation from singers on other wintering grounds, whales may sing often enough, either while migrating or while on a common summer feeding ground, to learn the changes made in the song over the season in the other wintering area, as well as to teach the whales from the other area the changes made in their own. If this were the case, the song, which would have grown more divergent in the two areas over the breeding season, would then converge on a norm before humpbacks return to the two wintering grounds. Several observations are consistent with this hypothesis. Tyack and Whitehead (1983) report having heard humpback song in open ocean, during migration, which is consistent with this hypothesis. Data from Lawton, Rice, Wolman, and Winn (1979) and from Darling and Jurasz (1983) show that the Southeast Alaskan feeding area contains whales from both the Hawaiian waters and the Mexican coastal waters. McSweeney et al. (in prep.) report having recorded some songs in the Alaska feeding area, though recordings from feeding areas appear to be rare despite much effort. However, Payne et al. (1983) find that most song changes occur within a winter season, rather than between them. This indicates that song changes occurring during migration and on the summer feeding grounds probably represent only part of the story. We hope that recording samples from full seasons in both areas will give further evidence to support or reject this possibility.

The second possibility is that singers may visit more than one wintering area in the same singing season and, when they arrive at the second area, introduce the song changes that are occurring in the first. If this were true, it might explain the discovery of Darling, Gibson, and Silber (1983) that most humpbacks appear to stay on the Hawaiian wintering grounds for no more than six weeks. This would also raise the possibility that there is a string of migrating humpbacks between the two wintering grounds who are singing at least occasionally and thereby maintaining a string of acoustic contact between the two areas.

The third possibility is that singers visit different wintering grounds in subsequent winters. Chittleborough (1960) reports that a male humpback whale marked with a discovery tag on the east coast of Australia (27°S, 153.5^{O}E) was "recaptured" approximately 11 months later on another wintering ground, the west coast of Australia ($25^{O}21'$S, $112^{O}26'$E). Darling and Jurasz (1983) report that out of 11 whales photographed in the Revillagigedos in 1979, 2 were photographed in Hawaii in other years. Even though the Darling and Jurasz sample is small, the results imply that there is probably considerable interchange between these two areas.

The above possibilities are not mutually exclusive and there may be a combination of these and other possibilities at work. More information on individual identification from photographs of fluke patterns and recording samples from full seasons in the two areas should help determine how and when humpbacks from these two areas mingle with each other.

However, the mechanism by which song information is shared is not our major concern here. We wish to focus instead on the fact that there is a strong potential for genetic exchange between humpback whales that are within acoustic range since such ranges are, of necessity, short. Data linking humpback song to reproductive activity (Tyack 1981) makes this seem even more likely. We don't know how extensive the genetic interchange is, but there seems to be a distinct possibility that humpbacks in Hawaii and the Revillagigedos are one population, and there is no evidence to the contrary. This same reasoning would also apply to humpbacks singing the same song in different wintering grounds in the North Atlantic and elsewhere.

Payne (1979) pointed out that humpback whale songs are a useful indicator of stocks. Winn and Thompson (1979) and Winn et al. (1981) also report "dialects" in humpback whale songs in different oceans but conclude that "while different populations in the same ocean basin have virtually identical song formats, it is important to note that they differ from other populations in other ocean basins." This is in basic agreement with our findings, though we disagree with their use of terms. Winn et al. do not define what they mean by the term "population". Our own research (and theirs) suggests that there is not significant enough isolation between humpbacks in the eastern North Pacific to refer to them as separate populations. The same situation seems to exist in the North Atlantic. Any claim of more than one population in either of these areas has yet to be supported.

The use of the term "dialects" by Winn et al. (1981) also seems in contradiction to common usage. The distinction between dialects and geographic variation in song is important. Geographic variation usually refers to variations between isolated populations or groups that are not in acoustic contact because of some geographic barrier, while dialects imply variations on a more local level which cannot be accounted for by geographic barriers (LeBoeuf and Peterson 1969, Nottebohm 1969, Grimes 1974, Conner 1982, Ford and Fisher 1983).

Our second major finding is that the North Atlantic and eastern North Pacific songs share some similarities in spite of their differences. These two groups of whales are almost certainly acoustically isolated. Several basic structural elements of song that they share are as follows: 1) songs from both oceans have an invariant theme order; 2) songs from both oceans are organized into the same basic components: themes, phrases, subphrases, units, and subunits; and 3) songs from both oceans are composed of sets of units which overlap to a great extent. While units themselves evolve to produce a graded series of sounds over time, it appears that there may be a finite number of kinds of these graded sets of units which can be put together to make up the song. This seems to be indicated by the fact that we have heard only one new kind of sound in the study of five years of recordings from the eastern North Pacific.

Winn and Winn (1978) state that humpback songs are organized into a fixed set of six themes. Although we agree that all sounds in humpback songs are organized into themes, we are unable to fit all themes from the Pacific Ocean study areas into songs with six themes such as they describe. We suggest that there may be a finite number of kinds of units, but that they can be put together in many different ways to make up different themes, and even songs with different numbers of themes, in different oceans. It does seem to be the case that there are themes in the different oceans with striking similarities. It will take a very large sample of songs from many isolated populations to determine whether the production of similar themes in the songs of different populations is coincidental or whether all humpbacks have similar preferences for certain combinations of units.

Given that there seems to be so much variability in song over long time periods, why should there be such strong selection for song constancy over enormous areas of ocean? Tyack (1981) has given evidence that it is female choice which drives the variability of humpback song. The apparent anomaly between songs that are constant over vast areas of ocean but variable over time may constitute an important clue to what it is that females are choosing for.

Because the analysis of the songs of birds has demonstrated that individuals within one population strictly adhere to one dialect (Marler and Hamilton 1966), it is not surprising that the analysis of humpback songs seems to have provided a powerful tool for delimiting stocks. The use of variability in pigmentation of humpbacks has been widely used as a stock indicator (see Glockner and Venus, 1983, for a review). Glockner and Venus cast serious doubts on the value of pigmentation patterns for stock determinations when they are used on any but the grossest level. They note, for example, that in some individuals the pigmentation of some of the body areas on which the technique depends changes significantly during the life of the whale. Until we have more direct biochemical or genetic evidence with which to characterize the gene pools of large numbers of humpback whales, "song pools" seem to be a particularly accessible way of assessing stocks. Indeed, the study of songs may currently constitute the best and most practical technique for stock assessment of this species.

Acknowledgements

Our primary thanks go to Mary Crowley and Angelica Theriot, who gave us copies of their 1977 humpback whale tapes from the Revillagigedo Islands. It was this generosity which triggered the rest of the research reported here.

Many others also provided recordings for which we are most grateful. They include: James Darling, Janet Heyman-Levine, Fred Levine, David Matilla, Katharine Payne, Judith Perkins, Peter Tyack, and Hal Whitehead. Captain George Nichols and the crew of the r/v *Regina Maris* (Expedition 13) were especially helpful in obtaining recordings in 1979 in the Revillagigedos. The Ocean Research and Education Society also provided us with tapes of the West Indies (1977).

Help with analysis of data was provided by Janet Heyman-Levine, Katharine Payne, Numi Spitzer, and Peter Tyack. For the statistical analysis, we also had help from Ronald Christensen, Eleanor Dorsey, and Katharine Payne.

James Bird, Eleanor Dorsey, Katharine Payne, Victoria Rowntree, and Peter Tyack read one or more drafts of the manuscript and offered valuable discussion and criticism.

Many kinds of valuable help (including, in some cases, funding) were provided by the following organizations: the Eppley Foundation, the Lahaina Restoration Foundation, the Mexican Navy, the National Geographic Society, the New

York Zoological Society, and the World Wildlife Fund - U.S., and by the following individuals: Capitan de Navio A. Esparza Rodriguez, Shirley Herpick, James Luckey, John and Walter McIlhenny, and Drake and Maureen Thomas. We are grateful to all of the above.

This report is Contribution #13 of the Ocean Research and Education Society, Gloucester, MA.

Literature Cited

Chittleborough, R.G.
1960. Marked humpback whale of known age. Nature 187: 4732.

Conner, D.A.
1982. Dialects versus geographic variation in mammalian vocalizations. Anim. Behav. 30: 297-298.

Darling, J.D., K.M. Gibson and G.K. Silber.
1983. Observations on the abundance and behavior of humpback whales (*Megaptera novaeangliae*) off West Maui, Hawaii 1977-79. *In* R. Payne (ed.), Communication and behavior of whales. AAAS Selected Symposia Series, p. 201-222. Westview Press, Boulder, Colo.

Darling, J.D. and C.M. Jurasz.
1983. Migratory destinations of North Pacific humpback whales (*Megaptera novaeangliae*). *In* R. Payne (ed.), Communication and behavior of whales. AAAS Selected Symposia Series, p. 359-368. Westview Press, Boulder, Colo.

Ford, J.K.B. and H.D. Fisher.
1983. Group-specific dialects of killer whales (*Orcinus orca*) in British Columbia. *In* R. Payne (ed.), Communication and behavior of whales. AAAS Selected Symposia Series, p. 129-161. Westview Press, Boulder, Colo.

Glockner, D.A.
1983. Determining sex of humpback whales (*Megaptera novaeangliae*) in their natural environment. *In* R. Payne (ed.), Communication and behavior of whales. AAAS Selected Symposia Series, p. 447-464. Westview Press, Boulder, Colo.

Glockner, D.A. and S.C. Venus.
1983. Identification, growth rate, and behavior of humpback whale, *Megaptera novaeangliae*, cows and calves in the waters off Maui, Hawaii, 1977-79. *In* R.

Payne (ed.), Communication and behavior of whales. AAAS Selected Symposia Series, p. 223-258. Westview Press, Boulder, Colo.

Grimes, L.G.
1974. Dialects and geographical variation in the song of the splendid sunbird *Nectarinia coccinigaster*. Ibis 116: 314-329.

Guinee, L.N., K. Chu, and E.M. Dorsey.
1983. Changes over time in the songs of known individual humpback whales (*Megaptera novaeangliae*). *In* R. Payne (ed.), Communication and behavior of whales. AAAS Selected Symposia Series, p. 59-80. Westview Press, Boulder, Colo.

Lawton, W., D. Rice, A. Wolman, and H. Winn.
1979. Occurrence of southeastern Alaska humpback whales, *Megaptera novaeangliae*, in Mexican coastal waters. Seattle: Abstracts from presentations of the third biennial conference on the biology of marine mammals, October 7-11, p. 35.

LeBoeuf, B.J. and R.S. Peterson.
1969. Dialects in elephant seals. Science (Wash. D.C.) 166: 1654-1656.

Levenson, C.
1972. Characteristics of sounds produced by humpback whales (*Megaptera novaeangliae*). Navocean Tech. Note No. 7700-6-72.

Marler, P. and W.J. Hamilton, III.
1966. Mechanisms of animal behavior. Wiley and Sons, New York. 771p.

McSweeney, D., W. Dolphin, and R. Payne.
In prep. Humpback whale songs recorded on summer feeding grounds.

Nottebohm, F.
1969. The song of the chingolo, *Zonotrichia capensis*, in Argentina: description and evaluation of a system of dialects. Condor 71: 299-315.

Payne, K. and R. Payne.
In prep. Large-scale changes over 17 years in songs of humpback whales in Bermuda. Z. Tierpsychol.

Payne, K., P. Tyack, and R. Payne.
1983. Progressive changes in the songs of humpback

whales (*Megaptera novaeangliae*): A detailed analysis of two seasons in Hawaii. *In* R. Payne (ed.), Communication and behavior of whales. AAAS Selected Symposia Series, p. 9-57. Westview Press, Boulder, Colo.

Payne, R.
1968. Among wild whales. The New York Zoological Society Newsletter, 6p.

1978. Behavior and vocalizations of humpback whales (*Megaptera* sp.). *In* K.S. Norris and R.R. Reeves (eds.), Report on a workshop on problems related to humpback whales (*Megaptera novaeangliae*) in Hawaii. U.S. Dep. Commer. NTIS PB-280 794, p. 56-78.

1979. Humpback whale songs as an indicator of "stocks". Seattle: Abstracts from presentations of the third biennial conference on the biology of marine mammals, October 7-11, p. 46.

Payne, R. and S. McVay.
1971. Songs of humpback whales. Science (Wash. D.C.) 173: 585-597.

Payne, R. and D. Webb.
1971. Orientation by means of long range acoustic signaling in baleen whales. Ann. N.Y. Acad. Sci. 188: 110-141.

Tyack, P.
1980. The function of song in humpback whales, *Megaptera novaeangliae*. *In* Abstracts of papers of the 146th national meeting 3-8 January, San Francisco, California, A. Herschman (ed.), Washington, D.C.: American Association for the Advancement of Science, AAAS Publication 80-2, p. 41.

1981. Interactions between singing Hawaiian humpback whales and conspecifics nearby. Behav. Ecol. Sociobiol. 8: 105-116.

Tyack, P. and H. Whitehead.
1983. Male competition in large groups of wintering humpback whales. Behaviour 83: 132-154.

Winn, H.E., W.L. Bischoff, and A.G. Taruski.
1973. Cytological sexing of cetacea. Mar. Biol. 23: 343-346.

Winn, H.E., P.J. Perkins, and T.C. Poulter.
   1971. Sounds of the humpback whale. Proceedings of the
   7th annual conference on biological sonar and diving
   mammals, Menlo Park, CA, p. 39-52.

Winn, H.E. and T.J. Thompson.
   1979. Comparisons of humpback whale sounds across the
   northern hemisphere. Seattle: Abstracts from presentations of the third biennial conference on the biology
   of marine mammals, October 7-11, p. 62.

Winn, H.E., T.J. Thompson, W.C. Cummings, J. Hain, J.
   Hudnall, H. Hays, and W.W. Steiner.
   1981. Song of the humpback whale -- population
   comparisons. Behav. Ecol. Sociobiol. 8:41-46.

Winn, H.E. and L.K. Winn.
   1978. The song of the humpback whale (*Megaptera
   novaeangliae*) in the West Indies. Mar. Biol. (Berl.) 47:
   97-114.

James D. Darling, Charles M. Jurasz

11. Migratory Destinations of North Pacific Humpback Whales (*Megaptera novaeangliae*)

Abstract

Previous research on the migration of the humpback whale shows that in the southern hemisphere most individuals follow a fairly direct north-south route between summer feeding and winter calving grounds. Our work suggests a different pattern in the North Pacific. Basing identification of individuals on their distinctive fluke markings, we have photographed the same seven whales in Hawaii (in winter) and in southeast Alaska (in summer) 4600 km away. This indicates a northeast-southwest axis of migration in the North Pacific and is in agreement with the only previous long range North Pacific tag returns for this species, of 8 whales that moved between the eastern Bering Sea/Aleutians and the waters south of Japan.

Two of the seven animals were photographed swimming side by side in Alaska and in the same channel in Hawaii the following winter. Both had calves in Hawaii. Six of the seven were photographed in Alaska and later in Hawaii, indicating a remarkable navigational ability for this species. An eighth individual was photographed first in Hawaii and two years later near Socorro Island -- a different calving ground 4800 km due east of Hawaii. This surprising result strongly supports the conclusions drawn by Payne and Guinee (1983) from the analysis of songs of humpback whales that the "populations" of eastern North Pacific humpbacks are either strongly overlapping or just one stock.

Introduction

During the summer, humpback whales (*Megaptera novaeangliae*) feed along the North Pacific rim from central

California to Japan, with higher concentrations in the areas of southeast Alaska, the Aleutian Islands and the Bering Sea (Rice 1974, 1977; Nemoto 1978). In the winter, this species migrates to three distinct assembly areas in the subtropical and tropical waters where calving and probably mating occur (Chittleborough 1965; Rice 1977). These three assembly areas are in the western North Pacific (from Taiwan north through the Ryukyu Islands to Kyushu, southern South Korea, and Honshu and back south through the Bonin Islands to the Marianas), in the central North Pacific (around the main Hawaiian Islands), and in the eastern North Pacific (off the west coast of Baja California and mainland Mexico and near the offshore Revillagigedo Islands) (Rice 1977).

Little is known about the migratory paths taken by humpbacks in the North Pacific. The only known connection between North Pacific summer and winter grounds is based on eight tagged whales which moved between the eastern Aleutians/Bering Sea and the Ryukyu and Bonin archipelago (a migration, in most cases, of more than 6000 km!) (Nishiwaki 1962; Ohsumi and Masaki 1975; Rice 1977).

By comparing color patterns on the dorsal surface of the flippers of Hawaiian and other eastern North Pacific whales, Herman and Antinoja (1977) inferred that the Hawaiian population is reproductively isolated from the eastern North Pacific stock and that it does not summer in the feeding grounds of that region. Our findings indicate that this inference is in error. Individual humpbacks can be identified by photographs of the black and white skin patterns on the underside of their flukes. Schevill and Backus (1960) first mentioned the value of these markings as a means of distinguishing individuals. The technique has been used since 1967 by Jurasz in Alaska and described in detail by Kraus and Katona (1977) and Katona, Baxter, Brazier, Kraus, Perkins, and Whitehead (1979). Using this technique, we have found whales wintering in Hawaii which were initially identified in southeast Alaska and vice versa. We have also found one whale which was identified in Hawaii one winter and in the Revillagigedo Islands two winters later. These are the first connections made between humpbacks in the Hawaiian and Mexican breeding areas and the North Pacific feeding grounds.

Methods and Results

The migration between humpback summering and wintering grounds reported here resulted from work in three areas. C.J. coordinated the photographic identification in southeast Alaska for Sea Search, Ltd. as part of a twelve-year study; J.D. compiled photographs from a three-year study in Hawaii; and the Ocean Research and Education Society provided photographs from the Revillagigedo Islands taken aboard the *Regina Maris*.

Photographs of the underside of flukes of 103 whales identified in southeast Alaska in July and August of 1977-78 were compared with photographs of 264 whales taken between Maui and Lanai in Hawaii in December-May of 1977-79, and with photographs of 11 individuals from the Revillagigedo Islands in Mexico obtained over a four week period from 27 February-25 March 1979.

Table 1 lists seven definite matches between whales identified in southeast Alaska in the summer and in Hawaii in the winter. Figure 1 (a and b) shows photographs of two of the seven tails matched. The whale which switched breeding grounds between Hawaii and the Revillagigedo Islands is also listed in Table 1 and shown in Figure 1(c).

Figure 2 shows the locations of matched whales. In Alaska, two of the whales were identified in Glacier Bay, the remainder in Frederick Sound, approximately 225 km to the

Table 1. Matched whales: Alaska-Hawaii-Mexico

Whale #	"Mark" date and location	"Recapture" date and location
15	18 August 1978 Frederick Sound, AK	31 March 1979 Auau Channel, HI
199	July 1977 Frederick Sound, AK	6 February 1979 Auau Channel, HI
3	19 January 1978 Auau Channel, HI	16 August 1978 Frederick Sound, AK
152	16 August 1978 Frederick Sound, AK	24 March 1979 Auau Channel, HI
109	20 August 1978 Frederick Sound, AK	11 April 1979 Auau Channel, HI
82	July 1977 Glacier Bay, AK	2 April 1978 Auau Channel, HI
107	July 1977 Glacier Bay, AK	21 February 1978 Auau Channel, HI
217	11 March 1977 Auau Channel, HI	1 March 1979 Socorro Is., Mex.

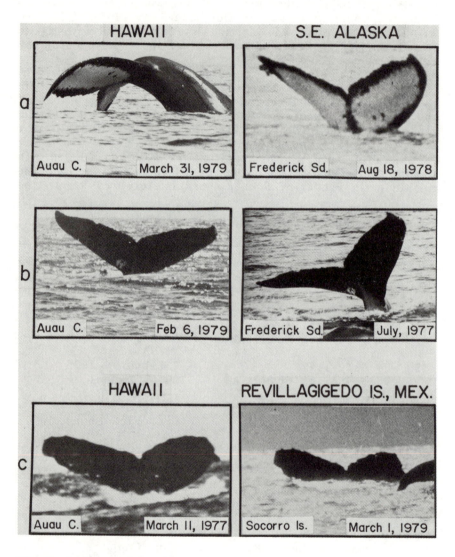

Figure 1. Examples of matches between Hawaii and southeast Alaska, and Hawaii and Mexico.

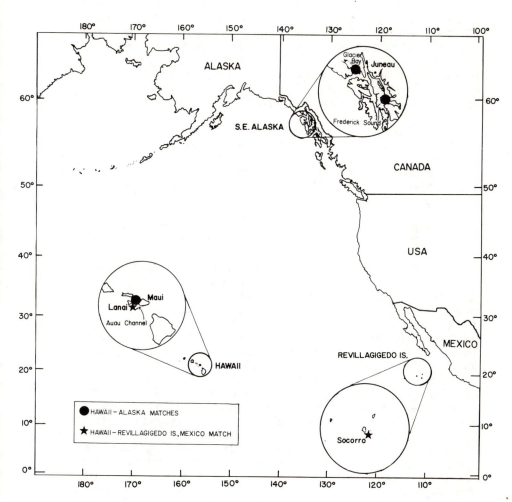

Figure 2. Locations where whales were photo-identified and later resighted.

southeast. It is of particular interest that both whales in Glacier Bay were seen side by side on several occasions in July 1977 by C.J. Both were photographed with calves in the Auau Channel the following winter. In fact, the Auau Channel between Maui and Lanai was the Hawaiian location for all eight matches reported here. Several matches have been made between whales in the Auau Channel and whales photographed by Daniel McSweeney (pers. comm.) in a second region of concentration along the west coast of Hawaii, and therefore one might also expect to find Alaskan whales there in the future.

The Alaska-Hawaii matches show that whales travel in both directions between Alaska and Hawaii in less than eight months. Whales photographed in Alaska during July-August were found in Hawaii the following February-April. One whale photographed in Hawaii in January was found in Alaska in August of the same year (Table 1).

Discussion

Studies conducted by Brown (1962), Dawbin (1966), and Chittleborough (1965) in the southwest Pacific-Australian and Antarctic areas show an apparent tendency for whales to move directly north and south between breeding areas and the nearest cold water feeding area. This tendency results in six primary humpback feeding areas in the southern hemisphere, each associated with a tropical breeding area (Mackintosh 1942). One might expect the North Pacific humpbacks to make similar north-south migrations, with Mexican whales feeding in southeast Alaska, Hawaiian whales in the central Aleutians and Bering Sea, and western Pacific whales in the northwest Pacific or western Aleutians. Apparently the case is not so simple. All the matches to date, including our own, show North Pacific humpback whales traveling southwest to northeast, rather than south to north, when migrating between their winter and summer grounds. Although coverage of the North Pacific Ocean in search of humpback whales marked either by Discovery tag or photographic identification is very spotty, what information we do have suggests a predominant migratory direction in the North Pacific different from the north-south directions observed in the southern hemisphere. Although their routes are not known, the number of matches between southeast Alaska and Hawaii and the fact that known individuals moved between these two areas every year for three years (1977-1979) suggests that migration between Hawaii and southeast Alaska is normal rather than exceptional.

Inasmuch as the Hawaiian Islands are the most geographically isolated chain in any ocean, the fact that whales are able to migrate to them from southeast Alaska indicates that this species has remarkable navigational ability.

A tendency for whales to consort together in winter and summer areas is suggested by 1) three whales photographed in the same period in Frederick Sound during summer 1978 which were photographed in the Auau Channel the following winter, and 2) two Glacier Bay whales seen side by side during summer 1977 that were both present in the Auau Channel the following winter. Both of these examples indicate social cohesion in this species.

Although data from the southern hemisphere make it surprising to find that humpbacks from southeast Alaska winter in Hawaii, it seems still more unusual to find an individual wintering in two different breeding areas (Hawaii and the Revillagigedo Islands) separated by approximately 4800 km of open ocean. This observation is not without precedent, however. Chittleborough (1960, 1962) reports two humpback whales which switched breeding grounds from the east to west coast of Australia, a distance of 5600 km by the shortest route; as well as a whale which was marked in the Fiji Islands wintering grounds and killed in the eastern Australian grounds, a distance of 3100 km. A Mexican-Hawaiian switch strongly supports the conclusions drawn by Payne and Guinee (1983) from the analysis of songs of humpack whales that the humpback "populations" of the eastern North Pacific are either strongly overlapping or just one stock.

The three southern hemisphere whales which switched breeding grounds were cases recovered after at least one intervening summer, suggesting that the transfer of grounds was accomplished by the whale choosing a different breeding area in different years. Examples of this kind indicate that genetic mixing probably occurs between adjacent breeding populations. The extent of this occurrence remains to be determined.

Perhaps as significant as the matches themselves is the demonstration given here of the value of photographic identification. We believe this method will provide more information about distribution, behavior, and population size of humpback whales than have the traditional Discovery marks, which can be read only by killing a whale. Similar matches have been made in the North Atlantic (Balcomb and Nichols 1978; Katona et al. 1979). In the Hawaiian studies in the seasons 1976-77 to 1978-79 (Darling, Gibson, and Silber 1983), 318 identifications were made. A total of 54 recoveries, excluding same day recoveries, were included in this number. The recovery rate is 16.9% and thus considerably higher than the 5.3% overall recovery rate for all humpback whales marked by Discovery tags in the southwest Pacific-Australia area (Rayner 1940; Dawbin 1959, 1964; Brown 1962; Chittleborough 1965; Ivashin 1973), or the

4.2% recovery of all humpbacks marked in the North Pacific (Rice 1977). The photographic technique also has the distinct advantage of allowing more than one recovery per whale.

Addendum

We have seen whale #199 (Figure 1b) on a third occasion, this time in Frederick Sound, Alaska on 24 August 1979. This is the first direct evidence of which we are aware, of an individual whale making a complete roundtrip between its migratory endpoints.

We also found in Hawaii on 1 February 1980 a second of the 11 whales photographed by Katharine Payne in Socorro waters on 11 March 1979. These findings, coupled with the evidence for nearly identical songs in Hawaiian and Mexican waters, force one to the conclusion that in spite of their great distance from one another, Hawaii and Socorro are occupied by a single population of whales.

Acknowledgements

We thank all those who made their photographs available to us, including Pat Eberhart, Bruce Wellman, and Mike Payne in Alaska; Kimberly Gibson, Deborah Glockner, Roy Nickerson, Gregory Silber, and Peter Tyack in Hawaii; and Katharine Payne and other members of the *Regina Maris* Expedition #13 in Mexico and Expedition #16 in Alaska. We gratefully acknowledge the use of Daniel McSweeney's and Rick Chandler's photographs from the island of Hawaii. Jim Bird helped in literature review and manuscript preparation. Linda Guinee helped in preparing the figures and Roger and Katharine Payne gave suggestions for the manuscript. Special thanks to Virginia, Susan, and Peter Jurasz for their invaluable assistance in the field and lab, and James C. Luckey and the Lahaina Restoration Foundation for support in Hawaii. The field study in Alaska was supported by the National Marine Fisheries Service, Northwest and Alaska Fisheries, Marine Mammal Division, and the National Park Service. The work in Hawaii was supported by the National Geographic Society, the New York Zoological Society and the World Wildlife Fund (grants to R. Payne), and the Vancouver Public Aquarium (1978 grant to J.D.).

Literature Cited

Balcomb, K.C. and G. Nichols.
    1978. Western North Atlantic humpback whales. Int. Whaling Comm. Rep. Comm. 28: 159-164.

Brown, S.G.
    1962. International co-operation in Antarctic whale

marking 1957 to 1960, and a review of the distribution of marked whales in the Antarctic. Nor. Hvalfangsttid. 51: 93-104.

Chittleborough, R.G.
1960. Australian catches of humpback whales 1959. C.S.I.R.O. Aust. Div. Fish. Oceanogr. Rep. No. 29: 16p.

1962. Australian catches of humpback whales 1961. C.S.I.R.O. Aust. Div. Fish. Oceanogr. Rep. No. 34: 13p.

1965. Dynamics of two populations of humpback whale, *Megaptera novaeangliae* (Borowski). Aust. J. Mar. Freshw. Res. 16: 33-128.

Darling, J.D., K.M. Gibson and G.K. Silber.
1983. Observations on the abundance and behavior of humpback whales *(Megaptera novaeangliae)* off West Maui, Hawaii, 1977-79. *In* R. Payne (editor), Communication and behavior of whales. AAAS Selected Symposia Series, p. 201-222. Westview Press, Boulder, Colo.

Dawbin, W.H.
1959. New Zealand and South Pacific whale marking and recoveries to the end of 1958. Nor. Hvalfangsttid. 48: 213-238.

1964. Movements of humpback whales marked in the South West Pacific Ocean 1952 to 1962. Nor. Hvalfangsttid. 53: 68-78.

1966. The seasonal migratory cycle of humpback whales. *In* K.S. Norris (editor), Whales, dolphins, and porpoises, p. 145-170. Univ. Calif. Press, Berkeley.

Herman, L.M. and R.C. Antinoja.
1977. Humpback whales in the Hawaiian breeding waters: Population and pod characteristics. Sci. Rep. Whales Res. Inst. Tokyo 29: 59-85.

Ivashin, M.V.
1973. Marking of whales in the Southern Hemisphere (Soviet materials). Int. Whaling Comm. Rep. Comm. 23: 174-191.

Katona, S., B. Baxter, O. Brazier, S. Kraus, J. Perkins, and H. Whitehead.
1979. Identification of humpback whales by fluke photographs. *In* H.E. Winn and B.L. Olla (editors), Behavior of marine animals - current perspectives in research, vol. 3: Cetaceans, p. 33-44. Plenum Press, N.Y.

Kraus, S. and S. Katona (editors).
1977. Humpback whales in the Western North Atlantic - A catalogue of identified individuals. College of the Atlantic, Bar Harbor, Maine, 26p.

Mackintosh, N.A.
1942. The southern stocks of whalebone whales. Discovery Rep. 22: 197-300.

Nemoto, T.
1978. Humpback whales observed within the continental shelf waters of the Eastern Bering sea. Sci. Rep. Whales Res. Inst. Tokyo 30: 245-247.

Nishiwaki, M.
1962. Ryukyuan whaling in 1961. Sci. Rep. Whales Res. Inst. Tokyo 16: 19-28.

Ohsumi, S. and Y. Masaki.
1975. Japanese whale marking in the North Pacific, 1963-1972. Bull. Far Seas Fish. Res. Lab. (Shimizu) 12: 171-219.

Payne, R. and L.N. Guinee.
1983. Humpback whale, *Megaptera novaeangliae*, songs as an indicator of stocks. *In* R. Payne (editor), Communication and behavior of whales. AAAS Selected Symposia Series, p. 333-358. Westwood Press, Boulder, Colo.

Rayner, G.W.
1940. Whale marking, progress and results to December 1939. Discovery Rep. 19: 245-284.

Rice, D.W.
1974. Whales and whale research in the Eastern North Pacific. *In* W.E. Schevill (editor), The whale problem, a status report, p. 170-195. Harvard Univ. Press, Cambridge, Mass.

1978. The humpback whale in the North Pacific: Distribution, exploitation, and numbers. *In* K.S. Norris and R.R. Reeves (editors), Report on a workshop on problems related to humpback whales *(Megaptera novaeangliae)* in Hawaii, p. 29-44. U.S. Dep. Commer., NTIS PB-280 794.

Schevill, W.E. and R.H. Backus.
1960. Daily patrol of a *Megaptera*. J. Mammal. 41: 279-281.

## Part 4

Research Techniques

Roger Payne, Oliver Brazier,
Eleanor M. Dorsey, Judith S. Perkins,
Victoria J. Rowntree, Alan Titus

12. External Features in Southern Right Whales *(Eubalaena australis)* and Their Use in Identifying Individuals

Abstract

Individual southern right whales *(Eubalaena australis)* can be recognized on the basis of several external features, the most important of which is the pattern of callosities (raised and thickened patches of skin) on the head. An analysis of these external features has been made, primarily from 16,000 aerial and shore-based photographs taken off Peninsula Valdés, Argentina, from 1971 through 1976. The callosities and several types of pigmentation patterns are described and are shown to be different for each individual and constant enough over time for individual recognition. A total of 484 individuals were identified at Peninsula Valdés between 1971 and 1976 (mortality rates are not known). Both similarities and differences are found between right whales in the western South Atlantic, the eastern South Atlantic, the North Atlantic, and the North Pacific. A technique is presented for calculating the chance that there exists in a population animals too similar to be distinguishable. It is argued that natural markings are better than visual tags as a means of identifying large populations of some wild animals.

Introduction

One of the distinguishing features of black right whales (genus *Eubalaena* Gray 1864) is the presence of unique structures called callosities, which are patches of thickened, cornified epidermis on the tops and sides of their heads. The function of these callosities has been the subject of speculation for years. During a study of southern black right whales, we noticed that the distribution, size, and

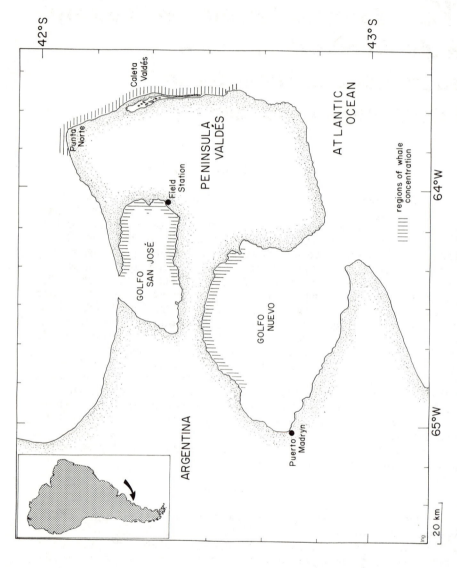

Figure 1. Map of Peninsula Valdés, Argentina showing principal areas frequented by right whales and the field station established for this study.

the callosities vary from whale to whale, as do some other external features. We guessed that these differences would allow us to identify individuals and would thus lay the foundation for modern behavioral studies of these whales.

Methods and Materials

Species Studied

The taxonomic status of right whales is not settled. It is accepted that the Greenland right whales or bowhead whales (*Balaena mysticetus* Linnaeus 1758), which are restricted to cold waters of the northern hemisphere, are a separate species from the black right whales (genus *Eubalaena* Gray 1864), which are found in temperate waters of both northern and southern hemispheres. Some authors place black right whales in the genus *Balaena* as one or more species (e.g., Eschricht and Reinhardt 1866; Rice 1977). More authors place them with the genus *Eubalaena*, with two to five species (e.g., Allen 1908; Lönnberg 1923).

The whales we have studied are black right whales from the southern hemisphere. We call them *Eubalaena australis*, in spite of the unsettled state of the taxonomy, because currently it is the most unambiguous reference to the black right whale that lives in the South Atlantic. We will refer to them as southern right whales in the remainder of the paper.

Study Areas

The main region of our study was the waters surrounding Peninsula Valdés, Argentina, a large cape which encloses two bays, Golfo San José and Golfo Nuevo. On the south shore of Golfo San José, the New York Zoological Society has built and maintained a field station from which our research is based (Figure 1). We have studied southern right whales in this region between June and December, the months when they are abundant there, from 1970 through 1981, but in this report we include the results only through 1976. These results are mostly from 52 airflights along the coast of Peninsula Valdés, from which we have analyzed nearly 16,000 photographs. Table 1 shows the monthly distribution of analyzed airflights for each year in Argentina.

In addition, one of us (R.P.) spent 10 days, in September 1974, in Cape Province, South Africa doing a brief survey of the southern right whales there for comparison with the Argentine whales. The survey consisted of two airflights, covering most of the coast of Cape Province from 13 miles north of Capetown to 25 miles east of Port Beaufort. Our techniques of collecting and analyzing aerial photographs were essentially the same as those described below for the work in Argentina.

Table 1. Monthly distribution of analysed airflights at Peninsula Valdés, 1971–1976.

Year	June	July	Aug	Sept	Oct	Nov	Dec	Total for Year
1971	0	0	3	5	7	0	0	15
1972	0	0	0	5	4	5	0	14
1973	1	1	1	2	2	2	1	10
1974	0	0	0	1	1	0	0	2
1975	0	0	0	2	0	1	1	4
1976	0	0	0	1	4	2	0	7
Total for Month	1	1	4	16	18	10	2	52

Aerial Techniques

In Argentina, we used a Cessna 180 airplane for observations from the air. We soon realized that the whales are concentrated near shore in three quite specific areas, so we focused our efforts in these areas (Figure 1). We almost always followed the coastline of the peninsula, staying less than two km out from the tide line, and flying at altitudes less than 200 m when searching for whales. When taking photographs for individual identification, we dropped down to between 65 m and 150 m (usually 100 m) while circling over each whale or group of whales. We found that it was important to take several pictures of each individual, striving particularly for motor drive sequences in which the shutter release was depressed for several frames. This removed the inevitable motion of the camera when pressing and releasing the shutter and made those pictures in the middle of the motor drive sequences steadier than the first and last frames of the same sequence.

We used a 300 mm f2.8 Topcon lens on a hand-held 35 mm, motor-driven Nikon camera, usually with a shutter speed of 1/500th of a second. In 1971-73, most photographs were made on black and white film (Kodak Plus-X), and in 1974-76 most were taken in color (Kodachrome or Ektachrome). Although color film is more difficult to work with when analyzing the data, we found that it made it easier to distinguish between apparent and real callosities (see below). Some photographs were taken with a gyrostabilizing unit (Kenyon Stabilizer, Model KS-6) attached to the base of the camera, but they were not of significantly better quality than those taken with a hand-held camera.

We directed the pilot to guide his approach to each whale so that the plane was closest to the whale when its head was closest to the plane, and when the plane was between the sun and the whale. We found it to be important to take photographs of the heads of whales from in front of them, not from behind, because much useful variability which occurs near the anterior end of the rostrum (the bonnet) is lost in a rear view. After every sequence of photographs, the photographer recorded the location, number, and behavior of the whales photographed and put a "blank" on the film to avoid ambiguity in later analysis. Usually, an additional observer counted the total number of whales seen in each group. The average flight time was 3.6 hours (range 1.5-6.5).

Analysis of Aerial Photographs

The process of identifying a whale was primarily based on the careful analysis of photographs of its head to determine the pattern of its callosities (see Figure 2 for an example). The pattern was often obscured partly or wholly by foam and/or specular reflections of sunlight. It was also subject to distortions by refraction from overlying waves when seen through the water. These problems could usually be overcome by taking a series of photographs of each whale and basing identifications on those features present in two or more photographs from different angles.

Once the callosity pattern was determined, it was compared to the current collection of photographs of known whales, which were organized into a "Head Catalog" containing the best single photograph of each individual. Individuals having similar bonnet outlines and numbers of rostral islands are grouped together in this catalog. When other distinguishing features, like white back marks, were present on the body of the whale, we used these as well in the process of identification. It could take an experienced person up to three hours, averaging about half an hour, to identify an unknown whale. Each identification was reviewed by a peer and any conflicting opinions had to be resolved before an identification was accepted. There is only one instance in our experience on which agreement has not been reached. Before a whale was considered to be a new individual, it was compared with every other whale in the catalog.

We graded the reliability of our identifications, using the letters A through D. Generally speaking, A was considered to be a whale identified with confidence and D was doubtful enough to be on the edge of usefulness, with B and C falling in between. Although these are subjective categories, we found them to be useful.

Once each whale was identified as a new or a familiar individual, the information in the accompanying field notes about location, group size, and behavior was recorded for it, as well as for each of its companions (if any).

Non-aerial Techniques of Observation

In addition to observations from a plane, we have been able to study the Argentine right whales to a greater or lesser extent from the shore, from small boats, by diving underwater, and by examining corpses that wash ashore.

Because the shore drops off very steeply in many parts of Peninsula Valdés, right whales often swim within less than one kilometer of shore, facilitating land-based observations. A particularly advantageous shore observation site was the top of a cliff, 46 m above mean low water, about one kilometer west of the field station, where we built an observation hut to shelter us from the frequent strong winds. From this site, we had more than a 180° view of Golfo San

Figure 2. Aerial photograph of a southern right whale showing a typical callosity pattern.

José and used a 1000 mm Century Precision Optics lens to photograph whales, which sometimes swam directly underneath the observation point.

We often used one or more Avon or Zodiac inflatable boats to approach whales closely on the water. By working from boats, we were able to obtain excellent above and below water close-up photographs of callosities and, in several cases, to collect by hand bits of loose callosity tissue, skin, and ectoparasites from whales that came within touching distance of boats or swimmers. We have had the opportunity of closely examining three right whale corpses that stranded at Peninsula Valdés: two male calves, probably stillborn (by their short length), that stranded in Golfo San José in August and October 1972, and one adult lactating female that stranded in Golfo Nuevo in October 1972.

Results

Reactions of Whales to Planes and Boats

We found the response of right whales to our airplane to be somewhat varied. Most whales, especially those in groups, demonstrated no obvious changes in their behavior. Calves playing with their mothers continued to play, courting groups continued their courtship, breaching or lobtailing whales continued those behaviors, and whales apparently asleep (motionless at the surface with infrequent shallow breaths and often a dry back) were usually not aroused by the plane. Whales that were traveling or "making-a-passage" were, however, usually very difficult to photograph; they would appear at the surface to breathe for only a few seconds after especially long periods underwater. They gave no indication, however, that they were avoiding the plane but rather were simply involved in a behavior that kept them below the surface most of the time.

Isolated individuals that were milling or still, in contrast, often appeared actively to avoid the plane, breathing only when the plane was not over them. Their avoidance seemed calm and deliberate as though they simply chose to be below water when a disturbance was nearby. There were also a very few individuals (probably less than 2%) that exhibited pronounced fright reactions: rapid swimming or diving as the plane came overhead and, very rarely, rapid swimming accompanied by defecation. Their behavior, as reported by observers on the shore, appeared soon to return to normal as the plane left the area. With the exception of these animals, we felt that small planes circling 100 m or more above the whales caused only minor interruptions of normal behavior, when they interrupted it at

Table 2. Sighting frequency of 484 identified right whales at Peninsula Valdés, 1971–1976. Repeated sightings are not necessarily in consecutive years.

# years photographed at Peninsula Valdés	1	2	3	4	5	6
# identified whales	216	135	81	38	13	1

all. For this reason we feel that our technique allows identification of right whale populations with a minimum of disturbance.

The reactions of individual whales to closely approaching boats powered by outboard motors was varied. Some stayed in the vicinity of the boat for several seconds or minutes after a slow approach to within touching distance; others made one or two passes on a stationary boat before leaving, and still others avoided boats altogether. We found that close encounters only followed slow approaches by the boat. On several occasions the whales made what we learned to be threat displays from later observations of interactions of individual right whales with their own and other species. However, in no instance was any threat carried as far as physical violence to the boat.

Identification of Individuals

From the 52 airflights in Argentina and nearly 16,000 photographs, we have recognized 484 different whales, that is, we got at least one photograph of each one with enough information to distinguish it from any other well-photographed whale. The number of whales identified per airflight ranged from 6 to 111. Table 2 shows the sighting frequency at Peninsula Valdés for the 484 identified whales, 268 of which we have seen in more than one year. In South Africa, from the two airflights and just over 400 photographs taken in 1974, we identified 30 different individuals. No whales were seen in both locations, suggesting that these are separate populations with little or no mixing.

During the airflights, we were able to distinguish two age categories of right whales: calves -- very small whales (clearly animals in their first year) accompanied by very large whales; and non-calves -- all other right whales. Within the category of non-calves, we could sometimes distinguish subadults from adults, a technique which required a boat or an obviously larger whale with which to compare lengths. Using airflight photographs we were often able to

Table 3. Number of southern right whales identified by year (Argentina airflights only).

	1971	1972	1973	1974	1975	1976	Total	Annual Mean
All whales	184	202	205	86	146	129	952	158.7
Mothers with calves	19	27	37	14	30	31	158	26.3
% of non-calves identified for the first time	100%	56%	34%	23%	27%	21%	-	-

calculate the ages of whales by measuring photographic negatives (Whitehead and Payne 1981).

Non-calves were much easier to identify than calves. In Argentina, we have encountered non-calves from the air a total of 1992 times and have been able to identify the whale in 1715 (86%) of these encounters. Many of these are, of course, the second, third...up to sixteenth identification of the same whale on different days. Calves, on the other hand, have been photographed on 568 occasions and identified in only 158 (28%) of these times. The reasons for the difficulty with calves are that the heads of young calves are small and are often covered with ectoparasites (cyamids) that obscure the callosity pattern by inhabiting bare skin as well as callosity tissue. Both problems disappear within the first few months of life.

We have also considered the accuracy with which we made our identifications. During the years 1971-75, we identified non-calves 1299 times. Of this total, we gave 81% of our identifications an A rating for reliability. Twelve percent were B rated, six percent C, and one percent D. Taken together with our previous data showing that we could identify 86% of all of the non-calf whales photographed, this indicates that we can identify with confidence (with an A rating for reliability) 70% of all the adults and subadult whales (i.e., non-calves) we photograph.

We have summarized the number of whales identified each year from airflights at Peninsula Valdés in Table 3. We identified a mean of 158.7 whales per year. Of these, an annual mean of 26.3 were identified mothers with calves. Since each mother had one calf, the total number of identified mothers, 158, is also a minimum figure for the number of calves born during these six years. We have also calculated the percent of non-calves identified each year which were seen that year for the first time. In 1971, of course, 100% of the non-calves were seen for the first time. That figure decreases with time and appears to be stabilizing somewhere between 20% and 30% in the last three years analyzed. In a later paper we will present the conclusions we can draw from our individual identification data about the population biology of these whales.

Catalog of Individuals

Our work on individual identification has resulted in a catalog of individual whales photographed at Peninsula Valdés. The main section of this catalog contains the best single photograph of each whale's head showing the distinctive features, the callosity patterns (see below). A second and much smaller section contains photographs of distinctive features on the backs of right whales -- pigment patterns and wounds (see below).

Callosities

Appropriateness of the term "callosity". Since the work of Matthews (1938), the term "callosity" has been used to describe the patches of thickened skin on the heads of right whales. However, prior to his work, several terms were used, with "excrescence" being the most common (Van Beneden 1868, Ridewood 1901, Lönnberg 1906, Allen 1916). We feel the term "excrescence" is inappropriate, inasmuch as in its primary meaning it indicates "an abnormal growth", whereas the growths on the heads of black right whales are normal. "Callus", on the other hand, refers to a host of tissue types which have in common that they are thickened, though not necessarily from wear. For example, the thickened pads of skin on the rumps of primates are called callosities, and in botany, "callus" refers to thickened formations on sieve areas and at the base of cuttings or below wounds. As Esau (1953, p. 271) explained, "These two unrelated formations bear the same name because both constitute thickened masses, a concept implied by the word 'callus'." Thus, adhering to the general biological meaning, we will call the normal, thickened patches of skin on the heads of *Eubalaena* callosities.

General description. Callosities of southern right whales vary from mostly smooth in fetuses with a molded or wrinkled appearance (Lönnberg 1906, Matthews 1938) to very rough in adults with tall, irregular projections and deep clefts. Some time after birth, they become roughened and pitted and almost completely covered with colonies of amphipod crustaceans of the family Cyamidae (whale lice). It is not known whether the nature of the whale-cyamid relationship is commensal, parasitic, or symbiotic, but the cyamids contribute significantly to the appearance of the callosities. We will discuss in some detail what causes callosities to appear light or dark because contrast with the darker skin is part of the way one detects the callosity outline in aerial photographs.

Color. When viewed from a distance, callosities appear to be principally white (Figure 2). However, as first suggested to us by W.E. Schevill, this could be due to the almost continuous layer of white cyamids covering each callosity. Our observations support his suggestion. Later we became aware of an article by Roussel de Vauzème (1834) who said that the chalk white cyamids occur in such prodigious quantities on the heads of right whales that one sees them a long way off at sea. On at least 20 occasions, we have had close looks at the callosities of whales that surfaced within less than two meters of our boat. In all cases, the callosity tissue visible in spaces devoid of cyamids was not white, but some shade of gray. We also have over 50 close-up color photographs of the heads of free-swimming right whales, and they indicate the same thing (see Payne 1972, 1976 for close-up color photographs). And once when we were able to dislodge a callosity projection from a live whale, the color of the tissue was dark gray.

We have dissected and examined callosities from all three right whale corpses described above. The callosity tissue from the two calves was very light gray (though skin near it was black), with several patches on the mandibles ("mandibular islands", see below) being white. On the adult corpse, one callosity had black tissue, but this color may have been a consequence of exposure after death. In all callosities that we dissected, the surface was darker than the underlying tissue, with the darker outer layer being no more than a few millimeters thick. Struthers (1887), who examined the corpse of a humpback whale, noted that "when the epidermis is off the *cutis vera* is at first white or cream colored ... then under exposure for some time to the air it becomes bluish, and on being scraped the cream color is restored." Our observations indicate that with the exception of white blazes (see below), all right whale skin, including callosities, darkens with time. We have several examples of this: 1) peeling skin is darker than the underlying layers it exposes; 2) when small calves are first seen, they are lighter gray than their mothers and they darken with age; 3) the rare calves that are white become gray as adults, although their ventral white blazes do not; and 4) scrapes on skin are lighter than surrounding skin. The dark surface color of callosities seems to be just another example of this phenomenon.

In all observations of living whales including partial albinos, we found callosity tissue to be lighter than the whales' living skin.

The color of callosities that one usually sees on live whales is a combination of the colors of the ectoparasites and of the callosity tissue itself. On adult southern right whales, cyamids of a white color are by far the most predominant ectoparasite on the callosities. There are often patches of yellow and of orange on the whales' heads as well. In the best of our photographs, these appear to be clusters of cyamids also, but we have not sampled from these colored patches.

Our observations fit exactly with those of Roussel de Vauzème (1834) who examined cyamids on freshly killed right whales from the South Atlantic which were floating alongside the whaling ship. That allowed him to observe what was presumably a more normal distribution of cyamids than would be found on the fully stranded specimens others have worked with. He noted three species of cyamids and described them as follows: 1) *Cyamus ovalis* - chalk white, living in great quantities on the callosities; 2) *C. erraticus* - wine-red, found at the base of callosities, on smooth skin between callosities, in the axes of flippers, in genital-anal grooves, and in wounds; and 3) *C. gracilis* - light yellow, like *C. ovalis* restricted to the callosities, but found there in jumbled masses ("pêle mêle"), not maintaining an orderly distribution as was true of *C. ovalis*.

The yellow, orange, and white organisms that we have seen are consistently located in different areas. White cyamids may completely cover a callosity and are most densely clustered on the sides and in shallow depressions. They are seldom found on very high projections. The orange color usually occurs separately from the white, either in isolated disc-shaped clusters on the smooth skin between callosities or in narrow bands (sometimes just a single row of animals) outlining the white cyamid concentrations. Orange is also seen in wounds and in other depressions on the body. An organism of this same orange color heavily infests the heads of most calves in their first few months of life, covering not just callosities, but usually the whole rostrum and sometimes most of the sides of the head as well. The yellow-colored organisms are often in distinct clusters, and are concentrated near the middle of the largest callosities and thus surrounded by white cyamids. Seen from a few meters away, callosities look somewhat like miniature snow-capped mountain ranges, studded with jagged black projections which, on closer inspection, turn out to be the bare peaks of callosity tissue projecting from out of the cyamid "snow" (like mountain peaks from which the snow has been blown). Occupying the snow basins like glaciers are the yellow cyamids, while fringing the mountains at their base are the orange cyamids.

A number of species of cyamids are reported to occur on right whales -- *Cyamus ovalis*, *C. gracilis*, *C. erraticus* (Roussel de Vauzème 1834, Leung 1965), *C. catodontis* (Best 1970), and *C. ceti* (Omura 1958). No color differentiation between these species is mentioned by Leung (1967) in his key to cyamids. Omura, Ohsumi, Nemoto, Nasu, and Kasuya (1969), however, noticed color variation in cyamids found on North Pacific black right whales. They report cyamids of a yellow-white color on the callosities and a yellow-brown color on "other body surfaces", without being clear about the species composition of these two colors.

On the beached right whales we have seen, cyamids were crawling all over the body. Other authors have erroneously assumed this indicates that cyamids are broadly distributed over the live right whales' bodies (Matthews 1938; Omura 1958). However, in our underwater observations of live right whales, only a few cyamids are seen anywhere on the body of a non-calf except for callosities, creases in the skin, wounds, or scars. (On calves, we find them on the sides of the head and elsewhere where skin is temporarily rough).

Distribution of callosities. The overall distribution of the callosities on a right whale's head is very similar to the overall distribution of facial hair on a human being -- we assume that the parallel is meaningless biologically. Callosities are distributed along the mandible, along the upper margin of the lower lip, along the narrow dorsal surface of the rostrum and over the eyes (Figures 2 and 3). The number, shape, size, and distribution of these callosities varies from whale to whale and is usually asymmetrical on the same individual. From our sample of 484 Argentine right whales and 30 South African right whales, we find that the areas in which callosities are likely to grow are limited (Figures 3 and 10). Within callosity growth regions many different callosity structures and placements are possible. In some regions all individuals have callosities, while in other regions callosities occur in only some individuals.

The following is a list of the callosities always present in the southern right whales we have seen (see Figure 3 for illustration):

1) Bonnet: a large callosity (usually the largest on the whale) at the tip of the rostrum. "Bonnet" was a term first used by early whalers who fancied it resembled a lady's bonnet. Although the term seems inappropriate, it is probably too fixed in common usage to be abandoned.

2) Coaming: a single, large callosity directly in front of the blowholes. This callosity, which is analogous to the bulge of tissue situated directly in front of the nostrils in most other mysticetes, appears to prevent water that is running along the head when the whale is at the surface from entering the nostrils, by deflecting it to each side. We call this callosity the coaming after its functional resemblance to coamings on boats (baffles placed on the decks to prevent water that is running along the deck from entering the cockpit and hatches).

3) Post-blowhole islands: one or two (rarely more) callosities directly caudal to the nostrils. (We have called all small isolated callosities "islands").

4) Chin callosities: callosities positioned anteriorly on each lateral surface of the lower jaw. They are thus roughly in the position of the beard in man.

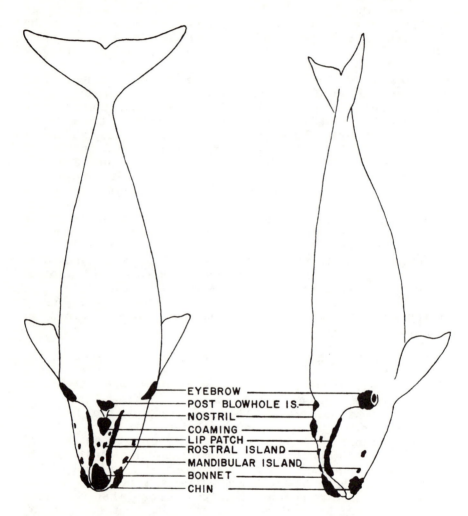

Figure 3. Diagrammatic view of a right whale showing the position and form of typical callosities. This is the same whale shown in Figure 2.

These are usually the largest callosities after the bonnet. In hundreds of opportunities we have never seen even the slightest indication of a callosity located medially on the lower jaw, such as was described by Clarke (1965) from a southern right whale off Chile.

5) Eyebrow callosities: large callosities lying directly over each eye. Our underwater photos of right whales showing the form of these callosities indicate that they are the callosities with the least variation in outline.

The following is a list of the callosities sometimes present in the southern right whales we have seen (Figure 3):

1) Mandibular and rostral islands: variable numbers of small callosities usually present in two areas: along the rami of the mandibles and on the rostrum between the bonnet and the coaming. Only in rare cases are they missing entirely. The rostral islands, along with the bonnet and the coaming, all occur between the nostrils and the tip of the snout. Collectively, these callosities comprise the whale's "moustache" -- a term first given to them by Van Beneden (1868), but not widely used since. We have never seen a continuous coverage of the rostrum in southern right whales (as is common in the northern right whale).

2) Nostril islands: small callosities lateral to the blowholes on either or both sides of the head. They are posterior to the anterior margin of the blowholes.

3) Lip callosities: callosities along the upper margin of the lower lips, ranging from a small spot to a long, broad strip covering almost the entire length of the upper margin of the lower lip.

A sampling of aerial photographs of Argentine right whale heads enables one to survey the individual variation in callosity patterns (Figure 4). The most obvious differences are in the extent of development of the lip callosities, in the number and arrangement of rostral islands, and in the size and shape of the bonnet. These callosities are all usually visible from the air when the whale blows. The post-blowhole islands, nostril islands, and coaming, though usually visible, are less variable from one whale to another and thus less useful for individual identification. Although there is much individual variation in the shape of the chin callosities and the number and arrangement of mandibular islands, these callosities are normally underwater and therefore obscured except in photographs taken on very calm days. Eyebrows

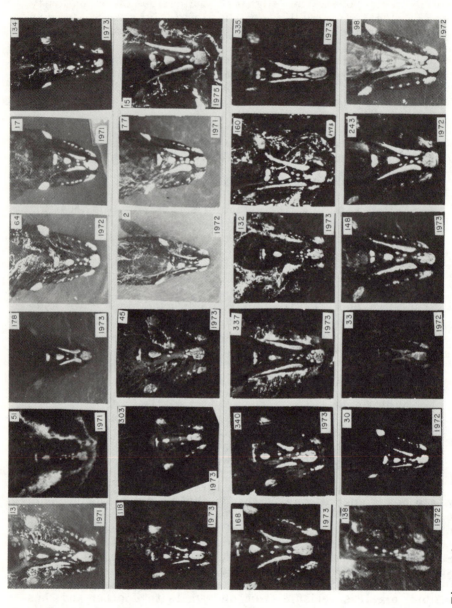

Figure 4. A sampling of Argentine right whale heads showing the individual variation in callosity patterns.

are never more than indistinct blurs through the water and are of no use in identifying individuals. Thus the principle features on which we have based our technique for distinguishing individual whales are the bonnet, the rostral islands, and the lip callosities.

In both populations of southern right whales which we studied, one of the features that is most obviously subject to individual variation is lip callosities. To study closely the individual variation of lip callosities, we selected for examination whales with good enough photographs of the lip area to be able to judge not only the presence or absence of lip callosities, but also the relative lengths of the callosities on each lower lip. The results are given in Table 4.

Two important conclusions can be drawn from these results. First, there is no significant difference between the Argentine and South African populations in the incidence of lip callosities: a chi-square test of independence between the three major categories of lip callosity distribution (no lip callosity, one lip callosity, and both lip callosities) and location resulted in χ^2 = 0.73, df = 2, p > 0.50. Second, there is a pronounced and quite surprising lateral asymmetry in the distribution of lip callosities in both populations. When only one lip callosity was present, 13 out of 14 whales had that callosity on the right lower lip. And when both lip callosities were present and unequal in length, 90 out of 91 whales had the longer callosity on the right side.

When lip callosities are unequal in length, the longer is rarely more than 50% longer, and when just one lip callosity is present, it is always relatively short. Short lip callosities are usually found about midway between the snout tip and the posterior margin of the post-blowhole callosities. Thus, the tendency to form lip callosities seems to be greatest in the center of the right lower lip and least at the anterior and posterior ends of the left lower lip.

Another interesting aspect of lip callosities that we have noticed from sea-level observations in Argentina is that lip callosities always have a relatively low profile, and lack, in older whales, the tall projections that most other callosities appear to acquire.

A second feature of the callosity pattern which is especially subject to individual variation is the rostral islands. Rostral islands exhibit a gradation in size along the rostrum, with the largest being nearest the anterior end and the smallest nearest the nostrils. To demonstrate this gradation, we selected good aerial photographs of 27 whales and measured the length of each rostral island along the longitudinal axis of the head and expressed that length as a percent of the length of the longest rostral island for each whale. We did a regression of these relative lengths against the position of each rostral island (n=142) along the rostrum

Table 4. Distribution of lip callosities in two populations of southern right whales. Only whales with good photographs of the lip callosity area were included in the survey.

Number of lip callosities	ARGENTINA		SOUTH AFRICA	
	# of whales	% of whales surveyed	# of whales	% of whales surveyed
None	38	20%	5	25%
One				
right only	11	6%	2	10%
left only	1	1%	0	0
Two				
right longer	83	44%	7	35%
left longer	0	0	1	5%
equal	55	29%	5	25%
Total whales surveyed	188	100%	20	100%

(Figure 5) and calculated a correlation coefficient of r=0.61. This is significantly different from r=0 at p<0.0001. So, although there is a lot of scatter around the regression line, the rostral islands do tend to be smaller posteriorly.

When we had the opportunity to see right whales close-up at sea level, we noticed that the posterior-most rostral islands were not only smaller in area, they were also less developed in height than the more anterior islands on the same animal. Furthermore, we sometimes noticed smooth swellings on the skin which lay on a caudal extension of the same arc on which the rostral islands occurred and at about their same average spacing. Although they lacked a thickened callus covering entirely, in many cases these had a single hair in their center. Such bumps in the skin thus appear to be incipient callosities. Every gradation existed between these bumps without roughened skin and fully developed callosities. Only bumps with roughened surfaces supported white cyamids.

We suppose that the very smallest callosities provide very little area suitable for attachment of cyamids. We expect that such islands are inferior habitat and will only be colonized when the infestation of cyamids is heavy. This would mean that fluctuations in the cyamid population could cause posterior rostral islands to appear and disappear or to vary between white and gray in aerial photographs of the same whale taken at different times. This is in fact just what we see, and we have learned to take into account this kind of variation in the visibility of small islands when identifying whales.

Growth and development. A right whale's callosities enlarge as the head grows, thus maintaining essentially the same relative size to the head throughout a whale's life, and their surfaces, which are smooth at birth, split and crack with age to form numerous clefts. This gives adult callosities an appearance reminiscent of deeply checked and furrowed tree bark, which develops in a similar fashion. The splits are sometimes several centimeters deep, thus penetrating the callosity tissue at least ten times deeper than the gray surface pigmentation penetrates the underlying white tissue in the corpses we have examined. Yet, all exposed surfaces of the callosities, including the faces of the deep cleft, are gray, suggesting that the tissue surface on the walls of the clefts became darker after the split exposed them.

In southern right whales, we have frequently seen long projections of the callosity tissue sticking several centimeters above the general level of the surrounding surface (misidentified as "barnacles" in Payne 1976, p. 335). Some projections are only slightly higher than the longest

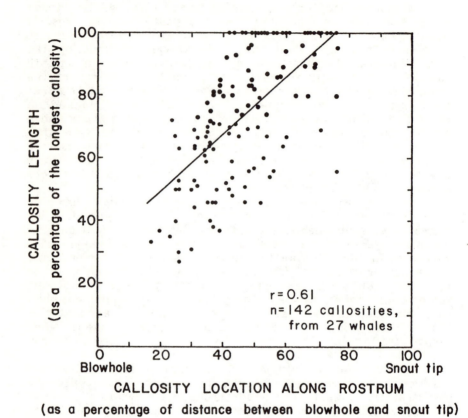

Figure 5. Length of rostral islands plotted against their position on the head. The correlation coefficient (r) is significantly different from zero (p<0.0001).

diameters of their base, but others are very tall, have narrow bases, and project so far beyond the callosity surface that they seem almost like thickened hair. A common feature of all such projections is that they are seen only on large adults, never on calves or young juveniles, and are most pronounced on whales which by their size, wrinkled skin, general inactivity, and overall appearance appear to be old. These projections are also free of cyamids which may be a clue to how they are formed.

Several authors have assumed that the cyamids cause the pitting of the callosity tissue in adults. Frederick Martens who was the first to describe cyamids, claimed that cyamids "bite out entire portions of the skin, giving an appearance to the skin, as if birds were picking at it." (Martens 1675 cited by Lütken 1873). Roussel de Vauzème (1834) concluded that the digestive anatomy of cyamids showed that they ate whale skin listing toothed mandibles, jaws armed with hooked claws, chewing apparatus of the stomach, and intestines without convolutions as circum-stantial evidence. He observed that the skin underneath cyamids was denuded of epidermis and corroded. Matthews (1938, p. 177) notes that myriads of cyamids "burrowed" in the callosities ". . . so that the deeper individuals were almost completely buried." Leung (1976) speculates that cyamids eat the cutaneous tissue of their hosts but offers no evidence in support. We have noted a high incidence of cyamids on the roughened skin, which was not itself callosity tissue of young calves, but we do not know whether the cyamids cause the roughness or only take advantage of pitting caused by another agent.

Our own examination of callosities leads us to believe that the smaller pits and depressions are probably excavated by cyamids. Part of the evidence concerns the projections of callosity tissue described above, which we propose are a consequence of simple fluid dynamic effects.

The places on the head in which cyamids concentrate demonstrate their preference for areas where water flowing past them is reduced in velocity. Thus they are likely to be concentrated in any depression as well as behind and between any obstacles, e.g., behind isolated callosities. Even when they cover callosity tissue densely, they rarely occupy the highest projections (Payne 1976, p. 334). However, when cyamids are attached to smooth skin surrounding a callosity, they are usually in dense clusters, which means only the animals on the perimeter are exposed to the full flow of water passing the whale. Furthermore, these clusters are only one layer deep (the cyamids are not clinging to each other; each is attached directly to the whale's surface). This must keep them well within the slower moving fluid boundary layers enveloping the whale.

Because flow rate past the whale will increase exponentially the further one gets from the surface (true of any submerged, moving object), callosity projections may extend into water layers that are moving too fast to be suitable habitat for cyamids. If cyamids are destroying callosity tissue, they could be expected to keep broad areas grazed flat as long as they kept up with callosity growth, but once a projection extended even a short distance above the surface, it might escape cropping by having its tip extend into fast-moving water. Such an escape into faster-moving layers of water would place the projection in a positive feedback situation in which any increase in length would improve the chances of it becoming still longer until it became long enough to be entirely exempt from cropping. In such a system, one would expect to get highly elongated projections of callosity tissue growing out of a level, grazed surface -- the condition we observe in older animals. These comments on cyamids eating callosity tissue do not presuppose any mechanism by which cyamids feed. In fact, the method of feeding in cyamids is a matter of considerable speculation.

Omura et al. (1969) suggest that callosities probably form around the bases of facial hairs. Callosities are not present in all areas where hairs are found, and hairs are not found on all callosity surfaces. However, in all right whale corpses we examined, hairs were widely scattered over the surface of the largest callosities, and the smallest callosities each had a single short, stiff hair located in a central depression.

The callosities on the calves also exhibit a peculiar feature visible in many photographs. A major part of the posterior portion of the bonnet is often found to consist of a bunch of discrete smaller callosities, each a roughly circular swelling touching the small callosities flanking it. In the center of each one of these circular swellings of "islands" there is a shallow depression (like a crater in a volcanic island) which is often darker than the surrounding tissue. From the center of this depression, a single hair emerges. In very young calves each discrete circular island in the bonnet appears to be almost entirely isolated. As the calf grows, the isolated regions also grow and make contact with each other on all sides until they fuse and form the single structure called the bonnet. The fusion does not disturb the outline of the bonnet enough to prevent recognition of the calf in future years.

Figure 6 shows the head of a calf (#191-72)[1] in which complete fusion of the bonnet has not yet taken place. The

---

[1] We designate a calf by giving it its mother's number followed by a hyphen and the year of its birth.

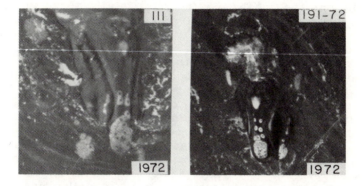

Figure 6. Two examples of bonnets showing fusion of small islands. The identification number of each whale is given in the upper right hand corner. A hyphenated number denotes a calf, the first digits being its mother's number, followed by two digits indicating its year of birth.

fact that it is formed of separate callosity islands can be
distinctly seen. Figure 6 also shows an adult whale (#111)
whose bonnet is fully developed. Even here one can still
discern the outline of the individual islands composing the
bonnet. Note the dark center of each of these islands. This
dark color appears to reflect that cyamids prefer a different
habitat. They are aggregated around the borders and sides of
isolated small callosities, the central peaks vacant. The
natural gray color of the callosity tissue, therefore, shows
through on the peaks, with the result that in aerial
photographs each small callosity looks like a white disc with
a dark center. Where several small callosities fuse to form a
larger one (such as the bonnet), there are sometimes a series
of dark spots that indicate the placement of the original
elements or islands which fused to form the larger callosity.

Constancy of callosities over time. In order to use
patterns of callosities to recognize whales over time, we
must know that the patterns are essentially unchanging with
time. Our evidence demonstrates that this is the case.

A small percent of the population of right whales in
Argentina has white or gray pigment patterns on the dorsal
surface that are usually visible when the callosities on the
head are exposed during breathing. These whales are thus
doubly marked, and each of the two types of marks serves as
a check on the constancy of the other. Figure 7 shows two
such whales photographed on two different occasions three
and four years apart. Both the distinctive white blazes and
the patterns of callosities have clearly remained the same
over time, and there can be no doubt that it is the same
individual whale photographed twice in each case.

A careful survey of 17 doubly marked whales (Table 5)
bears out the constancy of callosities demonstrated in Figure
7. To make Table 5, we took all of the doubly marked
whales for which we have photographs showing both the
callosity pattern and the dorsal mark well in at least two
different years. We examined both types of distinctive
marks for changes over time. The double marking assures
that we are looking at the same individual and that a
significant change in one mark could be detected. Table 5
lists the years in which each whale was seen and the results
of the survey. (The dorsal marks themselves are discussed in
a later section of this paper).

In surveying the callosities, we found it necessary to
make a distinction between changes in the distribution of
callosities and changes in their color. In more than half of
the whales, we noted variations in color between white and
gray in one or two callosities, variations of a sort already
described for posterior rostral islands and probably due to
changes in the level of infestation by cyamids. These color
changes were in some cases reversed later and in no case

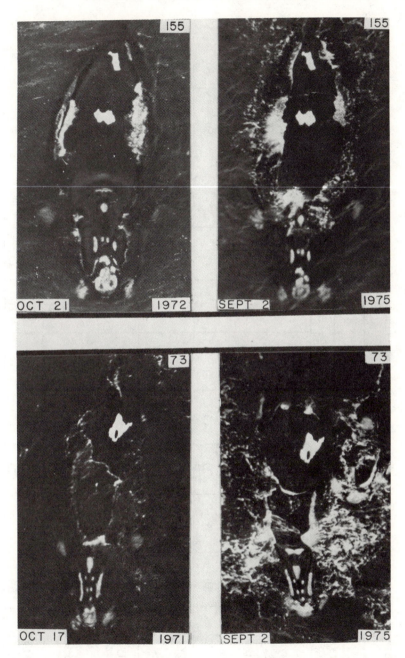

Figure 7. Photographs over time of two doubly marked whales. Neither the callosity pattern nor the blaze has changed in either whale.

Table 5. Constancy of distinctive features over time in doubly marked whales.

Type of dorsal marking	Whale I.D. #	1971	1972	1973	1974	1975	1976	call. change	call. color change	dorsal mark change
white blaze	54	CD						–	–	–
	73	CD						–	–	–
	79	CD		CD				–	1x	–
	155		CD	CD		CD		–	2,3x	–
	162		CD	CD		D		–	1x	–
	278		CD	CD		CD		–	3,4x	–
	304	D	CD	CD			CD	–	–	–
small white mark	19	CD	CD	CD			CD	–	1x	–
	44	CD	CD	CD	C	C	D	–	1x	–
	94	CD	CD	CD				–	–	–
gray blaze	48	CD	CD	CD	CD		CD	–	1x	–
	154	CD	CD	CD			CD	–	–	–
wound	81	CD	CD	CD	D		CD	–	5x	6x
	111	CD	CD			CD		–	3x	7x
	125		CD			CD		–	–	–
partial albino	355	CD	CD	CD	CD	CD	CD	8?	8?	–
	154–72		CD	CD	D	CD	CD	9–	9–	10x

[1] posterior rostral island

[2] peninsula island

[3] lip callosity

[4] rostral island

[5] bonnet

[6] absent in 1971 and '72, white in 1973 and '74, gray in 1976.

[7] edges of wound blurred in 1972 only

[8] left lip callosity a long gray strip in 1971 and a short white spot in 1974 -- could represent a decrease in callosity tissue or in gray sloughing skin.

[9] Because this whale's skin color was mostly white when a calf and then gray, rather then black, when older, its callosities were more difficult to distinguish. All callosity features that we could see, however, showed no change over time.

[10] white as a calf, gray in later years.

rendered the whale unrecognizable by its callosities alone. This is partly because there was enough information in the remaining callosities and partly because the callosity that was changing color could still be recognized as callosity tissue distinct from black skin.

We found no definite changes in distribution of callosity tissue over time, just one questionable change where the pictures were not good enough to tell whether an increase in grayness represented an increase in callosity tissue or in gray skin exposed by peeling.[2] This examination of doubly marked whales indicates that the distribution of the callosities of most or all whales is constant over at least six years and that while minor changes in appearance do occur, these do not hinder recognition of individuals by callosity pattern alone.

Figure 8 demonstrates the minor variations that occur in the appearance of whales' callosities due to three factors: (1) changes in the angles between the camera, the whale, and the sun; (2) the obscuring and distorting effect of overlying water with foam and waves; and (3) actual changes in the color of callosities. It also demonstrates that these minor variations do not interfere with individual identification.

In Figure 8, whale #119 is pictured on four different days in 1972. While the bonnet, the rostral, and the lip callosities are visible each time, all three mandibular islands on the whale's right side show up on only two of the four days. Whale #81 is pictured in four different years. The bonnet has two anterior projections which are obvious in the last year, 1976, but which show the effect of different angles of view in pictures taken during 1973 and 1972. In 1971, the left projection appears to be missing. Close examination of very good photographs in that year reveals a gray color where it is white in subsequent years, indicating that the callosity tissue itself has not changed. What is missing is the cyamid cover over that part of the bonnet. Whale #120 is similarly represented by photographs in four different years and again shows a consistent callosity pattern over time with one minor change: the rostral island nearest the coaming is gray instead of white in 1976 and in a poor photograph would appear to be missing altogether (which stresses the importance of good photographs to this technique). However, there is enough

---

[2]The height of most callosities is sufficient to allow one to distinguish between gray patches of skin and small callosities without a cyamid covering. However, callosities of very low relief such as lip patches and the smallest and most caudal of rostral islands may sometimes be confused with gray skin in poor photographs. In practice this seldom presents problems in identification since there is usually enough other information to confirm identity.

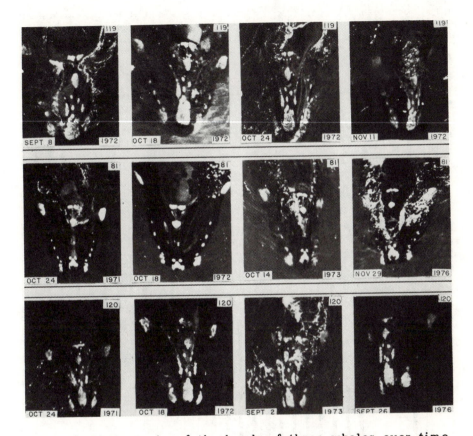

Figure 8. Photographs of the heads of three whales over time showing minor changes in the appearance of an essentially unchanging callosity pattern.

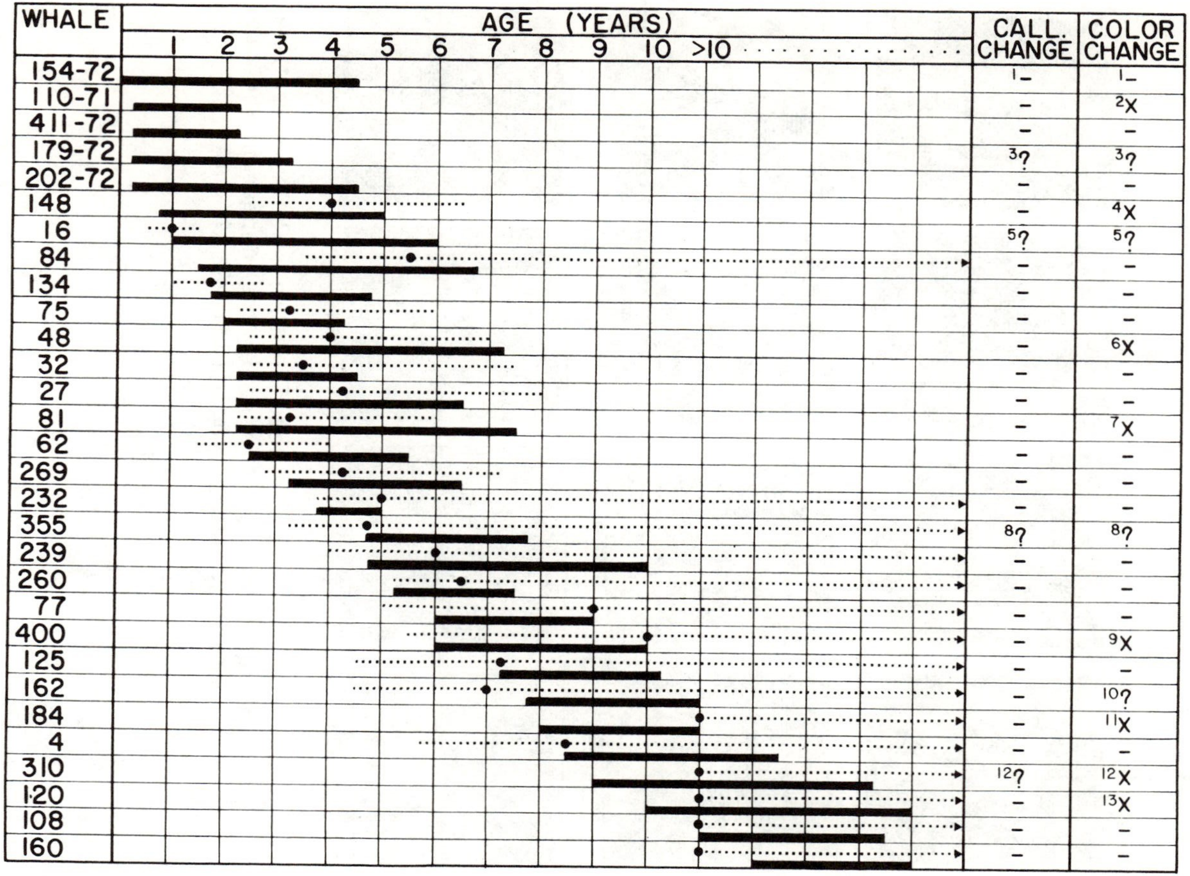
WHALE
AGE (YEARS)
1 2 3 4 5 6 7 8 9 10 >10
CALL. CHANGE
COLOR CHANGE
154-72
110-71
411-72
179-72
202-72
148
16
84
134
75
48
32
27
81
62
269
232
355
239
260
77
400
125
162
184
4
310
120
108
160
1–
3?
5?
8?
12?
2X
4X
6X
7X
9X
10?
11X
12X
13X

― AGE SPANNED BY GOOD CALLOSITY PHOTOGRAPHS   — NO DETECTABLE CHANGE

● MEAN AGE WHEN MEASURED   ? QUESTIONABLE CHANGE

...... 95 % CONFIDENCE INTERVAL FOR MEASUREMENT   X DEFINITE CHANGE

Figure 9. Survey of callosity changes with age in southern right whales. The first seven whales were seen as calves, so we know their year of birth unambiguously (hence the lack of dotted lines and mean age dots for them).

1 See Table 5, footnote 9.
2 Nostril islands whiter in one year.
3 Nostril islands darker or absent in the last year.
4 Much of bonnet darkens.
5 Anterior notch in bonnet may have grown and lip callosity may have whitened in last year.
6 Small lip callosity alternates between white and gray.
7 Part of bonnet whitens (see Figure 7).
8 See Table 5, footnote 8.
9 Most of bonnet whitens.
10 Part of bonnet and of one lip callosity may be less fused in one year.
11 Rostral islands darken.
12 Anterior notch in bonnet becomes more distinct and also coaming acquires a gray patch.
13 Posterior rostral island darkens (see Figure 7).

information in the rest of the callosities to identify the whale in spite of the apparent change.

Having established that major changes in callosities over time do not occur as far as we can tell, we looked to see whether minor changes are concentrated in any period in a whale's life. We have developed methods to estimate the age of whales from photographs (Whitehead and Payne 1981). This technique, briefly, is to drive a boat that is carrying a circular disc of known diameter alongside a whale and to take an aerial photograph of the boat and whale next to each other. The length of the whale is then measured from the photograph using the maximum diameter of the disc as a scale. Since right whales are growing during the first 10 years of their lives, by knowing their length we can make an estimate of their age.

In order to get some idea of how callosity changes are distributed throughout a whale's life, we have selected individuals of different ages for which we also had good callosity photographs over time. Figure 9 lists these whales with the ages, when they were seen, estimated by photographic measurements or, for calves, by assuming that they were born on July first of their calf year (birth also occurs in months other than July). We have samples of whales of all ages up to and beyond ten years, where our ability to discriminate age stops. Out of 30 whales, 4 showed only questionable changes in distribution of callosity tissue (all of which could have been changes in cyamid cover), and 14 whales had possible or definite changes in color of callosities. There seems to be no age-related concentration of either type of change.

Our failure to find any age class within which major changes in the callosity pattern occurs strongly suggests that the callosity pattern remains well enough fixed throughout life to make individual recognition by this means a practical technique.

Number of theoretically possible callosity patterns. Another requirement that must be met if we are to be able to use callosity patterns to distinguish individuals is that we must be sure that there are at least as many patterns possible as there are whales to be identified. For this purpose, we have tried to estimate the numbers of different patterns that could exist at the fineness of detail we are using to make our determinations of identity. That fineness of detail is quite coarse, and when identifying whales, we ignore much of the detail visible in the callosities. It is difficult to estimate just how much detail we do use, and it probably varies from whale to whale. However, we can approach the problem from a different angle. We can determine the "grain size" at which we can no longer tell heads apart and then assume that we must be examining patterns in somewhat more detail (i.e., at finer grain). If we base our calculations of the number of possible callosity patterns on grain size that is too coarse to distinguish between heads, we will get a conservative estimate of possible patterns.

In order to determine grain size, we took good quality aerial photographs of right whale heads and degraded them to a point where three observers, all experienced in identifying right whales' callosity patterns from photographs, were unable to identify them from among a sample of 450 heads. The degradation process consisted of projecting the photograph on a grid and fully blackening in every square of the grid that was touched by any portion of any callosity (Figure 10). The grid degrades the information in the head not only by eliminating details of outlines, but also by eliminating information on relief provided by shadows. We found that the experienced observers could no longer consistently identify whale heads that were degraded to a grain size of 50 squares by 50 squares. (The length of the side of each square was fixed at 1/50 the distance between the post-blowhole callosities and the tip of the snout).

To calculate the amount of information in a callosity pattern degraded to this degree, we first eliminated areas on the head which are never covered with callosities. The blackened areas in Figure 11 (arrived at by inspection of all the heads) constitute all points on which we have ever seen callosities. We next eliminated the areas on the head that always appear to support callosities (small areas at the location of the bonnet, chin, coaming, and post-blowhole callosity). When the resulting picture is superimposed on a 50 by 50 grid, the total number of squares occupied by optional callosity growth areas is about 1,000. If each square can be either black or white, then there are about 2^{1000} (or 10^{301}) possible combinations of patterns -- a number fixing the upper limit to the number of different degraded callosity patterns that could be recognized from close analysis of good photographs.

Number of observed kinds of callosity patterns that can be distinguished in practice. In the above argument, we have assumed that every square area of callosity growth can be covered with callosity tissue or not, independently of every other area. In fact, however, we have never observed many sorts of theoretically possible callosity arrangements, and it is very likely that many of them do not occur. In addition, we have assumed that we are using only the best photographs, with each head photographed from a certain angle, without tilt. But often we wish to work from less good photographs or from photographs showing the head tilted at a greater or lesser angle to the film plane. When working with heads at an angle, we have to make mental transformations to overcome perspective distortions, and these transformations

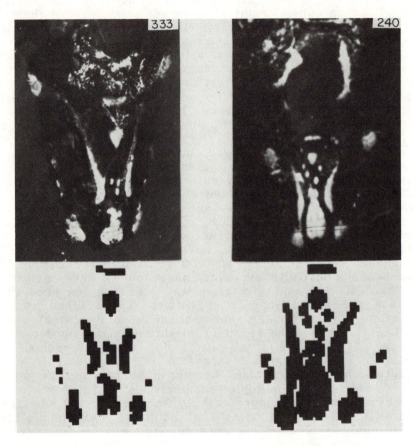

Figure 10. Photographs of two heads compared with their informationally degraded diagrams.

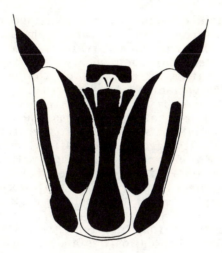

Figure 11. Diagram of the head of a right whale showing in
black areas where we have observed callosities.

introduce uncertainties. As a result, although a very large number of patterns exist at the grain size required to identify individuals, we will be unable in practice to detect many of them. To get a more realistic estimate of the amount of variability available to us, therefore, we tried to calculate the number of patterns that could occur, based on the variation we have observed in Argentina and South Africa, and the number of those that we could expect to detect given the limitations to the data under the conditions that we normally face.

To make this calculation, we examined our catalog of head photographs from Argentina and from South Africa for the number of distinct forms (values) that each kind of callosity (character) has been observed to take. It was at once apparent that some portions of the callosity pattern show more useful variability than others. For instance, the outline of the bonnet is a character that can have a great variety of shapes or values, whereas the outline of the coaming is relatively simple. We therefore took each callosity or callosity group and estimated its variability independently. We were only concerned with the number of values for each character which could be recognized in practice, so we tried to pick values that were unambiguous.

As an example, let us consider lip callosities. Judging conservatively, we felt that we could distinguish unambiguously between four different lengths of lip callosity, in addition to absent lip callosities and to those consisting of a single spot. Each of the four lengths could be either continuous or broken in increasing number of distinguishable places according to length (2, 3, 4, and 5 places for short, medium, long, and very long lip callosities, respectively). This gives the following number of distinguishable configurations for each length: absent (1), spot (1), short (4), medium (8), long (16), and very long (32). Finally, each configuration of the latter four lengths could be two distinguishable widths, giving a total of 122 distinguishable types of lip callosity. Each whale has two sites for lip callosity placement (left side and right side), so that if all lengths could co-occur, the number of recognizable lip callosity patterns would be 122 x 122 or 14,884.

This may be an overestimate for three reasons. First, we have never observed dramatically unequal lip callosities, like a spot on one side and a long or very long callosity on the other. Second, we have never observed more than three breaks in a single lip callosity. Third, we have almost never observed lip callosities of unequal length where the left lip callosity is the longer (Table 4). To be conservative, then, we have subtracted the combinations and configurations we have never or almost never seen to get a corrected number of expected and recognizable patterns equal to 7,215.

We performed similar calculations on the observed variations in the other five callosity areas, with the following number of recognizable configurations for each: bonnet (208), rostral islands (14,810), coaming (12), blowhole islands (11), and post-blowhole islands (37). If we assume that the six characters sort independently and that they occur with equal frequency, then the number of possible overall head patterns is the product of the number of recognizable variations in all six areas multiplied together, which equals 1.09×10^{14}. It is likely that neither independent assortment nor equal frequency holds throughout. However, it seems unlikely that adjusting for these assumptions would force us to change our conclusion that the variability is vastly more than is needed to allow recognition of every whale that has ever existed at one time. Even the most optimistic estimate of the present world population of right whales falls short of 10^4.

Look-Alike Whales

We have now calculated 1) how many different callosity patterns could exist at the level of detail we need to make our determinations and 2) how many different patterns we could recognize in practice with our technique. We still do not know the distribution of different recognizable patterns in the population. The reason we would like to know this is to be sure that there is not one pattern (or even several patterns) that is repeated.

Pennycuick (1978) offers a powerful technique based on information theory by which the reliability of identification can be calculated in terms of the probability that a pattern of natural markings will be duplicated in a population of given size. However, it might be wise to include a warning about applying his technique to field data. His technique involves determination of the frequency of occurrence of the values of all characters used to identify individuals of the species. When making this determination on animals in the wild, one is forced to assume that if an animal with a specific pattern is seen twice, that it is the same animal both times. But n sightings of the same identifying marks could represent anything between 1 and n animals of identical appearance. And inasmuch as Pennycuick's technique is employed in order to avoid confusing one's conclusions with animals of identical appearance, his technique seems to contain a circular fallacy. However, the fallacy can be avoided and Pennycuick's technique used if one marks some of the animals in the population and calculates the frequencies of characters only from them. When this is impossible, one is probably restricted to using corpses for calculating character frequencies.

We have developed a different approach to this problem. We offer a technique to estimate how many pairs of identical animals are likely to exist in a given population. Though applied here to right whales, this technique should be useful in research on many other animal species.

Let us assume that we have in a population several pairs of whales so identical in appearance that they cannot be distinguished by any means. We might think of these whales as perfect 'copies' or identical twins (though of course they need not necessarily be related). If such copies should exist, there are at least two types of events which could demonstrate their existence. One such event would be the photographing of a set of identical whales in the same photographic frame. A second such event would be the photographing of one whale in one place and its 'copy' in a different place, separated by a distance too great for a right whale to swim during the time between the photographs. One could also compare photographs taken simultaneously by different observers working in different areas, but in the case of right whales, this would require two planes, which has been a prohibitive expense for us.

If we assume that we have seen about 90% of the adult population visiting Peninsula Valdés, then there are about 550 right whales that pass through Peninsula Valdés waters. Following the usual form of statistical argument, we assume further, as a null hypothesis, that 25 sets of identical twins exist in this population. Assuming random distributions of whales, we have calculated the probability of not finding even one set of these twins by either of the first two methods mentioned above in all the aerial photographic data we have gathered and examined over six years. This probability turns out to lie between 0.03 and 0.04 (see Appendix 1 for derivation). Since the probability we obtained is less than the traditional level of 0.05, we shall reject the null hypothesis. In fact, in spite of two more years of census photographs beyond the data on which the above calculation was based, no identical twins have been detected.

If we accept random distribution of whales and independence of airflights, the conclusion from the above argument can be stated in another way. We can say that, even if sets of identical twins do exist in the population, we are at least 96% certain that there are fewer than 25 sets of twins in our population of 550 animals. Since 25 twins (50 whales) would constitute 9% of 550, we can feel confident that any conclusions we draw about the overall population based on our technique of recognizing whales will apply to at least 91% of the population. This is not as high a value as we would like, but as time passes, we will deal with a higher percentage of repeat sightings, meaning that as long as we do not detect any whales that look alike, the number of identical pairs of whales that could exist in our population will keep decreasing (the confidence we can have in our conclusions will increase).

We developed the above technique long after we began collecting data and, not surprisingly, found we had overlooked numerous chances to test for the occurrence of twins. For example, during our first two years of taking data, we often failed to note the exact time we photographed each group, meaning that we lost those data for use in intergroup comparisons in the same bay, since we could not calculate the time available to the whales to swim the distance between the groups. The census path we chose to fly could also have been improved had we the twin argument in mind. On censuses that are simple counts, one tries to cover the territory so that animals which have already been seen won't have time to move into the plane's path again as it returns from searching elsewhere. However, since we census by identifying individuals, we sometimes flew back over areas in efforts to improve photographic coverage of a whale we felt we might have missed. In the future, we should be able to avoid these problems and thus to reduce our estimate of the likelihood of identical whales existing in our population.

Other Distinctive Features

Dorsal Marks

A small percentage of our identified whales have, in addition to their callosities, distinctive markings on their dorsal surfaces that are visible in some photographs when the heads are also in view from above. As discussed earlier, these doubly marked whales provide a valuable check on the constancy of the callosity pattern over time, and two have been illustrated in Figure 7 as examples of the permanence of callosity patterns. We distinguish five kinds of marks. Although two types, peeling skin and partial albinism, affect the whole body, the back is most easily examined, hence their inclusion here as dorsal marks. Of the five types, three are essentially permanent and therefore useful for individual recognition throughout the whale's life. The fourth, wounds, may be useful over a few years. The fifth type, peeling skin, changes rapidly and is helpful for following individuals only within a sequence of photographs taken on the same day. We will discuss each type of dorsal marking in turn, omitting from consideration markings on the dorsal side of the flukes, because they are usually obscured by overlying water.

White blaze. The two dorsal markings illustrated in Figure 7 are examples of what we call white blazes. These are areas where the epidermis is presumably devoid of dark

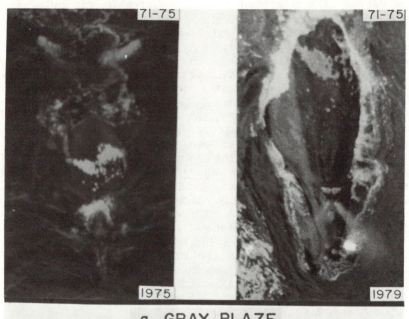

a. GRAY BLAZE

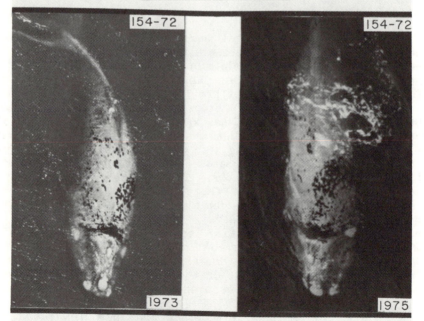

b. PARTIAL ALBINISM

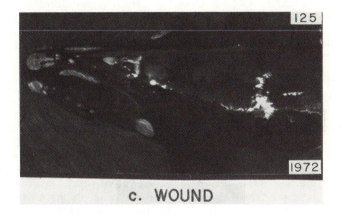

c. WOUND

Figure 12. Other distinctive features. a) Gray blaze. b) Partial albinism. c) Wound.

pigmentation. The skin is smooth and without special contour in the region of these white blazes, and the edges of the blazes are very distinct. Table 5 shows that we have found no evidence for any change in the shapes of white blazes over time. This kind of blaze appears not to darken with age but does appear to grow with a whale. We have measured the length of the blaze as a proportion of the total body length on one whale, #304-76, both as a two-to-three-month-old calf and 13 months later, and we found the proportions to be identical.

Out of the 484 identified whales in the Argentine population, nine, or 1.9%, have white dorsal marks that are clearly the white type of blaze. Another seven whales have white marks so small that we can't tell from our photographs whether they are blazes or small wounds (discussed below), so the rate of occurrence of white dorsal blazes at Peninsula Valdés is probably somewhere near 2%.

Gray blaze. A second kind of blaze occurs in these right whales, one which, unlike white blazes, darkens with age. It is also different from the white blazes in outline and in size. It is a brownish-gray color in adults and is not as easily visible as a white blaze (in poor photographs it may resemble peeling skin but the constancy of the pattern over time confirms its identity as a gray blaze). The outline of a gray blaze is much more complex than that of a white blaze: there are many more indentations and projections and small separate islands of gray and of black, so the ratio of perimeter area would be much higher in gray blazes than in white blazes. Compare the example of a gray blaze in Figure 12a with the white blazes in Figure 7. The smallest gray blaze we have seen is as large as the largest dorsal white blaze (the white blazes in Figure 7 are among the largest), while the largest gray blaze covers perhaps half of the dorsal surface.

We suspect that all of these gray blazes started out as white marks in the calf year. Four calves with white dorsal marks of the size and complexity of gray blazes have been photographed both as calves and as juveniles one or two years later. In three of these four, the dorsal mark was clearly brownish-gray after the first year. The marks of the fourth appeared still white, but it has only been seen as a yearling, and we expect that the marks will darken in the next year or two. Because we know that one gray-blazed whale is a full adult (#154, a female who has had at least two calves), we believe that the darkening of these blazes with age stops before they become as black as normal skin.

Aside from darkening from white to gray with age, the gray blazes do not change over time. The shape is constant (Table 5), and, as with white blazes, the size relative to the whale's length (measured in one calf, #403-75, at one to two months of age and again 13 months later) appears to remain unchanged as well.

In the Argentine population, we have seen seven whales, 1.4% of our identified individuals, with this type of dorsal mark, although two of these have been seen only as calves and thus with a white rather than with a gray color to the blaze. We have also seen this type of blaze on the ventral side of two Argentine whales, both of which have gray blazes on the dorsum as well.

Partial albinism.   A third type of body marking consists of a nearly complete lack of black pigmentation over the entire body.   In the first months of life, such whales appear to be pure white with scattered black markings on their bodies (Figure 12b).   In older whales, the background is no longer white, but is instead a gray or brownish-gray color similar to the color of gray blazes.   We have never seen adults whose entire bodies were as white as the bodies of the partial albino calves and we have watched one partial albino calf (#154-72) gradually darken as it matured, to the same brownish-gray color we find in adults.   It was quite gray by its second year, 1973, and has become darker since then.   We conclude that partial albinos are born a white color that darkens with age.   In the Argentine population, we have seen seven partial albinos, 1.4% of our identified individuals.

These whales stand out, even underwater, as noticeably paler than other whales.   Their black pigmentation is restricted to a complex pattern scattered sparsely in the white or gray background.   This pattern varies from individual to individual and does not change over time (Table 5).   There seems always to be a rather dense accumulation of black pigment in a narrow transverse band, posterior to the blowholes and extending partly down the sides of the head, but in no case as far as the eyes.   Best (1970) described three partial albino calves from South Africa, noting that in all three cases there was a black area on the posterior part of the rostrum or rostrum base.   From the photographs he includes, it is clear that in all three cases the black pigmentaton lies in the same region in which it is found in the Argentine partial albinos.

Wounds.   The fourth type of distinctive dorsal mark that is useful for individual identication over time is a wound.   We saw no apparent wounds in the South African whales, but in the Argentine population, nine whales or 1.9% of the population have what are clearly wounds, though the occurrence may be somewhat higher because there are seven whales with white marks so small that we can't tell whether they are blazes or wounds.   Wounds usually consist of raised or depressed areas of skin and are colored white or orange.

The colors appear to be due to cyamids, though the white could also be exposed blubber. The outlines of wounds are often more blurred than those of unpigmented blazes, and the cyamid-infested part is often surrounded by the gray color exposed by sloughing skin (see below). Figure 12c shows an example of a wound.

In three cases the wound is a single narrow scratch running longitudinally along the back, in a different position on each whale. In the one instance in which we can see surface contours as well, the scratch occurs within a shallow groove. The scratches are usually white and may be surrounded by gray-colored skin. We have watched a scratch over time in just one whale, #81. In 1971 and 1972, the whale was photographed with no distinctive dorsal mark. The scratch appeared in 1973 and was quite distinctly white. In 1974 and 1976, #81 was photographed again, and the scratch, although still visible, had become grayer and less distinctive. It is possible that with time a small wound, like a scratch, will heal to invisibility and will thus be of no further use as an identifying mark.

Skin mottling. Another kind of marking is usually distinctive from one individual to another but is constantly changing with time. This is the pattern of gray mottling on the black skin that is apparently caused by widespread irregular shedding of the outer layer of skin. Figure 13 shows two extreme conditions of gray mottling in the same individual 15 days apart -- almost all black and almost all gray. All stages in between occur as well. The gray appears to be the new light gray skin revealed when old black skin sloughs off. It is a bluish-gray that is distinguishable from the brownish-gray of permanent gray blazes and of partial albino non-calves. The rate of change in the mottling varies: individuals photographed on consecutive days may have almost exactly the same pattern of mottling down to very fine detail or may have changed completely with no trace of the earlier pattern. Where we have seen rapid change in consecutive days, the change has been an increase of gray areas rather than loss of gray. Because of the possibility of rapid change, the pattern of gray skin can be useful for individual recognition only within a single airflight by helping to follow a whale in frames where its callosity pattern is not discernible. Best (1970) has published two photographs which show skin mottling in South African whales very clearly.

Ventral Blazes

There is one other feature, located on the ventral surface, that is useful for individual identification of southern right whales. Most southern right whales have one

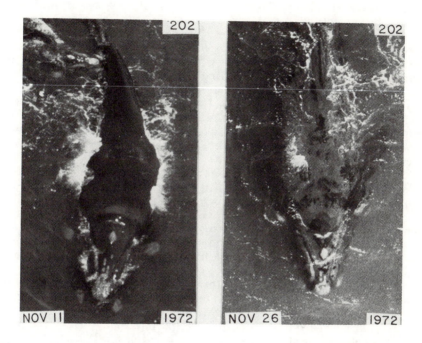

Figure 13. The change in the pattern of skin mottling due to sloughing skin on a whale photographed twice, 15 days apart.

or more white ventral blazes that range in size from small spots the size of larger rostral islands to enormous blazes covering a third or more of the ventral surface. Small single blazes invariably surround the umbilical scar and are usually elongated longitudinally. Larger blazes include the umbilical area as well as more extensive portions of the belly, throat, and/or lower flanks. They may be grossly asymmetrical and often have islands and peninsulas of white pigmentation trailing from them (particularly from the rear margins). In one peculiar case (Figure 14), a whale with a medium-sized ventral blaze has within the white pigmentation a strange, black zig-zag line surrounding the umbilical scar at a distance and in a form that looks just like the perimeter of a typical small ventral blaze. It suggests that at least two different underlying biochemical processes control development of white pigmentation skin cells and that where these processes overlap and interact, the end result is somehow normal black pigmentation.

Even small blazes may have a wealth of configurational information in the outline, as we observed in close-up examinations of nearby whales and of a corpse of a mature (lactating) female. In our aerial photographs of small, ventral blazes, however, most of the useful detail is too small to be seen. Larger blazes, however, are very complex and distinctive, making positive identifications possible from just a part of the ventral blaze. Figure 13 presents an example of a medium-sized ventral blaze and also demonstrates the constancy of the pattern over time. There is only a one year time span covered in this figure, but we have seen spans up to seven years with no visible change in pattern. This and the fact that even small calves have ventral blazes with distinct outlines leads us to believe that there is no change in these blazes over time. We have not yet, however, followed the ventral blaze of a calf as it matured.

We also believe that ventral blazes are unique to each individual, based on their variety and on the amount of information in even the smaller blazes. We cannot do a rigorous test of this, however, because the ventral blaze is almost always the only visible distinctive feature in a ventral view of a whale. In fact, more often than not, we cannot connect a ventral view to a dorsal view, showing the callosity pattern and thus the major source of information about each whale. In order to estimate the incidence of ventral blazes in the Argentine population, we examined photographs taken in 1971 alone and scored for presence or absence of a blaze only those ventral photographs of whales showing the area around the umbilicus. Out of 31 ventral surfaces that fit this criterion, 29 or 93.5% had a white blaze.

It is interesting to compare this figure with the rate of occurrence of white blazes on the dorsum which is only about 1.9%. While the latter type of blaze could be an aberration

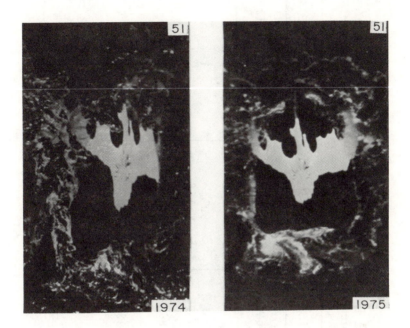

Figure 14. Example of a white ventral blaze showing constancy in the outline over a year. This whale has not been identified by any other distinctive features.

Table 6. Geographical variation in the incidence of external features of right whales. N is the number of whales that could be scored for presence or absence of each feature.

Feature	SOUTH AFRICA		ARGENTINA	NORTH PACIFIC
	Best (1970)	Our data	Our data	Omura (1958) Omura et al (1969) Rowntree et al (1980)
Lip callosities	86% N=7	75% N=20	80% N=188	18% N=12
Post-blowhole callosities	100% N=5	100% N=30	100% N=475	69% N=13
Dorsal markings (blazes and partial albinism)	11% N=75	23% N=30	5% N=484	7% N=14
Ventral blazes	100% N=2	100% N=2	94% N=31	86% N=14

of no adaptive significance, it would seem that blazes on the belly have been selected for and serve some function. Yablokov (1969) discusses coloration of this sort in several species of whales and suggests its possible use in species recognition. We suggest that it could also be useful to the whale in the same way it is to humans, for individual recognition when only the belly is visible.

Geographical Variation in the Incidence of External Features

We have compared our observations of callosities and other distinguishing features in the Argentine right whales with observations from other populations of right whales. Between the eastern and western South Atlantic and the North Pacific, we can make quantitative comparisons (Table 6), but only qualitative comparisons are possible with the North Atlantic because of the paucity of published data.

The data in Table 6 about the eastern South Atlantic right whales was collected off South Africa by us and by Best (1970). Our observed frequencies agree very closely with those of Best for most external features considered, but Best did not distinguish between individual whales. As a result, there may be overlap between individuals in his sample and in ours, or even overlap between individuals within his own data. For example, it is likely that some of the whales in his 1968 and 1969 census flights were the same individuals. For statistical comparisons with other areas, we have used only our own data from South Africa. In the North Pacific, we have data from three sources: Omura (1958), Omura et al. (1969), and Rowntree, Darling, Silber, and Ferrari (1980).

We found no differences in the incidence of any callosities between the South African right whales and those off Argentina. We therefore combined our data from both places and compared the South Atlantic to the North Pacific. In two features we found significant differences between the two oceans. The incidence of lip callosities in the North Pacific is only about one-fifth the incidence in the South Atlantic, a highly significant difference (t-test for comparing two percentages: $t=4.5688$, $df=218$, $p<0.001$). (Best noted a similar difference in incidence of lip callosities between these two areas). The incidence of post-blowhole callosities is also significantly lower in the North Pacific (t-test for comparing percentages: $t=4.1864$, $df=516$, $p<0.001$).

In the North Atlantic an unusual callosity configuration occurs. Andrews (1909) describes a whale from Long Island, New York and notes that there were callosities covering the head continuously from the snout to the blowhole. A similar configuration can be seen in photographs of three right whales, two from Florida and a third from Cape Cod published by Leatherwood, Caldwell and Winn (1976). In all these cases, the bonnet, maxillary islands, and coaming are fused into one continuous callosity. We have never seen this sort of continuous callosity coverage in Argentine or South African whales in spite of our large sample. Other published photographs (Sergeant, Mansfield, and Beck 1970; Anonymous 1976) show callosity patterns on western North Atlantic right whales which are basically the same as those from the South Atlantic. There is a photograph in Watkins and Schevill (1976) which shows a callosity pattern from New England waters intermediate between these two types (wherein the bonnet, although elongated, is not in fact continuous with the coaming). We have a single Argentine right whale with a bonnet extending caudally about as far as the bonnet in the Watkins and Schevill photograph, but it seems to be an unusual pattern.

The incidence of dorsal markings varies geographically in a different way from the incidence of callosities. As noted earlier, we have seen five kinds of dorsal marks: white blazes, gray blazes, wounds, partial albinos, and marks that are too small to identify as either wounds or blazes. Of the 39 whales in the Argentine population that had dorsal markings visible from the air, 7 were partial albinos, 9 had white blazes, 7 had gray blazes, 9 had wounds, and 7 had unidentifiable white marks. Of the 30 whales we photographed in South African waters, 2 were partial albinos, 3 had white blazes, and 2 had gray blazes. There were none in the other two categories. Neither Best (1970) nor Omura et al. (1969) break down dorsal markings into these categories, so we have combined pigmented dorsal markings in Table 6. However, Omura et al. (1969) discuss wounds separately, and Best (1970) describes partial albino calves.

Using only our own data from Argentina and South Africa, we have compared the two populations in the incidence of all pigmented back markings. South Africa has a significantly higher incidence (t-test for comparing percentages: $t=3.0358$, $df=517$, $0.001<p<0.01$). Even if we consider the unidentifiable white marks in the Argentine whales to be blazes, the result is similar ($t=2.6984$, $df=517$, $0.001<p<0.01$). Although Best reported a lower frequency of these markings in South Africa than we did, the difference is not significant statistically. It may be due to his overlooking gray blazes, which can look like sloughing skin. Another possible difference between the eastern and western South Atlantic populations is the incidence of wounds. The difference is not significant statistically, but it may be of biological significance, as discussed below.

In the North Pacific, there is a single report of a white dorsal blaze (Rowntree et al. 1980) and none of a gray blaze or a partial albino. The incidence of these dorsal marks in the North Pacific is not significantly different from that in Argentina.

Ventral blazes are also examples of white pigmentation, and we found that there was no significant difference between the percentage of whales that have them in the Argentine and North Pacific populations. We have not, as yet, measured possible differences in the average size of blazes in these populations. Samples from South Africa and from the North Atlantic are too small to test statistically.

Discussion

Geographic Variation

As noted at the outset, the taxonomic status of right whales is unresolved. Our results show that, with the exception of continuous callosity coverage on the heads of some North Atlantic right whales, the differences which exist between populations are not unique to any population but are statistically significant.

To review our results: there is a highly significant difference between North Pacific and South Atlantic populations in the incidence of lip and post-blowhole callosities. There is no significant difference between these populations in the incidence of dorsal markings. The sites for callosity growth in North Atlantic right whales include areas of skin on which we never saw callosity development in the South Atlantic, and for which it has not been reported in the North Pacific.

Right whales living in the North Pacific tend to have the fewest head regions supporting callosity growth while those living in the North Atlantic have the most. (This latter point has yet to be investigated with samples from the North Atlantic large enough to make statistical comparison possible). The degree of isolation of right whales in the North Atlantic would appear to be greatest inasmuch as some North Atlantic individuals have a feature in their callosity patterns (continuous coverage between snout tip and blowholes) that appears to be found nowhere else. Until more data from the North Atlantic becomes available, it cannot be argued that the missing post-blowhole callosities in some North Pacific right whales is a unique condition.

What these differences may indicate regarding the taxonomic status of right whales is for others to decide.

The difference in the incidence of dorsal pigmentation and partial albinism between eastern and western South Atlantic right whales and the fact that we have not seen the same individual in both areas, combined with the otherwise complete correspondence of callosity growth sites on the head, seems to indicate that although the whales in these two areas are closely related, they constitute distinct stocks with little or no interchange. If theory is correct, then the two stocks probably have different feeding destinations. This might explain why Argentine right whales have harpoon wounds whereas South African whales do not.

Of course it is not sure that the large wounds that we have observed in Argentina are caused by harpoons. They might with equal likelihood be the result of collisions with boats -- the white line directed towards the central portion of the wound being made by the keel near the bow rather than by the shaft, barb, or rope of a harpoon. But how then can we explain the fact that there are not such wounds in the South African population; and that in all cases where they do occur, the wounds are found in the same place on the back and in the same orientation to the whale's body?

It is well known (Clarke 1965) that right whales feed near South Georgia and we know of one example of what is almost certainly an Argentine right whale that was seen feeding there. There is still whaling going on in the area.

The occurrence of "harpoon" wounds only in the Argentine population also raises the intriguing possibility that the Argentine right whales, along with those seen off the coast of Tristan da Cunha (Best and Roscoe 1974) constitute a single stock, separate from the South African stock. We have made this speculation because there has been at least twice in the recent past, illegal hunting of right whales near Tristan da Cunha. One eyewitness to this was Mr. P.A. Day, then Administrator of Tristan da Cunha. His personal diary for 4 November 1963 includes this comment:

> "A fine day, only marred by violation of territorial waters by Russian whalers whom we saw killing whales within half a mile of the settlement. From reports while the island was unoccupied, (during a period of volcanic eruptions), they pretty well killed all the whale around here, and obviously are unaware anyone is here."

On the following day Mr. Day writes:

> "A beautiful day, whalers still active, steaming close in shore, towing whales out to a factory ship on the horizon. I have asked the Colonial Office to make diplomatic representations to the Russians."

Although the whales were not identified in his diary, Mr. Day has confirmed (pers. comm.) that they were right whales.

A second incident occurred 4 years later and was witnessed by Mr. Glen Norton, at the time Factory Manager for Tristan Investment (Pty.) Limited. He reports it as follows (pers. comm.):

"In March 1967, in Tristan da Cunha, an islander reported to the administrator, Mr. Brian Watkins, that he had seen a whale catcher operating off the southeast sector of the Island. The Administrator requested me to take a boat around the Island and if possible photograph any activity. We set out at 0900 in an easterly direction, keeping close inshore, and about noon, on rounding Sandy Point, sighted a Russian catcher operating off Lyon Point. We approached at low speed, taking photos from time to time. When we were about a half mile off, were sighted by the catcher, which immediately proceeded to the open sea. In doing so, all lines attached to the whale pulled out. We approached the whale as close as we dared, and photographed gaping wounds on its back, to which sea birds were paying attention. The whale was spouting pink frothy water to a height of about a foot, whilst moving very slowly in a westerly direction. By this time my film was used up, so we returned to the settlement.

The following day a Norwegian catcher (Capt. Carlsen) arrives off the settlement to drop mail. I discussed the incident with him, describing the whale, which he identified as a southern right. (Further evidence that it was this species which is often seen very close to shore comes from Mr. Norton's map showing that the catcher boat was 300 meters from the shore when he first saw it). He passed on the information to their factory ship, which was lying about ten miles n/w of the Island...

Months later we were advised; and I quote, 'that the Russians apologize for accidentally straying in Tristan waters'".

Reports such as these could explain the harpoon wounds on Argentine right whales. If the Argentine and Tristan da Cunha right whales are one stock, they may have the same behavior demonstrated by Darling and Jurasz (1983) and Payne and Guinee (1983) of visiting more than one breeding ground. It is a point we plan to investigate further.

Recognition by Natural Markings

In recent years the use of natural markings as a means to distinguish between individual baleen whales has been succesfully employed with at least two other species. Hatler and Darling (1974) showed that the oddly shaped patches of pigmentation covering the bodies of gray whales (*Eschrichtius robustus*) remained constant over at least three years. Katona et al. (1979) demonstrated the same for color patterns on the undersurfaces of humpback whale (*Megaptera novaeangliae*) flukes (about 900 have been identified by this means in the North Atlantic and about 400 in the North Pacific). In spite of these successes as well as our own with right whales (as this goes to press we can recognize 557), one repeatedly sees statements on recognition by natural markings such as the following by Kear (1978):

> "The method is particularly useful for intensive studies involving relativly small populations and a few thoroughly trained observers, but it breaks down when...populations (are) large...(natural markings) cannot substitute for conventional marking when every individual of a sizable population must be distinguishable with a high level of certainty. Nor is the method convenient when many observers are involved, as in zoos and in extensive schemes of marking and recovery."

This objection is simply not true. Natural markings have been very successfully used to study large populations of whales with many different observers, often untrained, providing the data. The study of humpback whales in the North Atlantic in which 900 animals are currently recognized is a particularly clear example. The study is a cooperative effort by dozens of observers, many of them tourists, who have contributed photographs of whale flukes accompanied by the date and location of the photograph.

Probably the single most important factor in the success of this cooperative project is the fact that a photograph is required to identify the whale, so one is not relying on the ability of inexperienced observers to interpret their own observations. A layman who takes a useful photograph of a natural marking contributes solid, enduring evidence which the researcher can interpret for himself, referring to it as often as necessary for confirmation. A person reporting the number seen on a tag, on the other hand, is not really providing evidence but only his interpretation of the evidence. There is no way to confirm his interpretation. A study relying on reports by others is fated to generate spurious sightings, which may take as much time or more time to investigate as it would have taken to make identifications from photographs in the first place.

A few examples may help to set this in perspective. 1) When, in the late 60's, one of us was installing acoustic tags on whales, he received in one week three sightings of a single successful implant from other people. One sighting was physically too far away for the whale to have swum the distance; the others were reported at the same time many miles apart. 2) P. Beamish (pers. comm.) had a similar experience with one of his early streamer tags deployed on a humpback whale. 3) B. Würsig (pers. comm.) spent several days following up a report that an orange tag he had put on the dorsal fin had been sighted near a distant island. When he finally located the eyewitness, he discovered that the orange object sighted was not on the dorsal fin but on the porpoise's head. A less careful observer might have used the report before verifying it, particularly since it indicated a plausible extension of range for Würsig's porpoises.

Another common objection to the use of natural markings for identifying animals is noted by Pennycuick (1978):

> "It is important to realise from the outset that there is no way to be absolutely certain that an individual with particular markings is the only one so marked in the population. To be certain of this, artificial markings must be used, and the user must be able to ensure that no two are the same."

Strictly speaking this is, of course, true. But it pertains to a degree of accuracy which may be entirely unnecessary for many studies and which, if it is ever attained, may be lost in the noise of the other kinds of observational errors likely to affect biological field studies. For the vast majority of studies, a 99% probability that a pattern of natural markings has no duplicate in the population will be better than necessary, for it is almost certainly better than the accuracy with which most field studies record the behavior of the species in question.

Furthermore, it must be realized that visual tags cannot be trusted absolutely either. Humans misread and misrecord numbers, and tags may be ambiguous if animals naturally produce marks which resemble symbols on the tag. For example, there was a report of a southern right whale with a number 10 on its flukes, when apparently no one was placing any such marks (P. Best pers. comm.). The number seems to have been a pattern of shedding skin that chanced to take the form of the number 10. Similarly, several right whales in the Argentine population have white blazes like alphanumeric symbols ("Dot J" and "Y Spot", for example) which look virtually identical to the white markings that have been produced by freeze branding the skin of toothed whales. Field biologists must live with uncertainties whether identification is based on tags or natural markings.

The natural markings found in right whales and other whales have several major practical advantages which may prove difficult, if not impossible, to duplicate with artificial tags. 1) They require no installation. 2) They are present at birth, in most cases. 3) They almost certainly last throughout the whale's life. 4) They don't interfere with locomotion or other natural behaviors. 5) They provide exceptional redundancy in confirming the identity of an individual. 6) They are rarely obscured from fouling marine growth. 8) They are much larger than any practical tag bearing the same information is ever likely to be and therefore can be read at a greater distance.

This last point needs to be stressed. The area over which the identifying marks of a right, gray, or humpback whale are spread is usually several square meters, but visual tags are limited to a much smaller size by the need to propel them for attachment to the animal. The smaller the size of a mark, the closer one must be to read it, the longer it will take on the average to obtain a satisfactory photograph or reading, and the smaller will be the proportion of successful identifications to sightings. The difficulty in reading artificial tags will increase as the number of animals tagged increases, for more information will be necessary on each tag to distinguish it from all others. If populations of whales comparable in number to those currently handled by natural markings are ever tagged, researchers who study them will have to deal with many of the same ambiguities of tag reading currently faced by researchers relying on natural markings.

Even when one considers smaller whales which have subtler natural markings, there are surprises in store when one considers the differences in the distances at which tags and natural marks can be read. There is an interesting example of this in Leatherwood, Caldwell, and Winn (1976). They have published a photograph of a free-swimming North Atlantic *Tursiops* with a spaghetti tag in place, as an example of what such a tag looks like in use. If we assume that they chose one of the better photographs available to them, then it is also a nice demonstration of just how difficult it is, under field conditions, to get a useful picture of something as small as a spaghetti tag. Were there not an arrow in the photo pointing out the tag, one might easily mistake it for turbulent water. This same photograph also shows several nicks on the trailing edge of the animal's dorsal fin. Though they too are somewhat blurred, the shapes and positions of the nicks can be clearly seen. It is from just such photographs (including some that were a lot worse) of just such nicks that individuals of this same species were identified in two excellent studies (Würsig and Würsig 1977; Shane 1977).

A final disadvantage of artificial tags concerns the way conspecifics may respond to them. It has been found in several species of terrestrial animals that the response of untagged animals to tagged conspecifics affects the social position and/or reproductive success of the wearer (Ramakka 1972; Burley 1981). If whales have similar reactions to their tagged fellows, it might be difficult to detect, but it would affect the behavior under study.

We have pursued the argument on the usefulness of natural markings to counteract the opinion that natural markings cannot specify an individual adequately for serious scientific studies. It should be apparent that this opinion is indefensible. Clearly, the degree of detail in callosity patterns is enough to distinguish between far more individual right whales than will ever exist at one time and with a reliability that may affect the final conclusions of a study less than would the disruption caused by installation of artificial tags. There are also practical ways to estimate reliability of identification and to measure empirically the maximum number of identical animals that exist in one's study population.

We realize that there are some species in which adequate natural markings amenable to practical use probably do not occur. However, it is also true that many apparently simple natural markings are found upon close examination to be sufficiently rich in information to specify individuals reliably. It is only when this does not hold that we would advocate applying some kind of artificial mark.

We do not wish to be misunderstood as opposing radiotelemetry or any other kinds of active tags. There are certainly many things that can be learned only through radio tags, and if tags that can be read electronically from distances of a kilometer or more (such as radio transponders giving back a pattern unique to each tag) ever become a reality, then major advances will be possible. We are in opposition, however, to shooting numbers onto animals that are already more than adequately marked by nature. We feel that, although natural marks take longer to read, it is a relatively minor price to pay for the major advantages offered by them.

Summary

1) Individual southern right whales, *Eubalaena australis*, were identified based on the pattern of callosities on the head and of other distinctive features on the body. 484 individuals were recognized in the vicinity of Peninsula Valdés, Argentina, from an analysis of 16,000 photographs taken on 52 airflights from 1971 through 1976 during the months of June through December. 30 individuals were recognized off Cape Province, South Africa from two airflights in 1974. No individuals were seen in both locations. Calves were identified at a lower rate than non-calves.

2) Callosities are described along with the covering of cyamid amphipods. Lip callosities are highly variable and show a distinct lateral asymmetry.

3) Although minor variations in appearance occur, the callosity patterns are unchanging with time.

4) The number of distinguishable callosity patterns possible, and the number of callosity patterns recognizable in practice are both estimated to be far in excess of the world population of right whales.

5) A technique is presented to calculate the likelihood that identical whales exist in a population.

6) Five types of distinctive dorsal marks are described, of which at least three are essentially unchanging with time. Distinctive ventral blazes are described as well.

7) North Pacific right whales have fewer lip callosities and post-blowhole callosities than South Atlantic right whales, but the same incidence of dorsal markings as the Argentine whales. North Atlantic right whales have a callosity configuration -- continuous rostrum coverage -- that does not occur in the South Atlantic and has not been reported from the North Pacific.

8) Argentine and South African right whales show no difference in callosity frequency, but differ significantly in incidence of pigmented dorsal marks (greater in South Africa). They thus appear to constitute separate stocks in the South Atlantic. The existence of harpoon wounds on the Argentine right whales alone, coupled with illegal hunting of right whales in Tristan da Cunha (of which eyewitness accounts are presented), suggest that the Argentine and Tristan da Cunha whales may constitute a single stock.

9) Arguments are presented for the advantages of using naturally occurring markings over visual tags for studying large populations of animals in the wild when persistent markings are present.

Acknowledgements

We gratefully acknowledge the help of many friends during all phrases of the research described herein. The following made especially large contributions: James Bird, Hugo Callejas, Kevin Chu, Nancy Cison, Christopher Clark, Jane Clark, Leslie Cowperthwaite, Ronald Christensen, Philip

DeNormandie, Carol Gould, James Gould, Nixon Griffis, Lisa Karnovsky, Lysa Struhsaker-Leland, Katharine Payne, Victor Solo, Peter Thomas, Hal Whitehead, and Bernd and Melany Würsig. Without their help, the whales would have remained incognito except to each other.

Peter Best and Peter Shaughnessy kindly arranged for a plane so we could make a photographic census in South Africa.

William Schevill and Peter Marler read the entire manuscript and Colin Pennycuick read sections of it. All made helpful suggestions for which we are grateful.

This research was supported by grants and contracts from the Marine Mammal Commission (Contract No. MM6AC017), the National Geographic and New York Zoological Societies and the World Wildlife Fund.

APPENDIX I

Derivation of the Probability
of Finding Identical Twins

The following calculations of the probability of detecting identical twins are based on several assumptions. We assume a population of 550 right whales, based on 484 identifications over 6 years, with no estimate of mortality. A larger population would reduce the likelihood of detecting twins, while a smaller population would increase it. We also assume the existence of 25 pairs of twins within the 550 whales, a number chosen arbitrarily. More twin pairs would increase the likelihood of detecting some, and fewer twin pairs would decrease that likelihood. If the variation in callosity patterns tended to cluster around a few common forms so that more than two animals had a given identical appearance, this would increase the likelihood of detecting at least one pair of them. The fact that we have not yet detected any twins suggests a lack of favored norms in callosity patterns.

We further assume random distribution of whales for the calculations below. We do not believe that their distribution is actually random, but we have no better model of distribution to use. We have no estimate of how the actual distribution of right whales would affect the probability of detecting identical twins, but we feel that our method provides a check against either of two extreme departures from randomness in the relative distribution of members of a twin set. If the two members of a twin set tended strongly to avoid each other, they would be more likely to occur in different bays at the same time, and the probability of their being thereby detected would increase. If, on the other hand, the two members tended strongly to seek each other's company, they would be more likely to be photographed in the same frame, and the probability of their being detected in that way would also increase.

Identical Twins in Different Bays

The first type of event which could demonstrate the existence of "identical twins" is the photographing on the same airflight of one whale in one bay and its twin in another bay. At Peninsula Valdés, there are three areas where right whales occur, each separated from one another by a greater distance than could be swum by a right whale during the

Compiled by Alan Titus and Eleanor M. Dorsey

duration of an airflight. For simplicity we will call these areas bays (although one is in fact a semi-protected lagoon behind fringing shoals). Some census flights included photographs of whales in all bays while others included only two bays. We shall deal with the two situations separately.

A. *Whales in two different bays.* We shall consider an airflight with photographs of "a" identified whales in one bay and "b" identified whales in another bay. We wish to determine the number of distinct events in which a twin set could be photographed, one in each bay, given the above conditions. We will represent this number by α. A first approximation to α is given by:

$$25 \times 2 \times \binom{548}{a-1} \times \binom{549-a}{b-1} \qquad (1)$$

The usual notation $\binom{n}{r}$ means "the number of combinations of n objects taken r at a time," which is given by the formula

$$\frac{n!}{r!\,(n-r)!}$$

The first factor is simply the number of twin sets we have postulated. The second factor, 2, is included because the first member of the twin set could be in either bay. Once we have specified the existence of one twin in the first bay, there remain (a - 1) unspecified whales that can be selected from the population of 550 whales diminished by the two specified twins. Thus, the third factor is $\binom{548}{a-1}$. The fourth factor is the number of ways in which the (b - 1) unspecified whales in the second bay could be selected from the population of 550 whales diminished by the "a" whales already selected for the first bay and also diminished by the specified twin in the second bay.

The expression, (1), shown above was called a first approximation because it gives too large a number. It does this because it counts twice each event when twins from two sets, one from each set, appear in each bay. The number of such events is approximated by:

$$\binom{25}{2} \times 4 \times \binom{546}{a-2} \times \binom{548-a}{b-2} \qquad (2)$$

The first factor is the number of ways that 2 sets of twins could be selected. The second factor, 4, is included because there are 4 ways in which the 2 sets of twins could be arranged such that one member of each set is in each bay.

The third factor is the number of ways of selecting the (a - 2) unspecified whales for the first bay from the 546 available whales. The fourth factor is the number of ways of selecting the (b - 2) unspecified whales for the second bay from the (548 - a) available whales. Therefore, our second order approximation to α is (1) - (2).

Let us consider now the events in which any given three sets of twins (labelled, for example, R, S, and T) are split between the two bays such that one twin from each set appears in each bay. These events will each be counted three times in (1), once with respect to twin set R, once for S, and once for T. Moreover, these events will each be counted three times also in (2), once with respect to the pairs of twin sets R and S, once for R and T, and once for S and T. Hence our second order approximation to α, namely (1) - (2), will not count any of the events when trios of twin sets are split between the two bays. An approximation to the number of such events is given by:

$$\binom{25}{3} \times 8 \times \binom{544}{a - 3} \times \binom{547 - a}{b - 3} \qquad (3)$$

following a logic very similar to that used to derive (2) above. Thus our third order of approximation to α is (1) - (2) + (3).

Continuing, let us consider the events in which any given four sets of twins are split between two bays such that one twin from each set appears in each bay. These events will each be counted four times in (1), $\binom{4}{2}$ times in (2) and $\binom{4}{3}$ times in (3). Therefore, these events will be counted twice in our third order approximation to α. The usual logic will show that such events are approximated by:

$$\binom{25}{4} \times 16 \times \binom{542}{a - 4} \times \binom{546 - a}{b - 4} \qquad (4)$$

This yields, as a fourth order approximation to α, the expression: (1) - (2) + (3) - (4).

This series could be developed further to obtain the 25th "approximation" which would, in fact, be the exact formula for α. However, in practice, the terms of the series decrease rapidly. Before showing this fact, let us recall that we are on the way to calculating a <u>probability</u> of observing a twin set in the two bays. To obtain this probability (or, to be strictly correct, a close approximation to it), we divide the fourth order approximation to α by $\binom{550}{a} \times \binom{550 - a}{b}$ which is simply the total number of possible events of "a" whales being photographed in one bay and "b" whales in another. It can easily be shown by applying the usual formula for combinations, that when each of the four expressions in our approximation is divided by the above denominator the resulting four terms are:

$1.65 \times 10^{-4} \times ab$ (5)
$1.33 \times 10^{-8} \times ab(a-1)(b-1)$ (6)
$6.83 \times 10^{-13} \times ab(a-1)(b-1)(a-2)(b-2)$ (7)
$2.54 \times 10^{-17} \times ab(a-1)(b-1)(a-2)(b-2)(a-3)(b-3)$ (8)

In order to make the most stringent test of the accuracy of successive approximations we choose actual values of a and b such that the ratio of each term in the series to the term following it is at a minimum. Empirically, we find these values are 48 and 40. For these values the expressions (5), (6), (7), and (8) take on the following values, respectively:

3.17×10^{-1}
4.68×10^{-2}
4.20×10^{-3}
2.60×10^{-4}

Clearly, the successive terms are diminishing by approximately an order of magnitude. For all other empirical values of a and b, successive terms will diminish even more rapidly. Therefore, the second order approximation, (5) - (6), to the probability of photographing a set of twins in the two bays, will be very close to the exact value. Moreover, since the third term is positive, this second order approximation will always yield a slight underestimate of the correct probability. Thus, the value obtained for the probability of _not_ photographing twins in the two bays will be a very small overestimate. In other words, any slight inaccuracy in the second order approximation would tend to lead, if anything, to _not_ rejecting our null hypothesis, that 25 sets of identified twins exist in the population.

Therefore, for each airflight with photographs from two bays, the second order approximation, (5) - (6), was subtracted from unity to obtain the probability of not photographing a set of twins in the two bays. We assumed each airflight to be an independent event and therefore multiplied these probabilities together, producing a value of 0.18347. This, then, is the probability, given the null hypothesis that 25 sets of twins exist in a population of 550 whales and, assuming random association between whales and independence of airflights, that none of the airflights, over six years, which covered two bays would have produced a photograph of a twin in one bay and its counterpart in the second bay.[1]

---

[1] Since writing this appendix, it has been pointed out to us that there exists an exact formula for α for the two bay case, which is the following: $\sum_{j=1}^{n} (-1)^{j+1} \, 2^j \, \binom{n}{j} \binom{N-2j}{a-j} \binom{N-a-j}{b-j}$

B. *Whales in three different bays.* We shall now consider an airflight with photographs of "a" whales in one bay, "b" whales in a second bay, and "c" whales in a third bay. The argument will follow similar lines to the two bay situation above but with somewhat more complicated formulae. We wish to determine β, the number of distinct events in which one member of a twin set could be photographed in a different bay. A first approximation to β is given by:

$$25 \times \left\{ \left[2 \times \binom{548}{a-1} \times \binom{549-a}{b-1} \times \binom{550-a-b}{c} \right] \right.$$
$$+ \left[2 \times \binom{548}{a-1} \times \binom{549-a}{c-1} \times \binom{550-a-c}{b} \right]$$
$$\left. + \left[2 \times \binom{548}{b-1} \times \binom{549-b}{c-1} \times \binom{550-b-c}{a} \right] \right\} \quad (9)$$

The first factor is simply the number of twin sets postulated in the null hypothesis. The first term inside square brackets represents the number of events where one twin could be photographed in the first bay and its counterpart in the second bay. The logic behind the first term is the same as that used for expression (1), except for the inclusion of the final factor which denotes the number of ways of selecting the "c" whales for the third bay from the population of 550 diminished by the "a" and "b" whales already selected for the first two bays. The second and third terms inside square brackets represent the events when twins are in the first and third, and second and third bays, respectively.

As before, expression (9) is only a first approximation to β since it will count twice each event when two sets of twins are photographed and each set is divided between any two of the three bays. For any given pair of twin sets, M and N, there are nine possible arrangements as shown by the following array:

M in bays 1 & 2	M in bays 1 & 2	M in bays 1 & 2
N in bays 1 & 2	N in bays 1 & 3	N in bays 2 & 3
M in bays 1 & 3	M in bays 1 & 3	M in bays 1 & 3
N in bays 1 & 2	N in bays 1 & 3	N in bays 2 & 3
M in bays 2 & 3	M in bays 2 & 3	M in bays 2 & 3
N in bays 1 & 2	N in bays 1 & 3	N in bays 2 & 3

The number of possible events represented by this array is given by expression (10). The first term in square brackets corresponds to row 1, column 1 in the array. The second

$$
\left[4 \times \binom{546}{a-2} \times \binom{548-a}{b-2} \times \binom{550-a-b}{c} \right] + 2 \times \left[4 \times \binom{546}{a-2} \times \binom{548-a}{b-1} \times \binom{549-a-b}{c-1} \right]
$$

$$
+ 2 \times \left[4 \times \binom{546}{a-1} \times \binom{548-a}{b-2} \times \binom{549-a-b}{c-1} \right] + \left[4 \times \binom{546}{a-2} \times \binom{548-a}{b} \times \binom{548-a-b}{c-2} \right]
$$

$$
+ 2 \times \left[4 \times \binom{546}{a-1} \times \binom{547-a}{b-1} \times \binom{548-a-b}{c-2} \right] + \left[4 \times \binom{546}{a} \times \binom{546-a}{b-2} \times \binom{548-a-b}{c-2} \right]
$$

(10)

term corresponds to row 1, column 2 and also to row 2, column 1: hence, the factor, 2, outside the square brackets. The second term will be used as an example to show the logic behind the formulae. We will use the situation depicted in row 1, column 2. The first factor, 4, is included because there are 4 possible configurations of the two sets of twins in the designated bays. In the next factor, only 546 whales are available for assignment to the first bay since the location of two sets of twins has already been specified. Likewise, there are only (a - 2) free places in bay 1 since 2 places are filled by one member of twin set M and one member of twin set N. In the next factor, the "b" places in bay 2 are diminished by 1, since one twin in set M is specified for bay 2. The population available for assignment to bay 2 is 550 whales diminished by the "a" whales already selected for bay 1 and further by the two twins not selected for bay 1. The final factor depicts the number of ways of selecting "c" whales less the single twin set N from the population of 550 whales diminished by the "a" and "b" whales already selected and further by the single twin designated for the third bay.

In order to determine the number of events when two sets of twins are each divided between two of three bays, we must multiply expression (10) by $\binom{25}{2}$, the number of possible combinations of pairs of twin sets. We shall define expression (11) in exactly this way.

Now we have, as a second order approximation to β, (9) - (11). In order to calculate the second order approximation to the probability of twins being photographed in different bays, we must divide this expression by:

$$\binom{550}{a} \times \binom{550 - a}{b} \times \binom{550 - a - b}{c}$$

This is the total number of ways of photographing "a", "b", and "c" whales in the three bays with no restrictions regarding twins. Algebraic manipulation will show that the first and second order terms of this probability are, respectively:

$$1.65 \times 10^{-4} \times (ab+ac+bc) \qquad (12)$$

and

$$1.33 \times 10^{-8} \times [a(a-1)b(b-1) + a(a-1)c(c-1) + b(b-1)c(c-1) + 2abc(a+b+c-3)] \qquad (13)$$

A third order term, analogous to that developed in the two bay case, has been derived, although the details of that derivation are too lengthy to be described here.

It can be shown that this third order term is given by:

$$6.83 \times 10^{-13} \times \{(a)(a-1)(a-2)(b)(b-1)(b-2)$$
$$+ (a)(a-1)(a-2)(c)(c-1)(c-2)$$
$$+ (b)(b-1)(b-2)(c)(c-1)(c-2)$$
$$+ 3(a)(b)(c) \times [(a-1)(a-2)(b-1)$$
$$+ (a-1)(a-2)(c-1) + (b-1)(b-2)(c-1)$$
$$+ (a-1)(b-1)(b-2) + (b-1)(c-1)(c-2)$$
$$+ (a-1)(c-1)(c-2) + 2(a-1)(b-1)(c-1)]\} \qquad (14)$$

Selecting actual values such that the ratio of the second order to the first order term is at a maximum, we find expressions (12), (13), and (14) take on the following values, respectively:

$$2.60 \times 10^{-1}$$
$$3.12 \times 10^{-2}$$
$$2.24 \times 10^{-3}$$

As in the two bay case, we find that the terms are diminishing by approximately an order of magnitude. Thus, we are again sure that the second order approximation not only is very close to the actual probability but also tends, if anything, to exert a force against rejecting the null hypothesis that 25 twin sets exist in the population.

Accordingly, for each airflight with photographs from three bays, the second order approximation, (12) – (13), was subtracted from unity to obtain the probability of not photographing a set of twins in two of the three bays. Taking each airflight to be an independent event, we multiply these probabilities together, yielding a probability of 0.20304. Finally, since the two bay situation and the three bay situation are independent events, we multiply the probability of one with the other and obtain a probability of not detecting a set of twins in different bays of 0.03725.

Identical Twins in the Same Photograph

The third type of event in which identical twins could be detected is the appearance of twins in the same photograph. Since the data include photographs with two, three, four, five, and six identified whales in the same photograph, separate probabilities were calculated for each situation.

A. Two whales in the same photograph. The number of such photographs is simply 25, that is, the number of twin sets postulated. The number of possible photographs of pairs of whales is simply $\binom{550}{2}$. There were 323 photographs of pairs of whales. Therefore the probability that none of these included a twin set is given by:

$$\left[1 - 25 / \binom{550}{2}\right]^{323}$$

B. *Three whales in the same photograph.* The number of such photographs which include a set of twins is 25 x 548, there being 548 ways of filling the third "slot", once it is stipulated that a twin set must fill two of the "slots". Since there are $\binom{550}{3}$ possible photographs of three whales and since there were in fact 57 such photographs, the probability that none of these included a twin set is given by:

$$\left[1 - \frac{25 \times 548}{\binom{550}{3}}\right]^{57}$$

C. *Four whales in the same photograph.* The first approximation to the number of such photographs which include a set of twins is $25 \times \binom{548}{2}$. This number is a little too large, since it counts twice the events in which two sets of twins are in a photograph. The number of such events is simply $\binom{25}{2}$. Following the usual procedure, then, we find that the probability that none of the 10 photographs of 4 whales included a twin set is:

$$\left[1 - \frac{25 \times \binom{548}{2} - \binom{25}{2}}{\binom{550}{4}}\right]^{10}$$

D. *Five whales in the same photograph.* The number of this type of photograph containing a least one set of twins is given by $25 \times \binom{548}{3} - \binom{25}{2} \times 546$, the second term being the correction for the events where two sets of twins appear together. Following the usual logic, the probability of not finding any sets of twins in these photographs (of which there are 3) is given by:

$$\left\{1 - \frac{\left[25 \times \binom{548}{3}\right] - \left[\binom{25}{2} \times 546\right]}{\binom{550}{5}}\right\}^{3}$$

   E. Six whales in the same photograph. The number of this type of photograph containing at least one set of twins is given by:

$$\left[25 \times \binom{548}{4}\right] - \left[\binom{25}{2} \times \binom{546}{2}\right] + \left[\binom{25}{3}\right]$$

The first and second terms are derived by the same logic as was used above. The third term is the final correction for the events when three sets of twins appear in one photograph. The probability, then, of not finding a set of twins in the two photographs of six whales is given by:

$$\left\{1 - \frac{\left[25 \times \binom{548}{4}\right] - \left[\binom{25}{2} \times \binom{546}{2}\right] + \left[\binom{25}{3}\right]}{\binom{550}{6}}\right\}^2$$

These five probabilities were all calculated and then multiplied together since they refer to independent events. The final probability of not finding any sets of twins in the same photographic frame, given the conditions of the null hypothesis, and given random associations between whales, works out to be 0.90331.

   If the photographing of twins in the same frame were completely independent of photographing them in separate bays, we could multiply the probability of each together and obtain the overall probability of not detecting twins by these methods (given the null hypothesis and our other assumptions) of 0.03365.

Literature Cited

Allen, G.M.
1916. The whalebone whales of New England. Mem. Boston
Soc. Nat. Hist. 8: 107-322.

Allen, J.A.
1908. The North Atlantic right whale and its near allies.
Bull. Am. Mus. Nat. Hist. 24: 277-329.

Andrews, R.C.
1909. Further notes on *Eubalaena glacialis* (Bonn.). Bull.
Am. Mus. Nat. Hist. 26: 273-275.

Anonymous.
1976. Whalewatchers rescue mother and calf. Mainstream
7: 31.

Best, P.B.
1970. Exploitation and recovery of right whales *Eubalaena
australis* off the Cape Province. S. Afr. Div. Sea Fish.,
Invest. Rep. 80, 20p.

Burley, N.
1981. Sex ratio manipulation and selection for
attractiveness. Science (Wash. D.C.) 211: 721-722.

Clarke, R.
1965. Southern right whales on the coast of Chile. Nor.
Hvalfangsttid. 54: 121-128.

Darling, J.D. and C.M. Jurasz.
1983. Migratory destinations of North Pacific humpback
whales (*Megaptera novaeangliae*). *In* R. Payne (ed.),
Communication and behavior of whales. AAAS Selected
Symposia Series, p. 359-368. Westview Press, Boulder,
Colo.

Esau, K.
1953. Plant anatomy. John Wiley & Sons, N.Y., 735p.

Eschricht, D.E. and J. Reinhardt.
1866. On the Greenland right-whale (*Balaena mysticetus*,
Linn.), with especial reference to its geographic
distribution and migrations in times past and present,
and to its external and internal characteristics. *In* W.H.
Flower (ed.), Recent memoirs on the Cetacea, p. 1-150.
The Ray Society, London. (Translated from K.
Videnskabernes Selskabs Skrifter 5, 1861)

Hatler, D.F. and J.D. Darling.
1974. Recent observations of the gray whale in British Columbia. Can. Field-Nat. 88: 449-459.

Katona, S., B. Baxter, O. Brazier, S. Kraus, J. Perkins, and H. Whitehead.
1979. Identification of humpback whales by fluke photographs. *In* H.E. Winn and B.L. Olla (eds.), Behavior of marine animals - current perspectives in research, vol. 3: Cetaceans, p. 33-44. Plenum Press, N.Y.

Kear, J.
1978. Recognition without marking. *In* B. Stonehouse (ed.), Animal marking - recognition marking of animals in research, p. 145. Macmillian Press, Ltd., London.

Leatherwood, S., D.K. Caldwell, and H.E. Winn.
1976. Whales, dolphins, and porpoises of the western North Atlantic - a guide to their identification. U.S. Dep. Commer., NOAA Tech. Rep., NMFS CIRC-396, 176p.

Leung, Y.M.
1965. A collection of whale-lice (Cyamidae: Amphipoda). Bull. South. Calif. Acad. Sci. 64: 132-143.

1967. An illustrated key to the species of whale-lice (Amphipoda, Cyamidae), ectoparasites of cetacea, with a guide to the literature. Crustaceana (Leiden) 12: 279-291.

1976. Life cycle of *Cyamus scammoni* (Amphipoda: Cyamidae), ectoparasite of gray whale, with a remark on the associated species. Sci. Rep. Whales Res. Inst. Tokyo 28: 153-160.

Lönnberg, E.
1906. Contributions to the fauna of South Georgia: I. Taxonomic and biological notes on vertebrates. K. Sven. Vetenskapsakad. Handl. 40: 1-104.

1923. Cetological notes. Ark. Zool. 15: 1-18.

Lütken, C.F.
1966. Contributions to the knowledge of the species of the genus *Cyamus* Latr., or whale lice. Fish. Res. Board Canada Transl. Ser. No. 642, 117p. (Translated from Kgl. Danske Vedensk. Selsk. Skr. 10, 1873).

Matthews, L.H.
    1938. Notes on the southern right whale, *Eubalaena australis*. Discovery Rep. 17: 169-182.

Omura, H.
    1958. North Pacific right whales. Sci. Rep. Whales Res. Inst. Tokyo 13: 1-52.

Omura, H., S. Ohsumi, T. Nemoto, K. Nasu, and T. Kasuya.
    1969. Black right whales in the North Pacific. Sci. Rep. Whales Res. Inst. Tokyo 21: 1-78.

Payne, R.S.
    1972. Swimming with Patagonia's right whales. Nat. Geogr. Mag. 142: 576-587.

    1976. At home with right whales. Nat. Geogr. Mag. 149: 322-339.

Payne, R. and L.N. Guinee.
    1983. Humpback whale (*Megaptera novaeangliae*) songs as an indicator of "stocks". *In* R. Payne (ed.), Communication and behavior of whales. AAAS Selected Symposia Series, p. 333-358. Westview Press, Boulder, Colo.

Pennycuick, C.J.
    1978. Identification using natural markings. *In* B. Stonehouse (ed.), Animal marking - recognition marking in animals in research, p. 147-159. MacMillian Press, Ltd., London.

Ramakka, J.M.
    1972. Effect of radio-tagging on breeding behavior of male woodcock. J. Wildl. Mgmt. 36: 1309-1312.

Rice, D.W.
    1977. A list of the marine mammals of the world. U.S. Dep. Commer., NOAA Tech. Rep. NMFS SSRF-711, 15p.

Ridewood, W.G.
    1901. On the structure of the horny excrescence, known as the "bonnet", of the southern right whale (*Balaena australis*). Proc. Zool. Soc. Lond. 1: 44-47.

Roussel de Vauzeme, A.
    1835. Memoire sur le *Cyamus ceti* (Latr.) de la classe des crustaces. Ann. Sci. nat. Zool. 1: 239-255, 257-265.

Rowntree, V. J. Darling, G. Silber, and M. Ferrari.
1980. Rare sighting of a right whale (*Eubalaena glacialis*) in Hawaii. Can. J. Zool. 58: 309-312.

Sergeant, D.E., A.W. Mansfield, and B. Beck.
1970. Inshore records of cetacea for eastern Canada, 1949-69. J. Fish. Res. Board Can. 27: 1903-1915.

Shane, S.H.
1980. Occurrence, movements, and distribution of bottlenose dolphins, *Tursiops truncatus*, in southern Texas. Fish. Bull., U.S. 78: 593-601.

Struthers, J.
1887. On some points in the anatomy of a *Megaptera longimana*. J. Anat. Physiol. 22(pt. 1): 109-127.

Van Beneden, P.J.
1868. Sur le bonnet et quelques organes d'un foetus des baleines de Groenland. (On the bonnet and several organs of a Greenland right whale fetus) Bull. Acad. R. Belg. 26: 186-195.

Watkins, W.A. and W.E. Schevill.
1976. Right whale feeding and baleen rattle. J. Mammal. 57: 58-66.

Whitehead, H. and R. Payne.
1981. New techniques for measuring whales from the air. U.S. Dept. Commer. NTIS PB81-161143, 36p.

Würsig, B. and M. Würsig.
1977. The photographic determination of group size, composition, and stability of coastal porpoises (*Tursiops truncatus*). Science (Wash. D.C.) 198: 755-756.

Yablokov, A.V.
1969. Types of colour of the cetacea. Dep. Fish. Res. Board Transl. Ser. No. 1239, 28p. (Translated from Bull. Moscow Soc. Nat. Biol. 68, 1963)

13. Determining the Sex of Humpback Whales *(Megaptera novaeangliae)* in Their Natural Environment

Abstract

A simple method for determining the sex of humpback whales in their natural environment is presented. It consists of noting in a ventral or lateral view of the abdomen of the whale the presence or absence of a hemispherical lobe posterior to the genital slit. The lobe is visible in females and absent in males. Through viewing underwater photographs of the abdomens of 96 individual humpback whales, 54 females and 42 males were sexed. Thirty-four known mothers and 33 calves were sexed. The sex ratio of calves was found to be 48.5% female, 51.5% male. Fourteen escorts and three singers were sexed. All were males. This morphological sexing technique has wide application in population studies of living humpback whales.

Introduction

Most modern behavioral studies of species in the wild rely on the knowledge of the sex of the individual observed. Recent studies of humpback whales (*Megaptera novaeangliae*) have profited from the technique of identification of individuals through photographs of the undersurfaces of their flukes (Schevill and Backus 1960; Katona, Baxter, Brazier, Kraus, Perkins, and Whitehead 1979). Such studies could benefit greatly from knowledge of the animal's sex.

In 1973, Winn, Bischoff, and Taruski developed a cytological technique for sexing living cetaceans. Their technique consisted of obtaining skin samples via a biopsy dart and analyzing them in the laboratory for sex chromatin bodies. Their method was effective but obviously complicated. A need still existed for a simpler method of determining the sex of a humpback whale in the wild.

Sexual dimorphism in length occurs in the humpback whale, the female averaging slightly longer than the male (Matthews 1937; Mackintosh 1942; Omura 1953; Chittleborough 1955a, 1955b; Nishiwaki 1959; Dawbin 1960). There are also differences between the sexes in the structure of the abdomen (Struthers 1887; True 1904; Lillie 1915; Matthews 1937). In 1904, True described the outline of the abdomen of the humpback whale as a series of depressions and elevations. He stated that the genital opening is surrounded by a thick wall which, in a female, ends in a hemspherical lobe. Matthews (1937), Angot (1951), and Mitchell (1973) measured the distance between the genital opening and the anus and found it to be almost 2.5 times greater in males than in females. They noted that the position of the genital opening is more anterior in males than in females. Each of these observations were made on corpses of whales obtained from the whaling industry, some killed specifically for the purpose of scientific study. From my underwater photographs of the ventral sides of humpback whales, I found that I was able to distinguish between females and males by applying True's (1904) description of the genital region. The hemisphical lobe at the end of the genital slit was clearly visible in females and absent in males. It is difficult, however, to obtain good photographs of the ventral side of a live whale. I felt that a field technique which relied on that might not find very broad application. It is much easier to photograph the lateral view of a whale while underwater and I found that such photographs show clearly the hemispherical lobe in the female and its absence in the male.

In this paper, I will demonstrate the distinction between the ventral outline of male and female humpback whales. I will show how to recognize the female's hemispherical lobe in lateral view and how to measure the angle between the posterior edge of the genital wall and the long axis of the whale's body.

I will also discuss the occurrence of sexual dimorphism in pigment patterns, the sex ratio of calves, and the different social behaviors of the two sexes.

Methods

I obtained 503 analyzable underwater photographs of humpback whales during a three year study, 1977-79 (Glockner and Venus 1983) and a one year study in 1980 (Glockner and Ferrari in prep.) conducted in Maui, Hawaii. These photographs included 62 ventral views and 441 lateral views of the whales.

Results

I placed the photographs into five categories: mothers, calves, yearlings, "escorts" (an adult accompanying a mother and calf, often seen swimming beneath the mother), and other adults. I was able to sex 6 different females and 4 different males from both the ventral view and the lateral view and an additional 48 different females and 38 different males from the lateral view. I had previously identified each of these whales as individuals by the identification technique outlined in Glockner and Venus (1983).

Figure 1 illustrates the external features of female and male humpback whales described by True (1904) and Matthews (1937). They are also seen clearly in two photographs in Figure 2. In the photograph showing the ventral side of a female whale, #3208, a known mother, one can see that the genital opening is surrounded by a thick ridge of raised skin (which I have called the genital wall) ending in a distinct hemispherical lobe. Mammary grooves are present on either side of the opening. The anus is located directly behind the lobe in a transverse groove. Caudal to this groove is a rounded area, another transverse groove, and then the keel-like carina, in that order. The male shown in Figure 2 is whale #1408, an escort. The genital slit is also surrounded by a genital wall but the hemispherical lobe is not present. The anus is situated further from the slit than in the female. The anus is also surrounded by a raised wall. This wall should not be mistaken for the hemispherical lobe anterior to the anus in the female. Although other authors have noted that mammary slits are visible in the male humpback, they are not in my experience visible in photographs under normal conditions in the field. In males the genital slit is located 10% of the body length anterior to the anus, while in females, it is only about 4% of the body length from the anus (Matthews 1937; Mitchell 1973).

A possibility exists that the hemispherical lobe is present only in lactating females. In order to see if this is true I examined photographs of calves. Figure 3 shows the ventral side of a female calf (#3208C9), and a male calf (#2217C8). It is apparent that the ventral view is the same as in the adult, the hemispherical lobe being clearly visible in the female but not in the male.

I also found that the hemispherical lobe was clearly visible in photographs showing lateral views of the same mother (#3208) and the same calf (#3208C9). This is illustrated in Figures 4 and 5. Lateral views of the same escort (#1408) and the same male calf (#2217C8) (Figures 4 and 5) show no lobe.

I analyzed photographs showing lateral views of the abdomens of 34 known individual "mothers" (adults accompanying calves). I found the characteristic hemi-

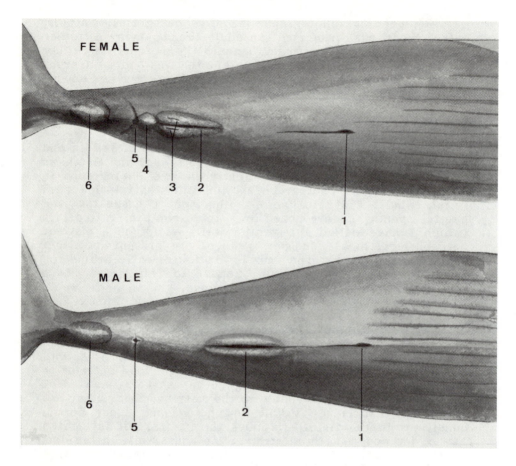

Figure 1. Diagram of the ventral side of a female and male
    humpback whale.
    1) the umbilicus
    2) the genital opening
    3) the mammary grooves, located on either side of
       the genital opening
    4) the hemispherical lobe
    5) the anus
    6) the carina

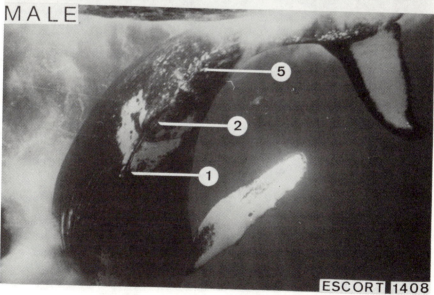

Figure 2. Comparison of the ventral view of the abdomen of a female and male humpback whale. (1-6 same as in Figure 1.)

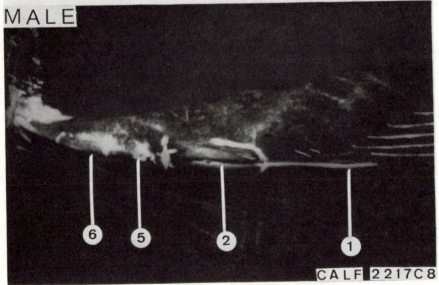

Figure 3. Comparison of the ventral view of a female and male calf (1-6 same as in Figure 1).

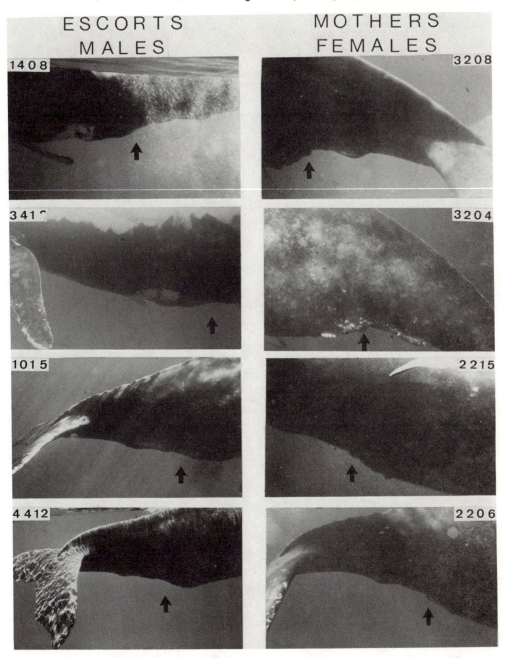

ESCORTS
MALES

MOTHERS
FEMALES

Figure 4. Comparison of the lateral view of the abdomen of
mothers and escorts. The arrows point to the
posterior end of the genital wall. The
hemispherical lobe can clearly be seen in the
female.

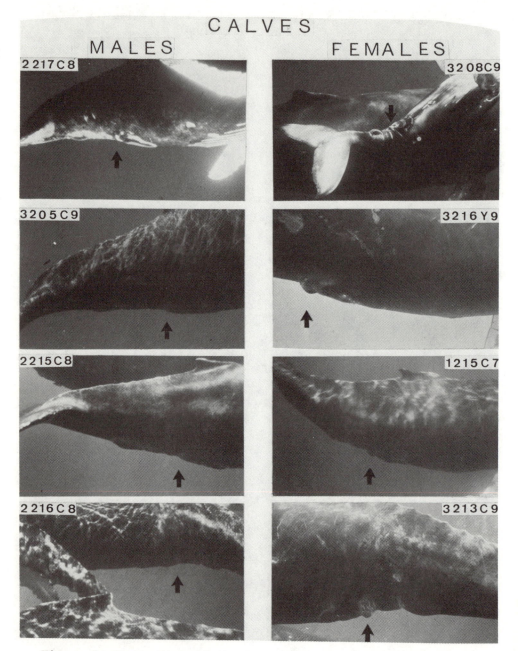

Figure 5. Comparison of the lateral view of the abdomen of female and male calves. Note: whale #3216Y9 is a yearling.

spherical lobe in every photograph. I assumed that the "mothers" were females.

In lateral views of 14 escorts, a lobe was not visible in any photograph. Figure 4 illustrates the lateral views of mothers and escorts. Figure 5 compares the lateral views of female and male calves.

In order to make a quantitative distinction between females and males, I measured the angle between the body axis and the posterior margin of the genital wall (which in profile views of females is the posterior margin of the hemispherical lobe). Figure 6 illustrates how this angle is determined. For this determination, I used only photographs in which I judged that the whale was perpendicular to the lens of the camera. I measured 17 females and 14 males. In females, the angle ranged from 41° to 67° (mean 51.9°, s.d. 8.68°). In males, the angle ranged from 19° to 32° (mean 24.3°, s.d. 2.80°). No overlap occurred between the angles measured for females and males, suggesting that measurements of this angle are a useful technique for determining sex.

On 22 February 1981, a humpback whale calf was found alive on Punaluu Beach, Oahu, Hawaii. The calf was transported to Sea Life Park, an oceanarium at Makapuu Point, Oahu and placed into a saltwater tank. I was able to photograph the underside of the whale from a lateral and ventral view. The calf was a male. Its genital slit was surrounded by a thick wall. Mammary slits were barely visible on either side of the genital slit. No hemispherical lobe was present. The anus, surrounded by a raised wall, was clearly visible. I measured the angle between the posterior margin of the genital wall and the body axis from a photograph of the lateral view of the calf. I repeated the measurement three times. I found the mean angle to be 20.3°, s.d. 1.35°.

Occurrence of Sexual Dimorphism in Pigment Patterns

The dorsal margin of the dorsal fin and of the caudal peduncle of a humpback whale (called here the dorsal ridge) is often white (Pike 1953). I found this pigmentation in all whales I photographed. However, as first suggested to me by Victoria Rowntree, the amount of white pigmentation in males was greater than in females.

In 90 animals I had good enough photographs to enable me to make a rough determination of the amount of white pigmentaiton on skin covering these areas. I examined my photographs, placing each whale in one of three arbitrary pigmentation categories: faintly, moderately, or prominently white-pigmented. Table 1 shows that in 76.2% of the adult males, the dorsal ridge was prominently pigmented, while in 81.1% of the adult females the dorsal ridge was faintly pigmented. A prominently pigmented dorsal ridge, appearing

Table 1. Extent of white pigmentation on the dorsal ridge.

	Faint		Moderate		Prominant		Total	
	N	%	N	%	N	%	N	%
Female Adults	30	81.1	3	8.1	4	10.8	37	100.0
Female Calves	7	46.6	4	26.7	4	26.7	15	100.0
Male Adults	1	4.8	4	19.0	16	76.2	21	100.0
Male Calves	1	5.9	4	23.5	12	70.6	17	100.0

like a white line, occurred in 70.6% of the male calves. For example, the male calf that was beached on Oahu 22 February 1981 had a prominently white-pigmented dorsal ridge. Of the female calves, 53.4% also had prominent to moderate pigmentation in this area. The larger percentage of prominent white pigmentation in female calves as against adults may suggest that humpback whales darken with age (Glockner and Venus 1983).

Sex Ratio of Calves

Very few data exist on sex ratios of commercially harvested whales in their first year of life in as much as calves and their mothers are given protection by nations signatory to the International Whaling Commission convention. Between 1977 and 1980, I was able to sex 33 live calves near Maui. I found that 16 (48.5%) were females and 17 (51.5%) were males. Though my sample is probably too low to be significant, these figures suggest a sex ratio skewed towards males which is in agreement with all but one study of fetal sex ratios of which I am aware. The exception (Matthews 1937) has a sample size of 23, however. These studies are summarized in Table 2 for purposes of comparison.

Sexes of Singers, Escorts,
Surface Traveling Groups, and Lone Individuals

On three occasions I approached and photographed adults which I was certain were alone and which were singing as determined by hearing songs so loud they appeared to come through one's whole body while swimming nearby and by the fact that the whale surfaced in the "breathing section" of the song (Payne, Tyack, and Payne 1983). In all three cases, I found the sex of the singers to be male.

One of the roles of the escorts towards mothers and calves has been postulated to be protective (Chittleborough 1953; Herman and Antinoja 1977; Glockner and Venus 1983). I was able to sex 14 escorts and found them all to be males.

Often a cow, calf, and escort will travel rapidly just below the surface while being followed by one or more other adults. On two occasions I was able to sex the pursuing adult. Each time it was a male. On several occasions I observed groups of adult whales, ranging in number from four to seven, traveling rapidly just below the surface. I was able to determine sexes of the whales in four cases. In three of these cases, the lead whale was a female. In these three observations, I was able to see that the lead individual was the most active and exhibited behaviors such as breaching and tail lifting. In the three cases mentioned above, the whale next in line behind the female was a male. From all my observations I have noted that whales in this position often

Table 2. Fetal sex ratios

Author	Area Sampled	Total in Sample	Female	Male
Matthews 1937	S. Georgia & Natal	23	13	10
Omura 1953	Antarctic	408	190	218
Chittleborough 1958	W. Coast Australia	249	121	128
Chittleborough 1958	Antarctic	1448	696	752
Totals		2128	1020	1108
Mean %		100	47.9	52.1

lift their heads above the surface at angles of approximately 45° with the surface of the water, a behavior which in gray whales (*Eschrichtius robustus*) has been suggested to be related to mating (Samaras 1974). I have also often seen these whales exhale streams of bubbles underneath the surface of the water. In my three cases where sexing was possible, the whale next in line behind the first male was also a male.

On the fourth occasion when I could determine sex of a group of adults traveling rapidly near the surface I found that none of the whales was lifting its head 45° above the surface. However, in this case the five lead whales were all males (the remaining two were unsexed).

On three occasions I was able to identify the sex of both members of a pair of humpback whales. In one case the pair was a male and a female. In the remaining cases both were males. On five occasions I have determined the sex of whales that were traveling alone. In one case, the lone individual was female and in four they were males.

Discussion

The work described here examined two techniques for determining the sex of humpback whales. The sexual dimorphism in pigmentation of the dorsal ridges indicates that in general males have more prominently white pigmented dorsal ridges than females. This may well result from scarring, as scars on humpbacks appear light on a dark surface (Lillie 1915) and males are known to fight (Darling, Gibson, and Silber 1983). However, a close percentage of male calves also exhibited the white pigmentation along their dorsal ridge. The white pigmentation on the dorsal ridge could prove useful in confirming sex but cannot be relied upon without other morphological or behavioral indications.

A more reliable indicator of sex in humpback whales of all ages is an underwater photograph showing the whale in profile view. I have found it is possible to obtain such photographs easily and inexpensively, suggesting that this technique has broad usefulness.

The sexing technique outlined here should provide valuable information for population and behavior studies of humpback whales and it may prove possible to adapt it to other cetacean species as well. Its great advantage over traditional techniques relying on corpses is that it makes possible population studies of whales without the necessity of killing them.

Summary

I was able to sex 96 whales, 54 females and 42 males, through photographs of the ventral and/or lateral sides of their genital region. These whales consisted of 34 known mothers, 14 escorts, 33 calves, 3 yearlings, and 12 other adults. I determined the sex ratio of calves to be 48.5% female, 51.5% male. I found the sex of all 14 escorts and 3 singers to be male.

Acknowledgements

I would like to thank Roger Payne of the New York Zoological Society for his suggestions, guidance, and tremendous contribution to the findings contained in this paper. I would also like to thank Katy Payne, Jim Bird, Vicky Rowntree, and Ellie Dorsey for their many suggestions and contributions. I would also like to thank Katy Payne for drawing the illustrations. I am especially grateful to Jim Bird for locating essential references and compiling a bibliography and to Vicky Rowntree for her invaluable assistance, suggestions, and photographic work. I am gratefully indebted to Roger and Katy Payne for so generously extending themselves and their laboratory to me. I am especially grateful to Mark Ferrari for his photographic work and for typing this manuscript. I would like to thank Gordon Lane of the Alexander Lindsay, Jr. Museum, Walnut Creek, CA for the use of their darkroom facilities. I would like to thank Roger Payne and William Schevill for reviewing this manuscript. My research was conducted under a federal permit issued by the National Marine Fisheries Service.

Appendix 1

Measurement of the angle formed between the body axis and genital wall involves the following steps:

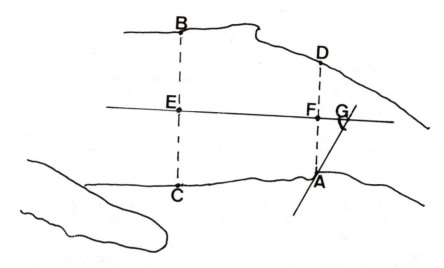

1) Locate point A, the intersection of the underside of the tail with the posterior edge of the genital wall in males or the posterior edge of the hemispherical lobe in females.

2) Locate point B, the point where the hump anterior to the dorsal fin emerges from the back.

3) Find point C. This point is on the underside of the abdomen below point B and is the closest point on the abdomen to point B -- it can be found by using a compass.

4) Draw line BC.

5) Locate point D by drawing a line through A parallel to line BC.

6) Find point E, the midpoint of BC, and point F, the midpoint of AD. Draw line EF, the line of the body axis.

7) Draw a line through point A along the straightest segment of the posterior edge of the genital wall and extend this line through the line of the body axis, creating point G at their intersection.

8) Measure angle FGA, the angle formed by the posterior edge of the genital wall with the body axis.

Literature Cited

Angot, M.
    1951. Rapport scientifique sur les expeditions baleinieres
    autour de Madagascar (saisons 1949 et 1950).
    (Scientific report on the whaling expeditions around
    Madagascar '1949 and 1950 seasons') Mem. Inst. Rech.
    Sci. Madagascar Ser. A. Biol. Anim. 5: 439-486.

Chittleborough, R.G.
    1953. Aerial observations on the humpback whale,
    Megaptera nodosa (Bonnaterre), with notes on other
    species. Aust. J. Mar. Freshw. Res. 4: 219-226.

    1955a. Aspects of reproduction in the male humpback
    whale, *Megaptera nodosa* (Bonnaterre). Aust. J. Mar.
    Freshw. Res. 6: 1-29.

    1955b. Puberty, physical maturity, and relative growth of
    the female humpback whale, *Megaptera nodosa*
    (Bonnaterre), on the Western Australian coast. Aust. J.
    Mar. Freshw. Res 6: 315-327.

    1958. The breeding cycle of the female humpback whale,
    Megaptera nodosa (Bonnaterre). Aust. J. Mar. Freshw.
    Res. 9: 1-18.

Darling, J.D., K.M. Gibson, and G.K. Silber.
    1983. Observations on the abundance and behavior of
    humpback whales (*Megaptera novaeangliae*) off West
    Maui, Hawaii 1977-79. *In* R. Payne (ed.),
    Communication and behavior of whales, AAAS Selected
    Symposia Series, p. 201-222. Westview Press, Boulder,
    Colo.

Dawbin, W.H.
    1960. An analysis of the New Zealand catches of
    humpback whales from 1947 to 1958. Norsk
    Hvalfangsttid., 49: 61-75.

Glockner, D.A. and S.C. Venus.
    1983. Identification, growth rate, and behavior of
    humpback whales (*Megaptera novaeangliae*) cows and
    calves in the waters off Maui, Hawaii 1977-79. *In* R.
    Payne (ed.) Communication and Behavior of Whales,
    AAAS Selected Symposia Series, p. 223-258. Westview
    Press, Boulder, Colo.

Glockner-Ferrari, D. and M.J. Ferrari.
    In prep. The breeding cycle of the female humpback
    whale, *Megaptera novaeangliae*.

Herman, L.M. and R.C. Antinoja.
    1977. Humpback whales in the Hawaiian breeding waters:
    Population and pod characteristics. Sci. Rep. Whales
    Res. Inst. Tokyo 29: 59-85.

Katona, S., B. Baxter, O. Brazier, S. Kraus, J. Perkins, and H.
Whitehead.
    1979. Identification of humpback whales by fluke
    photographs. *In* H.E. Winn and B.L. Olla (eds.),
    Behavior of marine animals - Current perspectives in
    research, Vol. 3: Cetaceans, p. 33-44. Plenum Press,
    N.Y.

Lillie, D.G.
    1915. Cetacea. British Antarctic ("Terra Nova")
    Expedition, 1910. Nat. Hist. Rep. Zool. (Lond.) 1: 85-
    124.

Mackintosh, N.A.
    1942. The southern stocks of whalebone whales. Dis-
    covery Rep. 22: 197-300.

Matthews, L.H.
    1937. The humpback whale, *Megaptera nodosa*. Discovery
    Rep. 17: 7-92.

Mitchell, E.
    1973. Draft report on humpback whales taken under
    special scientific permit by eastern Canadian land
    stations 1969-1971. Int. Whaling Comm. Rep. Comm.
    23: 138-154.

Nishiwaki, M.
    1959. Humpback whales in Ryukyuan waters. Sci. Rep.
    Whales Res. Inst. Tokyo 14: 49-87.

Omura, H.
    1953. Biological study on humpback whales in the
    Antarctic whaling Areas IV and V. Sci. Rep. Whales
    Res. Inst. Tokyo 8: 81-102.

Payne, K., P. Tyack, and R. Payne.
    1983. Progressive changes in the songs of humpback
    whales (*Megaptera novaeangliae*): A detailed analysis
    of two seasons in Hawaii. *In* R. Payne (ed.),
    Communication and behavior of whales, AAAS Selected
    Symposia Series, p. 9-57. Westview Press, Boulder,
    Colo.

Pike, G.C.
    1953. Colour pattern of humpback whales from the coast

of British Columbia. J. Fish. Res. Board Can. 10: 320-325.

Samaras, W.F.
1974. Reproductive behavior of the gray whale *Eschrichtius robustus*, in Baja California. Bull. South. Calif. Acad. Sci. 73: 57-64.

Schevill, W.E. and R.H. Backus.
1960. Daily patrol of a *Megaptera*. J. Mammal. 41: 279-281.

Struthers, J.
1887. On some points in the anatomy of a *Megaptera longimana*. J. Anat. Physiol. 22 (pt 1): 109-125.

True, F.W.
1904. The whalebone whales of the western North Atlantic. Smithson. Contrib. Knowl. 33, 332p.

Winn, H.E., W.L. Bischoff, and A.G. Taruski.
1973. Cytological sexing of Cetaceans. Mar. Biol. (Berl.) 23: 343-346.

## Part 5

Bibliography

14. An Annotated Bibliography of the Published Literature on the Humpback Whale *(Megaptera novaeangliae)* and the Right Whale *(Eubalaena glacialis/australis),* 1864–1980

Abstract

An annotated bibliography of the published literature on the humpback whale and the right whale, from 1864 to November 1980 is presented. The bibliography is divided into two parts, humpback whales and right whales, each arranged alphabetically by first author. Each part has the following indexes: author, subject, geographic location, and journal. Annotations are in key word form with subject key words followed by geographic location key words. Criteria for inclusion in the bibliography are: 1) humpback and/or right whales as the major topic of concern, or 2) a significant contribution to the knowledge of these species. A sample entry showing the citation format is presented as well as a list of the major reference sources and libraries consulted.

Introduction

The following annotated bibliography is an attempt to list all material on the humpback whale *(Megaptera novaeangliae)* and the right whale *(Eubalaena glacialis/Eubalaena australis)* published in any language from 1864 to November 1980. The year 1864 was chosen as the starting date because this was the year in which the Zoological Record, an invaluable guide to the literature, commenced publication. It is hoped that this bibliography will serve a useful function in the area of cetacean study by bringing together the diverse publications on these species of whales.

Most of the entries included in this bibliography are of a technical nature; however, popular papers are included when they add significant information to the study of the whale species involved. All subjects, including biochemistry, paleontology, and systematics, are presented. References include journal articles, technical reports, published letters, book chapters, and published abstracts of papers presented at scientific meetings.

Every effort was made to see all the references included in this bibliography (see the list of libraries visited). A small number of papers were not seen and these are marked with an asterisk (*) after the entry number. The source which referenced the paper is also given. In order to compile this bibliography, a variety of catalogues, indexes, and abstracts were used (see list of the major reference sources). Because of the time lag between publication and the indexing of a publication, a wide variety of journals were scanned every month. Letters were written to many scientists who have published work concerning humpback and right whales. Included in these letters was a list of their published papers concerning the species involved. They were asked to check this list of references for accuracy and completeness and to add papers that were overlooked. The response was gratifying and their help is greatly appreciated.

The following are the general criteria for inclusion of a reference in the bibliography: 1) The paper must have as its main topic of concern the humpback whale and/or the right whale, or 2) a contribution which is deemed significant to the study of these species of whales. This latter point is necessarily a subjective one, however it must be included if the bibliography is to serve its intended purpose: to provide a useful tool for the in-depth survey of humpback and right whale literature.

The main part of the bibliography is presented in two sections, the first on humpback whales and the second on right whales, both organized by first author. An "A" after the entry number refers to a paper that was added after the final draft had been typed. Works with no attributed authorship are listed under "anonymous" in the main part of the bibliography. Other access points include a co-author index (which includes the first author), a subject index based on the subject headings developed for the library of Roger Payne, a geographic index, and journal index. The journal index is included for the librarian who would want to access his or her library's strength in the cetacean field. Entries include the standard information as well as the number of plates, tables, figures, references, and the language of publication.

Foreign language titles are presented in their original language (with the exception of Chinese and Japanese which are presented in English) followed by an English translation which is given in parentheses. Russian language titles are transliterated using the *American National Standard System for the Romanization of Slavic Cyrillic Characters* (ANSI Z39.24-1976). In some cases only the English language title of a foreign language paper is known. This title is written in English with parentheses if a translation was not examined. If the entry is of a translated paper, the title is written in English without parentheses. Journal titles are abbreviated in accordance with *The World List of Scientific Periodicals published in the years 1900-1960*, 4th edition.

Annotations are in the form of key words based on the terms used in the subject index. Key words are listed alphabetically with subject key words first, followed by key words pertaining to geographic location. In cases in which further explanation is needed, a short written annotation is provided. If an abstract has been published in one of the major reference works such as Biological Abstracts or Chemical Abstracts, then a reference is made to this before the key word annotations. A list of abbreviations used in describing these reference works as well as other symbols and abbreviations used throughout this bibliography follows this introduction. Access to English translations is noted for foreign language entries, if available. In cases in which translations or English language abstracts have not been found, every effort has been made to annotate these entries on the basis of tables, figures, plates, references, authorship, and/or title translation.

I have endeavored to present a comprehensive look at the literature of the humpback and right whales within the scope defined in this introduction. However, some references may have been inadvertently neglected. From those using this bibliography, I would appreciate any additions or corrections.

Acknowledgements

I would like to thank the many scientists who responded to my letters of inquiry regarding their published papers on humpback and right whales. I would like to thank in particular Dr. James Mead who made his library, and the libraries of the National Museum of Natural History, available to me. My sincere gratitude goes to Dr. Roger Payne for giving me, over the last four years, the opportunity to complete this work. His guidance has been instrumental in its completion. Thanks go to Dr. James Mead, William E. Schevill, and Eleanor M. Dorsey for critically reading the manuscript and offering additions and many useful comments. I would like to thank Linda Guinee for final manuscript preparation and for typing the final draft. Last but not least, I would like to thank my wife Mary Margaret, whose moral support, library skills, and grammatical guidance have, in many ways, made this bibliography a reality.

Major Sources Consulted

Academy of Natural Sciences of Philadelphia - Catalogue of the Library
American Museum of Natural History, New York - Catalogue of the Library
Antarctic Bibliography
Aquatic Sciences and Fisheries Abstracts
Bernice Bishop Museum - Catalogue of the Library
Biological Abstracts
Biological and Agricultural Index
Chemical Abstracts
Conference Papers Index
Current Contents - Agriculture, Biology and Environmental Sciences; Biochemistry; Life Sciences
Current Programs
Department of the Interior - Catalogue of the Library
Directory of Published Proceedings
FAO Current Bibliography for Aquatic Sciences and Fisheries
Fisheries and Marine Service Translation Series (Canada)
Index Medicus
Index to Scientific and Technical Proceedings
Joint Publications Research Service (Transdex)
The John Crerar Library - Catalogue of the Library
Marine Biological Laboratories - Catalogue of the Library
Marine Biological Laboratories - Oceanographic Index 1946-1971 Subject Cumulation
Marine Mammal Information
Marine Mammal News
Marine Science Contents Tables
Monthly Government Publications Office Publications List
Museum of Comparative Zoology - Catalogue of the Library
Oceanic Abstracts
Oceanic Citation Journal
Polar Record
Reader's Guide to Periodical Literature
Received or Planned Current Fisheries, Oceanographic, and Atmospheric Translations. National Oceanic and Atmospheric Administration, National Marine Fisheries Service
Recent Polar Literature
Reports of the International Whaling Commission
Royal Society Catalogue of Scientific Papers
Science Citation Index
Social Science Index
Technical Translations - U.S. Dept. of Commerce; superceded by Translations Register Index, National Translations Center
Translated Tables of Contents of Current Foreign Fisheries, Oceanographic, and Atmospheric Publications. National

Oceanic and Atmospheric Administration, National Marine Fisheries Service
Wildlife Abstracts
Wildlife Review
Zoological Record

Major Bibliographies Consulted

Brown, R.J.
1975, January. Whales - A bibliography with abstracts - 1964 to October 1974. Springfield, VA: NTIS, PS-75/169/36A, 98p.

Gilbert, J.R., T.J. Quinn II, and L.L. Eberhardt.
1976, September. An annotated bibliography of census procedures for marine mammals. Washington: U.S. Marine Mammal Commission, 160p.

Gold, J.P.
1981, July. Marine mammals: A selected bibliography. Washington: U.S. Marine Mammal Commission, 91 p.

Harrison, E.A.
1976, January. Auditory perception in cetacea - a bibliography with abstracts - 1964 to January 1976. Springfield, VA: NTIS, PS-76-0008/36A, 75p.

Heizer, R.F.
1968, January. A bibliography of aboriginal whaling. J. Soc. Biblphy. nat. Hist., 4(part 7): 344-362.

Jenkins, J.T.
1948, November 3. Bibliography of whaling. J. Soc. Biblphy. nat. Hist., 2(part 4): 71-166.

Magnolia, L.R.
1977. Whales, whaling, and whaling research; A selected bibliography. Cold Spring Harbor, NY: The Whaling Museum, publication #WM-1, 91p.

Pederson, T. and J.T. Ruud.
1946. A bibliography of whales and whaling - selected papers from the Norwegian research work 1860-1945. Hvalrad. Skr., No. 30: 32p.

Severinghaus, N.C. and M.K. Nerini.
1977, January. An annotated bibliography on marine mammals of Alaska. Seattle: U.S. Dept. of Commerce, NOAA, NMFS, Marine Mammal Division, Northwest and Alaska Fisheries Center Processed Report, 125p.

Squire, I.L.
1964, December. A bibliography of cetacea - literature published between 1949 and 1963. China Lake, California: U.S. Naval Ordnance Test Station, NAVWEPS Report #8645, NOTS TP 3686, 118p.

Major Libraries Visited

Boston Public Library
Cabot Science Library - Harvard University
Countway Library of Medicine - Harvard University
Lindgren Library - Massachusetts Institute of Technology
Museum of Comparative Zoology Library - Harvard University
National Marine Fisheries Service Library - Woods Hole
Remington Kellogg Library of Marine Mammalogy - Smithsonian Institution
Biological Laboratories Library - Woods Hole
Science Library - Massachusetts Institute of Technology

Abbreviations and Symbols

Antarctic (1-6)	The 1-6 represents the six 'Areas' into which the waters south of 40°S are divided:
	Area 1 - 120 W to 60 W Area 2 - 60 W to 0 Area 3 - 0 to 70 E Area 4 - 70 E to 130 E Area 5 - 130 E to 170 W Area 6 - 170 W to 120 W
ArB	Arctic Bibliography
ASFA	Aquatic Sciences and Fisheries Abstracts
BA	Biological Abstracts
CA	Chemical Abstracts
CPI	Conference Papers Index
C.S.I.R.O.	Commonwealth Scientific and Industrial Research Organization
IEEE	Institute for Electrical and Electronic Engineers
JPRS	Joint Publications Research Service

MMC   Marine Mammal Commission

MTS   Marine Technical Society

NTIS   National Technical Information Ser-
                        vice

WR   Wildlife Review

ZR   Zoological Record

Humpback Whale Bibliography

1. Anonymous.
    1951, January. Pelagic catch of humpback whales in the Antarctic. Norsk Hvalfangsttid., 40(1): 11-12. (English and Norwegian)

    Whaling - Catch Statistics; Antarctic

2. ---------
    1952, February. Pelagic hunting of humpback whales in the Antarctic. Norsk Hvalfangsttid., 41(2): 77-78. (English and Norwegian)

    Whaling - Catch Statistics; Antarctic

    A correction to this report is given in Norsk Hvalfangsttid., 41: 3 (March, 1952), p. 134.

3. ---------
    1953, February. The catch of humpback whales. Norsk Hvalfangsttid., 42(2): 95-96. (English and Norwegian)

    Whaling - Catch Statistics; Antarctic

4. ---------
    1955, September. Twin embryos found in Moreton I. whale. Fish. Newsl., 14(9): 3, 1 unnumbered figure.

    Anatomy - Reproductive; Fetus; Whaling - Catch Statistics; Pacific - Western South: Australian Waters

5. ---------
    1956, January. 1955 whaling was worth £1,953,140. Fish. Newsl., 15(1): 3, 1 unnumbered table.

    Tagging/Marking; Whaling - Catch Statistics; Whaling - Industry Statistics; Pacific - Western South: Australian Waters

6. ---------
    1956, June. Norwegians want to take more humpbacks. Fish. Newsl., 15(6): 5.

    Whaling - Catch Statistics; Whaling - Quotas; Antarctic

7.   - - - - - - - - -
      1956, September. Eastern whaling over. Fish.
      Newsl., 15(9): 3.

      Whaling - Catch Statistics; Whaling -
      Industry Statistics; Pacific - Western South:
      Australian Waters   .

8.   - - - - - - - - -
      1956, September. Humpback proposal referred to
      comm'tee. Fish Newsl., 15(9): 5.

      International Whaling Commission; Whaling -
      Quotas; Antarctic

9.   - - - - - - - - -
      1957, January. 1956 whaling was worth £2,232,620.
      Fish. Newsl., 16(1): 5, 1 unnumbered table.

      Tagging/Marking; Whaling - Catch Statistics;
      Whaling - Industry Statistics; Pacific -
      Western South: Australian Waters

10. - - - - - - - - -
      1957, June. Want extention of humpback season.
      Fish. Newsl., 16(6): 3.

      International Whaling Commission; Whaling -
      Quotas; Antarctic

11. - - - - - - - - -
      1957, September. Australia's lead on humpbacks.
      Fish. Newsl., 16(9): 3.

      International Whaling Commission; Whaling -
      Quotas; Antarctic

12. - - - - - - - - -
      1957, September. Coastal humpback catch at
      optimum. Fish. Newsl., 16(9): 5.

      Stock Assessment; Whaling - Catch Statistics;
      Pacific - Western South: Australian Waters

13. - - - - - - - - -
      1958, April. Whaling in Australia in 1957. Norsk
      Hvalfangsttid., 47(4): 183-185, 1 unnumbered
      table.

14. - - - - - - - - -
      1958, July. Report on humpback stocks. Fish.

Newsl., 17(7): 7.

Stock Assessment; Whaling - Catch Statistics; Pacific - Western South: Australian Waters

15. ---------
    1958, August. Commission rejects humpback extension. Fish. Newsl., 17(8): 13, 28.

    International Whaling Commission; Whaling - Shore Stations

16. ---------
    1959, February. Whale marking and recoveries. Fish. Newsl., 18(2): 5-7, 1 table, 1 figure.

    Migration; Tagging/Marking; Pacific - Western South: Australian Waters

17. ---------
    1959, April. Australian whaling quotas for 1959. Fish. Newsl., 18(4): 7.

    Whaling - Quotas; Whaling - Shore Stations; Pacific - Western South: Australian Waters

18. ---------
    1959, June. Want more humpbacks. Fish. Newsl., 18(6): 23.

    Whaling - Catch Statistics; Whaling - Quotas; Antarctic

19. ---------
    1959, October. WA whaling slump. Fish. Newsl., 18(10): 7, 23.

    Tagging/Marking; Whaling - Catch Statistics; Whaling - Quotas; Antarctic (4,5); Indian - Eastern South: Australian Waters; Pacific - Western South: Australian Waters

20. ---------
    1960, February. Some humpback mingling possible. Fish. Newsl., 19(2): 13, 1 unnumbered figure.

    Distribution; Migration; Population Dynamics; Tagging/Marking; Indian - Eastern South; Australian Waters; Pacific - Western South: Australian Waters

21. ----------
        1960, March. Whaling production reflects catch
        fall. Fish. Newsl., 19(3): 7, 29, 1 unnumbered
        table.

        Tagging/Marking; Whaling - Catch Statistics;
        Whaling - Quotas; Pacific - Western South:
        Australian Waters

22. ----------
        1960, August. IWC cuts season in the Antarctic.
        Fish. Newsl., 19(8): 7.

        International Whaling Commission; Whaling -
        Quotas; Antarctic (4); Pacific - Western
        South: Australian Waters

23. ----------
        1961, January. Japanese humpback agreement.
        Fish. Newsl., 20(1): 11.

        Whaling - Quotas; Antarctic (4); Pacific -
        Western South: Australian Waters

24. ----------
        1961, February. Australia's 1960 whaling season.
        Fish. Newsl., 20(2): 13, 15, 1 unnumbered
        table, 2 unnumbered figures.

        Tagging/Marking; Whaling - Catch Statistics;
        Whaling - Industry Statistics; Pacific -
        Western South: Australian Waters

25. ----------
        1962, February. Australia's 1961 whaling season.
        Fish. Newsl., 21(2): 15, 26, 1 unnumbered
        table.

        Whaling - Catch Statistics; Antarctic (4,5);
        Indian - Eastern South: Australian Waters;
        Pacific - Western South: Australian Waters

26. ----------
        1963, August. IWC bans taking of southern
        humpbacks. Fish. Newsl., 22(8): 19.

        International Whaling Commission; Whaling -
        Catch Statistics; Whaling - Quotas; Whaling -
        Shore Stations; Antarctic; Pacific - Western
        South: Australian Waters

27. ----------
       1969, October 18. Singing whales. Nature, 224
       (5216): 217.

       Behavior - Acoustic; Vocalization

28.* ----------
       1973. Humpbacks fewer than estimated. Mar.
       Resour. Dig. Mar. Biol. Dig., 5(3): 5.

       Stock Assessment; Atlantic - Western North:
       Caribbean Waters

       Source: ZR 1973

29. ----------
       1973, July. Humpback whale population down.
       Bioscience, 23(7): 442.

       Stock Assessment; Atlantic - Western North:
       Caribbean Waters

30.* ----------
       1975. Humpback whale makes comeback. Mar.
       Resour. Dig. Mar. Biol. Dig., 6(12): 6.

       Source: ZR 1975

31. ----------
       1980, January-February. Tongan song of the
       humpbacks. Oceans, 13(1): 67.

       Behavior - Acoustic; Vocalization; Pacific -
       Western South: Tongan Waters

32. Adams, J.E.
       1975, September-October. Primitive whaling in the
       West Indies. Sea Frontiers, 21(5): 303-313.

       Whaling - Whaling Material; Atlantic -
       Western North: Caribbean Waters

33. Allen, G.A.
       1916, September. The whalebone whales of New
       England. Mem. Boston Soc. nat. Hist., 8(2):
       107-322 + 9 plates (29 figures); 99
       references.

       Anatomy - External; Coloration, Hair; Ana-
       tomy - Muscles; Anatomy - Skeletal; Behavior
       - Reproductive; Behavior - Respiratory; Be-

havior - Species Interaction: Killer Whale;
Behavior - Surface Dependent: Breach,
Lobtail; Distribution; Food; Parasites -
Ectoparasites; Whaling - Shore Stations;
Atlantic - Western North: US Waters

34. Andrews, R.C.
1909. Observations on the habits of the finback
and humpback whales of the Eastern North
Pacific. Bull. Am. Mus. nat. Hist., 26(paper
14): 213-226 + 11 plates (30 figures).

Behavior - Diving; Behavior - Feeding;
Behavior - Respiratory; Behavior - Surface
Dependent: Breach, Lobtail; Food; Photo-
graphs - Breach, Feeding, Lobtail, Sounding;
Physiology - Lactation; Pacific - Eastern
North: Alaskan Waters

35. ---------
1921, June 3. A remarkable case of external hind
limbs in a humpback whale. Am. Mus. Nov.,
No. 9: 6p., 4 figures, 2 references.

Anatomy - Skeletal; Photographs - Hind Limb;
Pacific - Eastern North: Canadian Waters

36. Angot, M.
1951. Rapport scientifique sur les expeditions
baleinieres autour de Madagascar (saisons
1949 et 1950). (Scientific report on the
whaling expeditions around Madagascar '1949
and 1959 seasons') Mem. Inst. Sci.
Madagascar, series A, 6(2): 439-486, 26
figures, 13 references. (French, English
summary)

BA 1952 (#19755)

Age Determination; Anatomy - External:
Coloration; Anatomy - Reproductive; Anatomy
- Skeletal; Fetus; Parasites - Ectoparasites;
Physiology - Lactation; Whaling - Catch
Statistics; Indian - Western South: Mada-
gascar Waters

37. Anthony, R.
1925, November 9. Sur un cerveau de foetus de
Mégaptère. (On a brain of the humpback
fetus) C.r. hebd. Séance. Acad. Sci., Paris,

181(19): 681-683, 3 unnumbered figures. (French)

Anatomy - Nervous; Fetus

38. Ash, C.E.
1953, July. Weights of Antarctic humpback whales. Norsk Hvalfangsttid., 42(7): 387-391, 1 table, 2 references. (English and Norwegian)

Weight; Whaling - Catch Statistics

39. ----------
1957, October. Weights and oil yields of Antarctic humpback whales. Norsk Hvalfangsttid., 46(10): 569-573, 2 tables, 6 references. (English and Norwegian)

Weight; Whaling - Catch Statistics; Antarctic (1,5)

40. Balcomb, K.C., III and G. Nichols.
1978. Western North Atlantic humpback whales. Rep. int. Whal. Commn. 28, Cambridge, SC/29/Doc 5, p. 159-164, 2 tables, 2 plates, 4 figures, 5 references.

Anatomy - External: Coloration; Behavior - Acoustic; Vocalization; Distribution; Individual Identification; Photographs - Individual ID by Dorsal Fin; Stock Assessment; Atlantic - Western North

Abstract of this paper appeared in: Proceedings (abstracts) of the second conference on the biology of marine mammals, December 12-15, 1977, San Diego, p. 4.

Also published as a cruise report of r/v *Regina Maris* Expedition #s1-3, Newfoundland /Labrador - Caribbean Sea, July 1976 - May 1977. Ocean Research and Education Society, Boston, MA, 1980.

41. ----------   ----------; and N.J. Haenel.
1980. (Preliminary report) Western North Atlantic humpback whale census near Hispaniola - Winter 1980. Boston: Ocean Research and Education Society, Expeditions #s20-22, 6p. + 1 table, 2 figures; 9 references.

Anatomy - External: Coloration; Behavior - Reproductive; Individual Identification; Stock Assessment; Atlantic - Western North: Caribbean Waters

42. ---------
    1980. (Abstract) Population estimates in Western North Atlantic. Boston: Humpback whales of the Western North Atlantic Workshop, November 17-21.

    Stock Assessment; Atlantic - Western North: Caribbean Waters

43. Bannister, J.L.
    1964. Australian whaling 1963 catch results and research. Sydney: C.S.I.R.O. Aust. Div. Fish. Oceanogr. Rep. No. 38, 13p. + 7 tables, 4 figures, and C.S.I.R.O. report list; 2 references.

    Tagging/Marking; Whaling - Catch Statistics; Indian - Eastern South: Australian Waters

44. --------- and A. de C. Baker.
    1967, July-August. Observations on food and feeding of baleen whales at Durban. Norsk. Hvalfangsttid., 56(4): 78-82, 2 tables, 23 references.

    Food; Physiology - Micturition; Indian - Western South: South African Waters

45. Beamish, P.
    1975. (Abstract) Biology and acoustics of a temporarily entrapped humpback whale. Santa Cruz: Conference on the biology and conservation of marine mammals - abstracts, December 4-7, p. 33.

    Behavior - Acoustic: Vocalization; Entrapments - Net; Atlantic - Western North: Canadian Waters

46. ---------
    1978, May. Evidence that a captive humpback whale (*Megaptera novaeangliae*) does not use sonar. Deep Sea Res., 25(5): 469-472 + 1 figure; 1 table, 13 references.

    BA 1979 v. 67 (#227)

Behavior - Acoustic: Echolocation, Vocalization; Entrapments - Net; Atlantic - Western North: Canadian Waters

Abstract of this paper appeared in: Proceedings (abstracts) of the second conference on the biology of marine mammals, December 12-15, 1977, San Diego, p. 72.

47. ---------
    1979. Behavior and significance of entrapped baleen whales. *In* H.E. Winn and B.L. Olla (eds.), Behavior of marine mammals - current perspectives in research - vol. 3: Cetaceans. New York: Plenum Press, p. 291-309, 1 table, 8 figures, 15 references.

    Behavior - Acoustics; Vocalization; Entrapments - Net; Tagging/Marking; Atlantic - Western North: Canadian Waters

48. Beauregard, H.
    1865, December 19. Note sur une Mégaptère échouée au Bruse près Toulon. (Note on a humpback stranded at Bruse near Toulon) C.R. Soc. Biol., 8th series, 2(43): 753. (French)

    Distribution; Entrapments - Net; Mediterranean Sea

49. Beck, A.B.
    1961, November. The copper levels of fetal and newly born marsupials and whales. Aust. J. Sci., 24(5): 245-246, 2 tables, 9 references.

    Fetus; Pollutants; Indian - Eastern South: Australian Waters

50. Beklemishev, C.W.
    1960, August 6. Southern atmospheric cyclones and the whale feeding grounds in the Antarctic. Nature, 187(4736): 530-531, 2 figures, 12 references.

    Food; Oceanography; Antarctic

51.* ---------
    1961. (The influence of atmospheric cyclones on the feeding ground of whales in Antarctica) Trudy Inst. Okeanogr., 51: 121-141.

BA 1962 (#17628)

Food; Oceanography; Antarctic

52. Bell-Marley, H.W.
    1909, February 15. Hunting the hump-back whale
    (*Megaptera longimana*) in Natal waters.
    Zoologist, 4th series, 13(146, paper 812): 54-
    63 + 1 plate.

    Anatomy - External: Scars; Behavior -
    Respiratory; Food; Parasites - Ectoparasites;
    Photographs - Anterior View; Indian -
    Western South: Natal Waters

53. Beneden, P.-J. van.
    1887, October. Histoire naturelle de la baleine a
    bosse (*Megaptera boops*). (Natural history of
    the hunch whale *'Megaptera boops'*) Mem.
    cour. Acad. r. Sci. Belg., 40: 42p., 12
    references. (French)

    Anatomy - External: Coloration; Anatomy -
    Skeletal; Classification; Distribution; Fetus;
    Food; Parasites - Ectoparasites; Photographs
    - List of Published Photographs

54. Bentley, P.J.
    1963. Composition of the urine of the fasting
    humpback whale (*Megaptera nodosa*). Comp.
    Biochem. Physiol., 10(3): 257-259, 1 table, 5
    references.

    BA 1964 (#41522)

    Biochemical Studies; Physiology - Micturition

55. Bernström, J.
    1966. Om en knölval, *Megaptera novaeangliae*
    (Borowski), strandad vid Kosteröama i
    Bohuslan 1735. (On the humpback whale,
    Megaptera novaeangliae 'Borowski', on the
    beach in Koster Islands in Bohuslan in 1735)
    Fauna Flora, 3-4: 97-101, 17 references.
    (Swedish)

    BA 1967 (#78109)

    Strandings; Atlantic - Eastern North: Scan-
    dinavian Waters

56.   Berzin, A.A. and A.A. Rovnin.
        1966. The distribution and migrations of whales in
        the northeastern part of the Pacific,
        Chuckchee and Bering Seas. Izv. tikhookean.
        nauchno-issled. Inst. rȳb. Khoz. Okeanogr.
        58: 179-207, 2 tables, 8 figures, 47
        references. (Russian)

        Distribution; Arctic: Chukchi Sea; Pacific -
        Eastern North

        Translation available from - Office of
        International Fisheries, Washington, D.C.
        20235

57.   ---------
        1978, July-August. Distribution and number of
        whales in the Pacific whose capture is
        prohibited. Sov. J. Mar. Biol., 4(4): 738-743,
        4 figures, 15 references.

        Distribution; Antarctic (1,4); Pacific -
        Eastern North: U.S. Waters

58.   Bootle, K.
        1966, December. What happened to the humpback
        whale? Fish. Newsl., 25(12): 17, 19, 21.

        Anatomy - External: Coloration; Distribution;
        Migration; Stock Assessment; Whaling -
        Quotas; Whaling - Shore Stations; Antarctic
        (4,5); Indian - Eastern South: Australian
        Waters; Pacific - Western South: Australian
        Waters

59.   Breathnach, A.S.
        1955, July. The surface features of the brain of
        the humpback whale (Megaptera novae-
        angliae). J. Anat., 89(3): 343-354 + 2 plates
        (4 figures); 5 figures, 24 references.

        BA 1956 (#22806)

        Anatomy - Nervous; Photographs - Brain
        Anatomy

60.   Brown, S.G.
        1956, December. Whale marks recently recovered.
        Norsk Hvalfangsttid., 45(12): 661-664, 1
        unnumbered table, 7 references.

Migration; Tagging/Marking; Antarctic (5); Pacific - Western South: Australian Waters

61. --------
1957, June 22. Migrations of humpback whales. Nature, 179(4573): 1287.

Distribution; Migration; Pacific - Western South: New Zealand Waters

62. --------
1957, October. Whale marks recovered during the Antarctic whaling season 1956/57. Norsk Hvalfangsttid., 46(10): 555-559, 1 unnumbered table.

Migration; Population Dynamics; Tagging/ Marking; Antarctic (5); Pacific - Western South: Tongan Waters

63. --------
1958, October. Whale marks recovered during the Antarctic whaling season 1957/58. Norsk Hvalfangsttid., 47(10): 503-507, 3 unnumbered tables.

Migration; Tagging/Marking; Antarctic (5); Pacific - Western South: Australian Waters

64. --------
1959, December. Whale marks recovered in the Antarctic seasons 1955/56, 1958/59, and in South Africa 1958 and 1959. Norsk Hvalfangsttid., 48(12): 609-616, 3 unnumbered tables, 4 references.

Migration; Population Dynamics; Tagging/ Marking; Antarctic (5,6); Pacific - Western South: Australian and Tongan Waters

65. --------
1976. Sightings of blue and humpback whales on the Icelandic whaling grounds 1969 to 1974. Rep. int. Whal. Commn. 26, London, SC/27/Doc 7, p. 297-299, 1 table.

Distribution; Atlantic - North: Icelandic Waters

66. Bryant, P.; G. Nichols, Jr.; T. Boettger; and K. Miller. 1980. Krill availability and the distribution of

humpback whales in Southeastern Alaska. Boston: Ocean Research and Education Society, expedition #16, 8p. + 1 figure; 20 references.

Behavior - Feeding; Distribution; Food; Pacific - Eastern North: Alaskan Waters

67. Budker, P.
1950. La question des Megapteres. (The question of humpbacks) Cybium, No. 5: 49-63, 4 tables, 1 figure. (French)

Distribution; Migration; Whaling - Catch Statistics; Atlantic - Eastern South: African Waters

68. ---------
1951, November 25. L'industrie baleinière au Gabon. (The whaling industry of Gabon) Bull. Soc. zool. Fr., 76(4): 271-276 + 2 plates (4 figures). (French)

Photographs - Ventral View; Whaling - Whaling Material; Atlantic - Eastern South: African Waters

69.* --------- and J. Collignon.
1952. Trois campagnes baleinières au Gabon: 1949-1950-1951. (Three whaling cruises to Gabon: 1949-1950-1951) Bull. Inst. Etud. centra., 3: 75-100. (French)

Source: Reference #73

70. ---------
1952, March. Quelques considèrations sur la campagne baleinière 1951 au Cape Lopez (Gabon). (Some considerations of the whaling cruises of 1951 at Cape Lopez 'Gabon') Mammalia, 16(1): 1-6, 2 figures, 4 references. (French)

Distribution; Migration; Whaling - Catch Statistics; Antarctic (2-5); Atlantic - Eastern South: African Waters

71. ---------
1953, September. Les campagnes baleinières 1949-1952 au Gabon (Note preliminaire). (The

whaling cruises 1949-1952 at Gabon 'preliminary note'). Mammalia, 17(3): 129-148, 7
tables, 4 charts, 3 figures, 4 references.
(French, English summary)

Stock Assessment; Whaling - Catch Statistics;
Atlantic - Eastern South: African Waters

72. ---------
1954, June. Whaling in French oversea territories.
Norsk Hvalfangsttid., 43(6): 320-326, 4
unnumbered tables, 3 references. (English and
Norwegian)

Whaling - Catch Statistics; Atlantic - Eastern
South: African Waters; Indian - Western
South: Madagascar Waters

73. --------- and C. Roux.
1968, November-December. The 1959 summer
whaling season at Cape Lopez (Gabon). Norsk
Hvalfangsttid., 57(6): 141-145, 3 figures, 7
references.

Distribution; Oceanography; Whaling - Catch
Statistics; Atlantic - Eastern South: African
Waters

74.* Budylenko, G.A.
1970. (Observations on the behavior of newborn
humpback whales) Trudȳ Atlant. nauchno-
issled. Inst. rȳb. Khoz. Okeanogr., 29: 231-
233. (Russian)

Source: ZR 1970

75. Casinos, A, S.; S. Filella; and J. Pelegri.
1977. Notas sobre cetaceos de las Aguas Ibericas.
I. Sobre un ejemplar de *Megaptera
novaeangliae* (Borowski, 1781) (Cetacean,
Balaenopteridae) capturado frente a las
costas Gallegas. (Notes on cetaceans of
Spanish waters. I. On an example of
Megaptera novaeangliae 'Borowski, 1781'
'Cetacean, Balaenopteridae' captured off the
Gallegas coast) Miscelanea zool., 4(1): 299-
303, 1 table, 2 figures, 5 references.
(Spanish, English summary)

BA 1979 v. 67 (#1356)

Anatomy - External: Coloration; Food; Parasites - Ectoparasites; Atlantic - Eastern
North: European Waters

76. Chittleborough, R.G.
1953, November. Aerial observations on the
humpback whale, *Megaptera nodosa* (Bonnaterre), with notes on other species. Aust. J.
mar. Freshwat. Res., 4(2): 219-226 + 2 plates
(4 figures), 6 tables, 2 figures, 3 references.

Behavior - Locomotion; Behavior - Species
Interaction: Killer Whale; Distribution;
Migration; Photographs - Adult/Calf, Sound-
ing; Indian - Eastern South: Australian
Waters

77. ----------
1954, March. Studies on the ovaries of the
humpback whale, *Megaptera nodosa* (Bonnaterre), on the Western Australian coast. Aust.
J. mar. Freshwat. Res., 5(1): 35-63 + 6
plates; 8 tables, 1 unnumbered appendix table,
9 figures, 10 references.

Anatomy - Reproductive; Fetus; Photographs
- Ovaries; Physiology - Lactation; Physiology
- Ovulation; Indian - Eastern South:
Australian Waters

78. ----------
1955, February. Aspects of reproduction in the
male humpback whale, *Megaptera nodosa*
(Bonna-terre). Aust. J. mar. Freshwat. Res.,
6(1): 1-29 + 1 plate (4 figures); 9 tables, 1
unnumbered appendix table, 12 figures, 33
references.

Anatomy - Reproductive; Photographs -
Testis; Indian - Eastern South: Australian
Waters

79. ----------
1955, October. Puberty, physical maturity, and
relative growth of the female humpback
whale, *Megaptera nodosa* (Bonnaterre), on the
Western Australian coast. Aust. J. mar.
Freshwat. Res., 6(3): 315-327, 6 tables, 3
figures, 12 references.

BA 1956 (#18304)

Anatomy - Reproductive; Growth Rates; Indian - Eastern South: Australian Waters

80.* ---------
1957. An analysis of recent catches of humpback whales from the stocks in Group IV and V. Sydney: C.S.I.R.O. Aust. Div. Fish. Oceanogr. Rep. No. 4: 1-17.

Whaling - Catch Statistics; Indian - Eastern South: Australian Waters; Pacific - Western South: Australian Waters

Source: Mackintosh, N.A. The stocks of whales. London: Fishing News (Books) Ltd., 1965, 232p.

81. --------- and K. Godfrey.
1957, May. A review of whale marking and some trials of a modified whale mark. Norsk Hvalfangsttid., 46(5): 238-248, 2 tables, 2 figures, 10 references. (English and Norwegian)

Tagging/Marking

82. ---------
1957, June. Breeding cycle of the female humpback whale. Fish. Newsl., 16(6): 5.

Anatomy - Reproductive

83. ---------
1957, July. Australian whale marking. Fish. Newsl., 16(7): 11, 1 table.

Tagging/Marking; Pacific - Western South: Australian Waters

84. ---------
1958. Australian catches of humpback whales 1957. Sydney: C.S.I.R.O. Aust. Div. Fish. Oceanogr. Rep. No. 17: 23p., 9 tables, 12 figures, C.S.I.R.O. report list, 8 references.

Growth Rates; Whaling - Catch Statistics; Indian - Eastern South: Australian Waters; Pacific - Western South: Australian Waters

85. ---------
1958, March. The breeding cycle of the female

humpback whale, *Megaptera nodosa* (Bonnaterre). Aust. J. mar. Freshwat. Res., 9(1): 1-
18 + 5 plates (10 figures); 4 tables, 4 figures,
23 references.

BA 1958 (#40372)

Anatomy - Reproductive; Fetus; Photographs
- Female Reproductive Cycle; Physiology -
Lactation; Indian - Eastern South: Australian
Waters

86.* ---------
1959. Australian humpback whales 1958. Sydney:
C.S.I.R.O. Aust. Div. Fish. Oceanogr. Rep.
No. 23: 21p.

Whaling - Catch Statistics; Indian - Eastern
South: Australian Waters; Pacific - Western
South: Australian Waters

Source: Mackintosh, N.A. The stocks of
whales. London: Fishing News (Books) Ltd.,
1965, 232p.

87. ---------
1959, February. Australian marking of humpback
whales. Norsk Hvalfangsttid., 48(2): 47-55, 3
tables, 1 figure, 12 references. (English and
Norwegian)

Distribution;   Migration;   Tagging/Marking;
Ant-arctic (4,5); Indian - Eastern South:
Australian Waters; Pacific - Western South:
Australian Waters

88. ---------
1959, October. Determination of age in the
humpback whale *Megaptera nodosa* (Bonnaterre). Aust. J. mar. Freshwat. Res., 10(2):
125-143 + 4 plates (7 figures); 7 tables, 9
figures, 21 references.

BA 1960 (#17690)

Age Determination; Anatomy - Ear; Photo-
graphs - Male Ear Plugs; Tagging/Marking;
Indian - Eastern South: Australian Waters;
Pacific - Western South: Australian Waters

Reprinted in Norsk Hvalfangsttid., 49: 1 (January, 1960), p. 12-26, 29-32, 35-38. (English and Norwegian)

89. ---------

1959, October. Intermingling of two populations of humpback whales. Norsk Hvalfangsttid., 48(10): 510-514, 517-521, 3 tables, 3 figures, 9 references. (English and Norwegian)

Distribution; Migration; Tagging/Marking; Whal-ing - Catch Statistics; Antarctic (4,5); Indian - Eastern South: Australian Waters; Pacific - Western South: Australian Waters

90. ---------

1960. Australian catches of humpback whales 1959. Sydney: C.S.I.R.O. Aust. Div. Fish. Oceanogr. Rep. No. 29: 16p. + 13 tables, 17 figures, C.S.I.R.O. report list, 13 references.

Age Determination; Tagging/Marking; Whaling - Catch Statistics; Antarctic (4,5); Indian - Eastern South: Australian Waters; Pacific - Western South: Australian Waters

91. ---------

1960, March. Apparent variations in the mean length of female humpback whales at puberty. Norsk Hvalfangsttid., 49(3): 120-124, 2 tables, 1 figure, 6 references. (English and Norwegian)

Growth Rates; Indian - Eastern South: Aust-ralian Waters; Pacific - Western South: Australian Waters

92. ---------

1960, April. Lengths of puberal female humpbacks. Fish. Newsl., 19(4): 13.

Growth Rates; Whaling - Catch Statistics; Indian - Eastern South: Australian Waters; Pacific - Western South: Australian Waters

93. ---------

1960, July 9. Marked humpback whale of known age. Nature, 187(4732): 164, 5 references.

BA 1960 (#68885)

Age Determination; Growth Rates; Migration; Tagging/Marking; Antarctic (4,5); Indian - Eastern South: Australian Waters; Pacific - Western South: Australian Waters

94.   ----------

1961. Australian catches of humpback whales 1960. Sydney: C.S.I.R.O. Aust. Div. Fish. Oceanogr. Rep. No. 31: 12p. + 17 tables, 10 figures, C.S.I.R.O. report list; 8 references.

Age Determination; Tagging/Marking; Whaling - Catch Statistics; Indian - Eastern South: Australian Waters; Pacific - Western South: Australian Waters

95.   ----------

1962. Australian catches of humpback whales 1961. Sydney: C.S.I.R.O. Aust. Div. Fish. Oceanogr. Rep. No. 34: 13p. + 24 tables, 19 figures, C.S.I.R.O. report list; 7 references.

Age Determination; Tagging/Marking; Whaling - Catch Statistics; Indian - Eastern South: Australian Waters; Pacific - Western South: Australian Waters

96.   ----------

1962. Exploitation of a population of the humpback whale. Fishg News int., 1(5): 38-41, 2 tables, 4 figures, 6 references.

Whaling - Catch Statistics; Indian - Eastern South: Australian Waters

97.   ----------

1963. Australian catches of humpback whales 1962. Sydney: C.S.I.R.O. Aust. Div. Fish. Oceanogr. Rep. No. 35: 5p. + 15 tables, 6 figures, C.S.I.R.O. report list; 1 reference.

Age Determination; Whaling - Catch Statistics; Indian - Eastern South: Australian Waters; Pacific - Western South: Australian Waters

98.   ----------

1965, March. Dynamics of two populations of the humpback whale, *Megaptera novaeangliae*

(Borowski). Aust. J. mar. Freshwat. Res., 16(1): 33-128, 44 tables, 36 figures, 57 references.

BA 1965 (#96970)

Age Determination; Anatomy - External: Coloration; Anatomy - Reproductive; Behavior - Locomotion; Distribution; Growth Rates; Migration; Population Dynamics; Stock Assessment; Tagging/Marking; Whaling - Catch Statistics; Antarctic (4,5); Indian - Eastern South: Australian Waters; Pacific - Western South: Australian Waters

99. ---------
1978. Australian humpback whaling. *In*: Whales and Whaling, Vol. 1 - Report of the independent inquiry conducted by the Hon. Sir Sydney Frost. Canberra: Australian Government Publishing Service, p. 276-287, 3 tables.

International Whaling Commission; Whaling - Catch Statistics; Whaling - Quotas; Antarctic (4,5); Indian - Eastern South: Australian Waters; Pacific - Western South: Australian Waters

100. Cocks, A.H.
1884, September-November. The finwhale fishery on the coast of Finmark. Zoologist, 3rd series, 8(93):366-370, 2 references. Continued in 3rd series, 8(94):417-424, 2 references; and 3rd series, 8(95): 455-465, 5 references.

Anatomy - External: Coloration; Behavior - Acoustics: Vocalization; Fetus; Parasites - Ectoparasites; Atlantic - Eastern North: Scandinavian Waters

101. Cope, E.D.
1865, September. Note on a species of hunchback whale. Proc. Acad. nat. Sci., Philad., p. 178-181, 3 references.

Anatomy - Skeletal

102. ---------
1867, March 19. On the Bahia finner. Proc. Acad. nat. Sci., Philad., p. 32.

Anatomy - Skeletal; Classification; Atlantic - Western South: Brazilian Waters

103.   ---------
1871, January-July. On *Megaptera bellicosa*. Proc. Am. phil. Soc., 12(86): 103-108, 8 figures, 4 references.

Anatomy - Skeletal; Behavior - Surface Dependent: Spyhop; Classification; Atlantic - Western North

103A. Couch, L.K.
1930, September. Humpback whale killed in Puget Sound, Washington. Murrelet, 11(3): 75.

Behavior - Object Interaction: Logs; Distribution; Pacific - Eastern North: U.S. Waters

104.   Dakin, W.J. and G. Rayner.
1940, May 18. (Letters) Humpback whales off West Australia. Nature, 145(3681): 783-784.

Migration; Tagging/Marking; Indian - Eastern South: Australian Waters

105.   ---------
1940, August 21. The migrations of Antarctic whales, and migrations along the coasts of Australia. Aust. J. Sci., 3(1): 4-5, 1 unnumbered figure.

Migration; Tagging/Marking; Indian - Eastern South: Australian Waters

106.   Dall, W. and D. Dunstan.
1957, January. *Euphausia superba* Dana from a humpback whale, *Megaptera nodosa* (Bonnaterre), caught off Southern Queensland. Norsk Hvalfangsttid., 46(1): 6-9, 3 figures, 13 references. (English and Norwegian)

Behavior - Feeding; Food; Photographs - *Euphausia superba*; Indian - Eastern South: Australian Waters; Pacific - Western South: Australian Waters

107.   Dall, W.H.
1872, December 16. On the parasites of the cetaceans of the north west coast of

America, with descriptions of new forms. Proc. Calif. Acad. Sci., 4(21): 299-301.

Parasites - Ectoparasites; Pacific - Eastern North: U.S. Waters

108.   Darling, J.D.
1978. The winter whales of Hawaii - part one: To Hawaii for the winter. Waters - Journal of the Vancouver Aquarium, 3(4th quarter): 9-18, 18 unnumbered figures, 7 references.

Behavior - Acoustic: Vocalization; Behavior - Locomotion; Behavior - Surface Dependent: Breach, Surface Active Behavior; Individual Identification; Photographs - Breach, Fluke Coloration Pattern, Individual ID by Dorsal Fin and Fluke, Surface Active Behavior; Pacific - North: Hawaiian Waters

109.   ---------; K.M. Gibson; and G.K. Silber.
1979. (Abstract) Observations on the abundance and behavior of humpback whale, *Megaptera novaeangliae*, off West Maui, Hawaii 1977-79. Seattle: Abstracts from presentations at the third biennial conference of the biology of marine mammals, October 7-11, p. 11.

Anatomy - External: Coloration; Individual Identification; Stock Assessment; Pacific - North: Hawaiian Waters

110.   ---------
1980. (Abstract) Migration and behavior of identified humpback whales (*Megaptera novaeangliae*). *In*: Abstracts of papers of the 146th national meeting 3-8 January 1980 San Francisco, California. A. Herschman (ed.), Washington, D.C.: American Association for the Advancement of Science, AAAS Publication 80-2, p. 40.

Anatomy - External: Coloration; Behavior - Agonistic; Behavior - Cow/Calf/Escort; Behav-ior - Surface Dependent: Tail Lash; Individual Identification; Migration; Pacific - Eastern North: Alaskan, Mexican Waters; Pacific - North: Hawaiian Waters

111.   Dawbin, W.H. and R.A. Falla.
1953. A contribution to the study of the humpback

whale based on observations at New Zealand shore stations. Proceedings of the 7th Pacific Science Congress, part 4, p. 373-382, 1 unnumbered figure.

Fetus; Food; Migration; Whaling - Catch Statistics; Whaling - Shore Stations; Pacific - Western South: New Zealand, Tongan Waters

112. ---------

1956, September. Whale marking in South Pacific waters. Norsk Hvalfangsttid., 45(9): 485-504, 507, 508, 1 table, 5 figures, 1 unnumbered figure, 5 references. (English and Norwegian)

Distribution; Tagging/Marking; Pacific - Western South: Australian, New Zealand, Tongan Waters

113. ---------

1956, October. The migrations of humpback whales which pass the New Zealand coast. Trans. Roy. Soc. N.Z., 84(part 1): 147-196, 3 tables, 6 figures, 49 references.

Behavior - Feeding; Behavior - Locomotion; Behavior - Reproductive; Behavior - Species Interaction: Petrels, Prions, Sea Birds, Sooty Shearwaters; Behavior - Surface Dependent: Mating Activity; Distribution; Entrapments - Net; Food; Migration; Oceanography; Whaling - Catch Statistics; Pacific - Western South: New Zealand Waters

114. ---------

1957, July 21. (Abstract) Changes in parasitic infections during the growth of humpback whales. Aust. J. Sci., 20(1): 19.

Parasites - Ectoparasites; Parasites - Endoparasites

Presented at monthly meeting of The Society for Experimental Biology of New South Wales, April 30, 1957.

115. ---------

1959, May. New Zealand and South Pacific whale marking and recoveries to the end of 1958. Norsk Hvalfangsttid., 48(5): 213-234, 236-

238, 3 tables, 2 figures, 9 references.
(English and Norwegian)

Age Determination; Behavior - Surface
Dependent: Breach, Flippering; Distribution;
Growth Rates; Migration; Tagging/Marking;
Antarctic (1-5); Pacific - Western South:
Australian, Fiji, New Zealand, Tongan Waters

116. ---------

1959, June 20. Evidence on growth-rates obtained
from two marked humpback whales. Nature,
183(4677): 1749-1750, 2 figures, 8 references.

Age Determination; Anatomy - Reproductive;
Growth Rates; Photographs - Male Ear Plugs;
Tagging/Marking; Pacific - Western South:
New Zealand Waters

117. ---------

1960. Problems of humpback whale migration in
the Pacific Ocean. Rep. Challenger Soc.,
3(12): 32.

Migration; Population Dynamics; Tagging/
Marking; Pacific - Western South: Fiji, New
Zealand, Tongan Waters

118. ---------

1960, February. An analysis of the New Zealand
catches of humpback whales from 1947-1958.
Norsk Hvalfangsttid., 49(2): 61-75, 5 tables,
8 figures, 4 references. (English and
Norwegian)

Anatomy - Reproductive; Migration; Population Dynamics; Whaling - Catch Statistics;
Pacific - Western South: New Zealand Waters

119. ---------

1964, March. Movements of humpback whales
marked in the South West Pacific Ocean 1952
to 1962. Norsk Hvalfangsttid., 53(3): 68, 70-
74, 76-78, 6 tables, 1 figure, 12 references.

Distribution; Migration; Population Dynamics;
Tagging/Marking; Pacific - Western South:
Australian, Fiji, New Zealand, Tongan Waters

120.   ---------
      1966. The seasonal migratory cycle of humpback whales. *In*: Whales, Dolphins, and Porpoises. K.S. Norris (ed.), Berkeley: University of California Press, p. 145-170, 1 table, 6 figures, 61 references.

      Behavior - Feeding; Behavior - Locomotion; Behavior - Species Interaction: Shearwater; Distribution; Migration; Oceanography; Population Dynamics; Tagging/Marking; Antarctic (4,5); Indian - Eastern South: Australian Waters; Pacific - Western South: Australian, New Zealand Waters

121.*   ---------
      1974. World humpback whale stocks. Report of a special meeting of the International Whaling Commission Scientific Committee, La Jolla, CA, December 3-13, SC/SP74/Doc 33.

      Stock Assessment

      Source: Rep. int. Whal. Commn 26, London, 1976

122.   Dawson, J.W.
      1883, March. On portions of the skeleton of a whale from gravel on the line of the Canada Pacific Railroad, near Smith's Falls, Ontario. Am. J. Sci., 3rd series, 25(147, paper 20): 200-202, 2 references.

      Anatomy - Skeletal; Evolution; Atlantic - Western North: Canadian Waters

      Also published in: The Canadian Naturalist and Quarterly Journal of Science, 10:7 (March 8, 1883), p. 385-387.

123.   Dempsey, E.W. and G.B. Wislocki.
      1941, June 25. The structure of the ovary of the humpback whale (*Megaptera nodosa*). Anat. Rec., 80(2): 243-257, 3 plates (14 figures), 1 figure, 8 references.

      BA 1941 (#21997)

      Anatomy - Circulatory; Anatomy - Reproductive; Photographs - Ovaries, Corpus Luteum

124.* Doroshenko, N.V.
1969. (On the distribution and migration of the humpbacked whale in the north-eastern parts of the Pacific Ocean, the Bering and Okhotsk Seas). *In*: (Marine Mammals) V.A. Arsen'ev; B.A. Zenkovich; and K.K. Chapskii (eds.), Moscow: Akad, Nauk SSSR Izdatel'stvo Nauka, p. 176-182. (Russian)

Source: ZR 1970

125. Dunn, D. and J. Darling.
1978. Humpback whales in British Columbia: A hope for the future. Waters - Journal of the Vancouver Aquarium, 3(4th quarter): 31-32, 1 unnumbered figure.

Behavior - Feeding; Behavior - Species Interaction: Killer Whale; Conservation; Food; Pacific - Eastern North: Canadian Waters

126. ---------
1978. Introducing *Megaptera novaeangliae*: The humpback whale. Waters - Journal of the Vancouver Aquarium. 3(4th quarter): 4-8, 8 unnumbered figures, 1 reference.

Basic information on the humpback whale

127. Dunstan, D.J.
1957, October. Caudal presentation at birth of a humpback whale, *Megaptera nodosa* (Bonnaterre). Norsk Hvalfangsttid., 46(10): 553-555, 2 figures, 4 references. (English and Norwegian)

Anatomy - Reproductive; Fetus; Photographs - Fetus

128.* Earle, S.A.
1979. (Abstract) Feeding behavior of humpback whales. Abstracts of papers from the proceedings of the 14th Pacific Science Congress, Khabarovsk, USSR, committee F, section FIII, p. 129.

Behavior - Feeding

Source: Kawamura, A. A review of food of Balaenopterid whales. Scient. Repts. Whales

Res. Inst., Tokyo, No. 32 (December, 1980), p. 155-197.

129. ---------
1979, January. Humpbacks: The gentle whales. Natn. geogr. Mag., 155(1): 2-17, 10 unnumbered figures.

Behavior - Feeding; Behavior - Species Interaction: Bottlenose Dolphin, Fish, Pilot Whale, Spinner Dolphin; Behavior - Surface Dependent: Breach; Conservation; Photographs - Barnacles, Breach, Feeding, Underwater View; Pacific - North: Hawaiian Waters; Pacific - Eastern North: Alaskan Waters

130. ---------
1979, September-October. Underwater encounters with whales. Pacific Discovery, 32(5): 1-9, 11 unnumbered figures.

Behavior - Acoustic: Vocalization; Behavior - Feeding; Behavior - Locomotion; Behavior - Reproductive; Behavior - Surface Dependent: Breach; Individual Identification; Photographs - Breach, Feeding, Underwater View; Pacific - North: Hawaiian Waters

131. ---------
1979. (Abstract) Quantitative sampling of krill (*Euphausia pacifica*) relative to feeding strategies of humpback whales (*Megaptera novaeangliae*) in Glacier Bay, Alaska. Seattle: Abstracts from presentations at the third biennial conference of the biology of marine mammals, October 7-11, p. 15.

Behavior - Feeding; Food; Pacific - Eastern North: Alaskan Waters

132. Edel, R.K. and H.E. Winn.
1978. Observations on underwater locomotion and flipper movement of the humpback whale *Megaptera novaeangliae*. Marine Biology, 48(3): 279-287, 4 figures, 21 references.

BA 1979 v. 67 (#26016)

Behavior - Feeding; Behavior - Locomotion; Entrapments - Net; Physiology - Thermo-

regulation; Atlantic - Western North:
Caribbean, Canadian Waters

Abstract of this paper appeared in:
Proceedings (abstracts) of the second
conference on the biology of marine mammals,
December 12-15, 1977, San Diego, p. 64.

133. Engle, E.T.
    1927, February. Notes on the sexual cycle of the
    Pacific cetacea of the genera *Megaptera* and
    Balaenoptera. J. Mammal., 8(1): 48-51, 1
    table, 1 reference.

    BA 1927 (#11847)

    Anatomy - Reproductive; Anatomy - Skeletal;
    Pacific - Eastern North: U.S. Waters

134. Erdman, D.S.; J. Harms; and M.M. Flores.
    1973, June 30. Cetacean records from the
    Northeastern Caribbean region. Cetology, No.
    17: 14p., 5 tables, 12 references.

    BA 1974 v. 57 (#24424)

    Distribution; Atlantic - Western North:
    Caribbean Waters

135. Forestell, P.H.; R. Antinoja; and L. Herman.
    1977. (Abstract) Organization and behaviors of
    humpback whales as a function of pod size.
    San Diego: Proceedings (abstracts) of the
    second conference on the biology of marine
    mammals, December 12-15, p. 27.

    Behavior - Species Interaction: Bird, Bottle-
    nose Dolphin, Fish, Tiger Shark; Behavior -
    Surface Dependent: Breach, Lobtail; Popu-
    lation Dynamics; Pacific - North: Hawaiian
    Waters

136. --------- and L.M. Herman.
    1979, January. (Abstract) Observations on
    humpback whale calves (*Megaptera novae-*
    angliae) in the Hawaiian winter assembly
    area. Pacif. Sci., 33(1): 120.

    Population Dynamics; Stock Assessment;
    Pacific - North: Hawaiian Waters

Paper delivered at the Third Annual Albert L. Tester Memorial Symposium, University of Hawaii, Honolulu, April 13-14, 1978.

137.   --------- ---------

1979. (Abstract) Behavior of "escorts" accompanying mother-calf pairs of humpback whales. Seattle: Abstracts from presentations at the third biennial conference of the biology of marine mammals, October 7-11, p. 20.

Behavior - Agonistic; Behavior - Cow/ Calf/Escort; Pacific - North: Hawaiian Waters

138.   Fruchtman, P.

1978, November-December. Cetacean encounter. Oceans, 11(6): 52-55, 4 unnumbered figures.

Behavior - Species Interaction: Narwhal; Entrapments - Ice; Atlantic - Western North: Canadian Waters

139.   Fujino, K.

1953, June. On the serological constitution of sei, fin, blue, and humpback whales. Scient. Repts. Whales Res. Inst., Tokyo, No. 8: 103-124, 13 tables, 56 references.

Anatomy - Circulatory; Biochemical Studies

140.   ---------

1960, November. Immunogenetic and marking approaches to identifying subpopulations of the North Pacific whales. Scient. Repts. Whales Res. Inst., Tokyo, No. 15: 85-142, 46 tables, 1 unnumbered appendix table, 9 figures, 50 references.

Anatomy - Circulatory; Biochemical Studies

141.   Gates, D.J.

1956, July. Whaling may yield over £2 million. Fish. Newsl., 15(7): 25.

Whaling - Catch Statistics; Whaling - Industry Statistics; Indian - Eastern South: Australian Waters; Pacific - Western South: Australian Waters

142. ----------
     1961, February. Australia's 1960 whaling season.
     Fish. Newsl., 20(2): 13, 15, 1 unnumbered
     table, 2 unnumbered figures.

     Tagging/Marking, Whaling - Catch Statistics;
     Whaling - Industry Statistics; Indian -
     Eastern South: Australian Waters; Pacific -
     Western South: Australian Waters

143. ----------
     1963, May. Australian whaling since war. Norsk
     Hvalfangsttid., 52(5): 123-127, 3 unnumbered
     tables.

     Whaling - Catch Statistics; Whaling -
     Industry Statistics; Whaling - Shore Stations;
     Indian - Eastern South: Australian Waters;
     Pacific - Western South: Australian Waters

     Also published in: Fish. Newsl., 22:2
     (February, 1963), p. 21-25.

144. ----------
     1964, February. Value of whale products falls.
     Fish. Newsl., 23(2): 16-17, 1 unnumbered
     table.

     Whaling - Industry Statistics

145. Gersh, I.
     1938, November 14. Note on the pineal gland of
     the humpback whale. J. Mammal., 19(4): 477-
     480, 1 unnumered table, 5 figures.

     Anatomy -Endocrine

146. Gervais, H.P.
     1883. Sur une nouvelle espèce du genre *Mégaptère*
     provenant de la baie de Bassora (Golfe
     Persique). (On a species of the genus
     Megaptera from the Bassora Bay 'Persian
     Gulf') C.r. Hebd. Séance. Acad. Sci., Paris,
     97(27): 1566-1569. (French)

     Anatomy - Skeletal; Classification; Distri-
     bution; Indian - Western North: Persian Gulf

     *Paper entitled: Sur une nouvelle espèce de
     Mégaptèra (*Megaptera indica*) provenant du
     Golfe Persique. (On a new species of

Megaptera '*Megaptera indica*' found in the
Persian Gulf) published in Nouv. Archs. Mus.
Hist. nat. Paris (1888), p. 199-218 + 3 plates.
(French)

Source: ZR 1889

147.   Glockner, D.
         1978. The winter whales of Hawaii - part two:
         Underwater with mothers and calves. Waters
         - Journal of the Vancouver Aquarium, 3(4th
         quarter): 19-25, 12 unnumbered figures, 6
         references.

         Anatomy - External: Coloration; Behavior -
         Cow/Calf/Escort; Individual Identification;
         Photographs - Underwater View; Pacific -
         North: Hawaiian Waters

148.   --------- and S.C. Venus.
         1979. (Abstract) Humpback whale, *Megaptera
         novaeangliae*, cows with calves identified off
         West Maui, Hawaii, 1977-79. Seattle:
         Abstracts from presentations at the third
         biennial conference of the biology of marine
         mammals, October 7-11, p. 24.

         Anatomy - External: Coloration; Individual
         Identification; Pacific - North: Hawaiian
         Waters

149.   Goodale, D.R.; H.E. Winn; et al.
         1980. (Abstract) Distribution of calves on
         Mouchoir, Silver and Navidad Banks. Boston:
         Humpback whales of the Western North
         Atlantic Workshop, November 17-21.

         Distribution; Population Dynamics; Atlantic -
         Western North: Caribbean Waters

150.   Goodall, T.B.
         1913, June 15. With the whalers at Durban, and a
         few notes on the anatomy of the humpback
         whale (*Megaptera boops*). Zoologist, 4th
         series, 17(864): 201-211 + 1 plate (2
         figures); 1 unnumbered figure, 2 references.

         Anatomy - Baleen; Anatomy - External: Col-
         oration; Anatomy - Eye; Anatomy -
         Reproductive; Parasites - Ectoparasites;
         Parasites - Endoparasites; Photographs - Male

Ventral View; Atlantic - Eastern South: South African Waters

151.   Gray, J.E.
1864, May 24. On the cetacea which have been observed in the seas surrounding the British Isles. Proc. zool. Soc. Lond., No. 14: 195-248, 2 unnumbered tables, 24 figures.

Anatomy - Skeletal; Classification; Atlantic - Eastern North: British Isles Waters

152.   ---------
1874, January. On the bladebones of *Balaena hectori* and *Megaptera novae-zelandiae*. Ann Mag. nat. Hist., 4th series, 13(73, paper 11): 56-58, 1 unnumbered figure.

Anatomy - Skeletal; Classification

153.   ---------
1874, February. On the Bermuda humpbacked whale of Dudley (*Balaena nodosa*, Bonnaterre, *Megaptera americana*, Gray, and *Megaptera bellicosa*, Cope.) Ann. Mag. nat. Hist., 4th series, 13(74): 186.

Anatomy - Skeletal; Behavior - Surface Dependent: Spyhop; Classification; Migration; Atlantic - Western North

154.* Grevé, C.
1904. Fossile und recente wale des Russischen Reichsgebietes. (Fossil and recent whales of the Russian territory of the Reich) Korresp Bl. NaturfVer. Riga, No. 47: 67-76. (German)

Anatomy - Skeletal

Source: Reference #183

155.* Grewingk, C.
1886. Neve funde subfossiler wirbeltierreste unserer provinzen. (New find of subfossil vertebrate in our province) Sber. naturf. Ges. Dor., 8: 143-144. (German)

Source: Reference #183

156.   Grzimek, R.
1980. Bei den buckelwalen um Newfundland. (On

the humpbacks of Newfoundland) Tier, 20(2): 7-13. (German)

Behavior - Locomotion; Behavior - Surface Dependent: Breach; Entrapments - Net; Photographs - Breach; Atlantic - Western North: Canadian Waters

157.   Guerin, C. and J.-C. Rage.
1970, June. Sur deux os de jubarte (Mammalia, Mysticeti). (Concerning two whale bones 'Mammalia, Mysticeti') Bull. Mems. Soc. Linn. Lyon, 39(6): 185-196, 1 table, 3 figures, 20 references. (French)

BA 1971 (#35940)

Anatomy - Skeletal

158.   Guldberg, G.A.
1903, December 15. Über die wanderungen verschiedener bartenwale. (On the various migration routes of the baleen whales) Biol. Zbl., 23(24): 803-816, 7 references. Continued in 1904, June 1. 24(11): 371-384, 2 references. (German)

Distribution; Migration

Part 2 contains humpback whale information

159.   ---------
1908. Eine missbildung bei den cetacean. (Deformity among the cetacea) Skr. Vidensk-Selsk. Christiania, No. 12: 1-7 + 1 plate; 5 references. (German)

Fetus

160.* Gustavson, H.
1967. (How a whale was caught in the Gulf of Finland) Eesti Loodus, 10: 173-174, 1 figure (Estonian, English summary)

Source: ZR 1967

Atlantic - Eastern North: Scandanavian Waters

161.   Haast, J. von.
1882. Notes on the skeleton of *Megaptera ialandii*

(*novaezealandiae*), Gray. Trans. Proc. N.Z. Inst., 15: 214-216, 2 unnumbered figures, 1 reference.

Anatomy - Skeletal; Classification

162. Hafner, G.W.; C.L. Hamilton; W.W. Steiner; T.J. Thompson; and H.E. Winn.
       1979, July. Signature information in the song of the humpback whale. J. acoust. Soc. Am., 66(1): 1-6, 5 tables, 3 figures, 14 references.

       Behavior - Acoustic: Vocalization; Atlantic - Western North: Caribbean Waters

       Abstract of this paper appeared in: Proceedings (abstracts) of the second conference on the biology of marine mammals, San Diego, December 12-15, 1977, p. 35.

162A. Hain, J.H.W.; G.R. Carter; S.D. Kraus; C.A. Mayo; and H.E. Winn.
       1980. (Abstract) Feeding behavior of the humpback whale, *Megaptera novaeangliae*, in the continental shelf waters of the Northeastern United States. Boston: Humpback whales of the Western North Atlantic Workshop, November 17-21.

       Behavior - Feeding; Food; Atlantic - Western North: U.S. Waters

163. Haldane, R.C.
       1905, April. Notes on whaling in Shetland, 1904. Annls. Scot. nat. Hist., No. 54: 65-72 + 1 plate; 1 unnumbered figure.

       Anatomy - External; Coloration; Behavior - Object Interaction: Rocks; Parasites - Ectoparasites; Atlantic - Eastern North: British Isles Waters

164. Hall, J.D.
       1979, December. A survey of cetaceans of Prince William Sound and adjacent vicinity - their numbers and seasonal movements. *In*: Environmental assessment of the Alaskan Continental Shelf - final report of principal investigators, Vol. 6. Biological studies. Boulder, CO: Outer Continental Shelf Environmental Assessment Program, Depart-

ment of Commerce, Department of Interior, p. 631-726, 3 appendices, 3 tables, 7 figures, 17 references.

Anatomy - External: Coloration; Distribution; Individual Identification; Photographs - Fluke Coloration Patterns; Tagging/Marking; Pacific - Eastern North: Alaskan Waters

165. Hanström, B.
        1944. Zur histologie und vergleichden anatomie der hypopjyse der cetaceen. (On the histology and comparative anatomy of the hypophysis of the cetacean) Acta Zool., 25(1-3): 1-25 + 1 plate; 15 figures, 35 references. (German)

        BA 1947 (#16216)

        Anatomy - Endocrine

        Translation avilable from: Central Library, C.S.I.R.O., P.O. Box 89, E. Melbourne, Victoria 3002, Australia. (Translation #2012)

166. Harboe, A. and A. Schrumpf.
        1952, August. The red blood cell diameter in blue and humpback whale. Norsk Hvalfangsttid., 41(8): 416-418, 3 references. (English and Norwegian)

        Anatomy - Circulatory

167. Hasegawa, Y. and Y. Matsushima.
        1968. Fossil vertebrae of humpback whale from alluvial deposits in Yokohama City. Bull. Kanagana Prefect. Mus. Nat. Sci., 1: 29-36 + 1 plate; 1 table, 1 figure, 1 unnumbered figure. (Japanese summary)

        Anatomy - Skeletal

168. Hector, J.
        1875, July. Notes on New Zealand whales. Trans. Proc. N.Z. Inst., 7(paper 35): 251-265 + 3 plates (23 figures); 2 references.

        Anatomy - Baleen; Anatomy - Skeletal; Pacific - Western South: New Zealand Waters

169.* Herman, L.M.
        1976. (Abstract) Observations on the humpback

whale population of Hawaii. Boulder, CO: Abstracts of papers presented at the Animal Behavior Society meetings, June, abstract #13.

Pacific - North: Hawaiian Waters

Source: Animal Behavior Society, pers. comm., 1981

170. ---------- and P. Forestell.
1977. (Abstract) The Hawaiian humpback whale: Behaviors. San Diego: Proceedings (abstracts) of the second conference on the biology of marine mammals, December 12-15, p. 29.

Behavior - Reproductive; Behavior - Surface Dependent: Breach, Surface Active; Pacific - North: Hawaiian Waters

171. ---------- and R.C. Antinoja.
1977, December. Humpback whales in the Hawaiian breeding waters: Population and pod characteristics. Scient. Repts. Whales Res. Inst., Tokyo, No. 29: 59-85, 8 tables, 4 figures, 59 references.

Anatomy - External: Coloration; Behavior - Cow/Calf/Escort; Behavior - Locomotion; Behavior - Species Interaction: Killer Whale; Distribution; Population Dynamics; Stock Assessment; Pacific - North: Hawaiian Waters

Abstract of this paper appeared in: Report on a workshop on problems related to humpback whales (*Megaptera novaeangliae*) in Hawaii. K.S. Norris and R.R. Reeves (eds.), Springfield, VA: NTIS, PB-280 794, Report No. MMC-77/03, April, 1978, appendix 6, p. 54.

172. ----------
1978, April. (Abstract) Humpback whales in the Hawaiian breeding waters. II. Behaviors. *In*: Report on a workshop on problems related to humpback whales (*Megaptera novaeangliae*) in Hawaii. K.S. Norris and R.R. Reeves (eds.), Springfield, VA: NTIS, PB-280 794, Report No. MMC-77/03, appendix 13, p. 79.

Behavior - Cow/Calf/Escort; Behavior - Reproductive; Behavior - Surface Dependent: Breach; Pacific - North: Hawaiian Waters

173. ---------
1979, January. Humpback whales in Hawaiian waters: A study in historical ecology. Pacif. Sci., 33(1): 1-15, 2 figures, 58 references.

Behavior - Surface Dependent: Breach, Flippering, Lobtail; Distribution; Harassment; Oceanography; Whaling - Shore Stations; Pacific - North: Hawaiian Waters

Preliminary paper appeared in: Report on a workshop on problems related to humpback whales (*Megaptera novaeangliae*) in Hawaii. K.S. Norris and R.R. Reeves (eds.), Springfield, VA: NTIS, PB-280 794, Report No. MMC-77/03, April, 1978, appendix 3, p. 25-28.

174. ---------; R.C. Antinoja; C.S. Baker; and R.S. Wells. 1979. (Abstract) Temporal and spacial distribution of humpback whales in Hawaii. Seattle: Abstracts from presentations at the third biennial conference on the biology of marine mammals, October 7-11, 1979, p. 28.

Distribution; Stock Assessment; Pacific - North: Hawaiian Waters

175. ---------; P.H. Forestall; and R.C. Antinoja. 1980, April. The 1976/77 migration of humpback whales into Hawaiian waters: Composite description. Springfield, VA: NTIS, PB80-162332, 55p., 10 tables, 3 appendix tables, 13 figures, 4 appendix figures, 8 references.

Conservation; Harassment; Migration; Population Dynamics; Stock Assessment; Pacific - North: Hawaiian Waters

176. ---------; C.S. Baker; P.H. Forestall; and R.C. Antinoja. 1980, May 31. Right whale *Balaena glacialis* sightings near Hawaii: A clue to the wintering grounds? Mar. Ecol. Prog. Ser., 2: 271-275, 2 figures, 24 references.

Anatomy - External: Callosities, Coloration; Behavior - Acoustic: Vocalization; Behavior - Species Interaction: Bottlenose Dolphin, Right Whale; Behavior - Surface Dependent: Flippering, Surface Active; Distribution; Photographs - Callosities, Flukes; Pacific - North: Hawaiian Waters

Preliminary report appeared as: Baker, C.S.; P.H. Forestall; R.C. Antinoja; and L.M. Herman. (Abstract) Interactions of the Hawaiian humpback whale, *Megaptera novae-angliae*, with the right whale, *Balaena glacialis*, and odontocete cetaceans. Seattle: Abstracts from presentations at the third biennial conference of the biology of marine mammals, October 7-11, 1979, p. 2.

177. Honigmann, H.
      1915. Das primordialkranium von *Megaptera nodosa* Bonnat. (The primordial cranium of *Megaptera nodosa* Bonnat.) Anat. Anz. Jena, 48: 113-127 + 3 figures. (German)

      Anatomy - Skeletal; Photographs - Primordial Cranium

178. ----------
      1917. Bau und entwicklung des knorpelschadels des vom buckelwal. (Structure and evolution of the skull cartilage in the humpback whale) Zoologica, Stuttg., 27(69): 85p. + 2 plates (4 figures); 28 figures, 70 references. (German)

      Anatomy - Skeletal

179. Hubbs, C.L.
      1965, November. Data on speed and underwater exhalation of a humpback whale accompanying ships. Hvalrad. Skr., No. 48: 42-44, 2 figures.

      Anatomy - External: Coloration; Behavior - Locomotion; Behavior - Lone Whale; Behavior - Respiratory; Pacific - Eastern North: Mexican Waters

180. Hudnall, J.
      1977, May. In the company of great whales. Audubon, 79(3): 62-73, 10 unnumbered figures.

Behavior - Surface Dependent: Breach, Sur-
face Active; Harassment; Photographs -
Breach, Surface Active Behavior, Underwater
Bubble Blow, Underwater View; Pacific -
North: Hawaiian Waters

181.   ---------
1978, March. A report on the general behavior of
humpback whales near Hawaii, and the need
for the creation of a whale park. Oceans,
11(3): 8-15, 12 unnumbered figures.

Anatomy - External: Coloration; Behavior -
Reproductive;   Conservation;   Harassment;
Indi-vidual   Identification;   Photographs   -
Under-water View; Pacific - North: Hawaiian
Waters

182.   Ichihara, T.
1959, September. Formation mechanism of the ear
plug in baleen whales in relation to the
glove-finger. Scient. Repts. Whales Res.
Inst., Tokyo, No. 14: 107-135 + 9 plates (18
figures); 10 figures, 33 references.

Anatomy - Ear

183.   ----------
1964, March. Prenatal development of the ear plug
in baleen whales. Scient. Repts. Whales Res.
Inst., Tokyo, No. 18: 29-48 + 5 plates (13
figures); 10 figures, 17 references.

Anatomy - Ear; Fetus

184.   Ingebrigtsen, A.
1929, May. Whales caught in the North Atlantic
and other seas. *In*: Whales and plankton in
the North Atlantic - a contribution to the
work of the whaling committee and of the
North Eastern Area Committee. Rapp. P.-v
Reun. Cons. perm. int. Explor. Mer., 56: 26p.,
5 unnumbered tables, 4 references.

Behavior   -   Feeding;   Distribution;   Fetus;
Migration; Atlantic - North

185.*   Ivashin, M.V.
1957. (Method of determination of the traces of
corpora lutea in pregnancy and ovulation in
humpback whales) Trudȳ vese. nauchno-issled.

Inst. morsk. rȳb. Khoz. Okeanogr., 33: 161-172, 5 tables, 5 figures, 13 references. (Russian)

ArB 1965 (#72485)

Anatomy - Reproductive

186.* ---------
1958. (The method of determination of the yellow marks in the flesh in pregnancy and ovulation in the humpback whale) *In*: (Biology and hunting of marine mammals) B.A. Zenkovich (ed.), Moscow: (All-Union Scientific Exploration Institute of Marine Fishery, Economics and Oceanography), 33: 161-172. (Russian)

Anatomy - Reproductive

Source: Card catalog of J.G. Mead's collection. Further information from: J.G. Mead, pers. comm., 1981

187. ---------
1958. O sistematitseskon polozhenii gorbatogo kita (*Megaptera nodosa islandii* Fischer) yuzhnogo polushariya. (On the systematic position of the humpback whale '*Megaptera nodosa islandii* Fischer' of the southern hemisphere) Byulleteni Sovetskoĭ Antarkticheskoĭ Zkspedidii, No. 3: 77. (Russian)

Anatomy - External: Coloration; Classification

Translated in: Soviet Antarctic Expedition Information Bulletin, 1 (1964), p. 145-146

188.* ---------
1959. (Breeding of the humpback whale in the southern hemisphere) Informatsionnyi sbornik VNIRO, No. 7.

Source: Gilbert, J.R.; T.J. Quinn II; and L.L. Eberhardt. An annotated bibliography of census procedures for marine mammals, September, 1976, 160p.

189. ---------
1960. O mnogoplodii, urodstvakh razvitiya i zmbrional' noĭ smertnosti u kitov. (On

multiparity and abnormalities of the development and embryo mortality in whales) Zool. Zh., 39(5): 755–758, 2 figures, 6 references. (Russian, English summary)

Fetus

190.   ----------

1961. O periodichnosti pitanya gorbatogo kita v yuzhnoi chasti Atlanticheskogo Okeana. (On the periodicity of feeding of the hunched whale in the southern part of the Atlantic Ocean) Byull. mosk. Obshch. Ispȳt. Prir. (Biology), 66(6): 110–115, 1 figure, 11 references. (Russian, English summary)

BA 1963 (#21039)

Behavior - Feeding; Antarctic (2,3)

Translation available from: JPRS #63-18408 - specify Ivashin

191.   ----------

1962. Mechenie gorbatykh kitov v yuzhnom polusharii. (Marked humpback whales in the southern hemisphere) Zool. Zh., 41(12): 1848–1858, 2 tables, 4 figures, 30 references. (Russian, English summary)

Behavior - Locomotion; Distribution; Migration; Tagging/Marking; Antarctic (4,5); Indian - Eastern South: Australian Waters; Pacific - Western South: Australian Waters

192.*   ----------

1971. (Cases of multiple pregnancy and deformities of embryos of mysticetes.) Trudȳ Atlant. nauchno-issled. Inst. rȳb. Khoz. Okeanogr., 39:75–84.

BA 1973 (#59782)

Anatomy - Reproductive; Fetus

193.   ----------

1971. Nekotorye rezol'taty mecheniya kitov, provedennogo s Sovetskikh sudov v yuzhnom polusharii. (Some results of whale marking carried out from Soviet ships in the southern hemisphere) Zool. Zh., 50(7): 1063–1078, 6 tables, 5 figures, 19 references. (Russian, English summary)

BA 1972 v. 53 (#24722)

Migration; Tagging/Marking; Antarctic

194. ----------
1973. Marking of whales in the Southern Hemisphere (Soviet materials). Rep. int. Whal. Commn 23, London, Appendix IV, Annex Q, p. 174-191, 5 tables, 1 appendix table, 6 references.

Tagging/Marking; Antarctic (4,5); Indian - Eastern South: Australian Waters; Pacific - Western South: Australian Waters

195. Japha, A.
1908. Zusammenstellung der in der ostsee bisher beobacheten wale. (Summary of the whales observed in the Baltic Sea) Schr. physokon. Ges. Konigsb., 49: 119-189, 1 unnumbered table, 119 references. (German)

Distribution; Atlantic - Eastern North: Baltic Sea

196. Jonsgård, Å.
1957, January. Some comments on W. Dall and B. Dunstan's article on the find of *Euphausia superba* in a humpback whale caught near South Queensland. Norsk Hvalfngsttid., 46(1): 10-12. (English and Norwegian)

Behavior - Feeding; Behavior - Locomotion; Food; Pacific - Western South: Australian Waters

197. ----------; J.T. Ruud; and P. Oynes.
1957, April. Is it desirable and justified to extend the open season for humpback whaling in the Antarctic? Norsk Hvalfangsttid., 46(4): 160-177, 10 tables, 5 references. (English and Norwegian)

Tagging/Marking; Whaling - Catch Statistics; Antarctic (1-5); Indian - Eastern South: Australian Waters; Pacific - Western South: Australian Waters

198. Jurasz, C.M. and V. Jurasz.
        1975. (Abstract) Bubble net feeding of the
        humpback. Santa Cruz: Conference on the
        biology and conservation of marine mammals
        abstracts, December 4-7, p. 44.

        Behavior - Feeding; Pacific - Eastern North:
        Alaskan Waters

199. --------- ---------
        1977. (Abstract) Censusing of humpback whales,
        Megaptera novaeangliae, by body charac-
        teristics. San Diego: Proceedings (abstracts)
        of the second conference on the biology of
        marine mammals, December 12-15, p. 54.

        Anatomy - External: Coloration; Individual
        Identification; Stock Assessment; Pacific -
        Eastern North: Alaskan Waters

200. --------- ---------
        1977. (Abstract) Vessel and humpback whale
        Megaptera novaeangliae, interactions. San
        Diego: Proceedings (abstracts) of the second
        conference on the biology of marine mammals,
        December 12-15, p. 64.

        Behavior - Locomotion; Behavior - Object
        Interaction: Boat; Behavior - Respiratory;
        Harassment; Pacific - Eastern North: Alaskan
        Waters

201. --------- ---------
        1978. Humpback whales in Southeastern Alaska.
        Alaska Geographic, 5(4): 116-127, 17
        unnumbered figures.

        Anatomy - External: Coloration; Behavior -
        Feeding; Behavior - Sleeping; Behavior -
        Surface Dependent: Fluke Up; Individual
        Identification; Photographs - Feeding, Fluke
        Coloration Pattern; Pacific - Eastern North:
        Alaskan Waters

202. --------- ---------
        1979, December. Feeding modes of the humpback
        whale, *Megaptera novaeangliae*, in Southeast
        Alaska. Scient. Repts. Whales Res. Inst.,
        Tokyo, No. 31: 69-83, 11 figures, 7
        references.

> Behavior - Feeding; Behavior - Species
> Interaction: Gull; Food; Photographs -
> Feeding; Pacific - Eastern North: Alaskan
> Waters

203. Katona, S.; B. Baxter; O. Brazer; S. Kraus; J. Perkins;
     and H. Whitehead.
     > 1979. Identification of humpback whales by fluke
     > photographs. *In*: Behavior of marine animals -
     > current perspectives in research, vol. 3:
     > Cetaceans. H.E. Winn and B.L. Olla (eds.),
     > New York: Plenum Press, p. 33-44, 28 figures,
     > 25 references.
     >
     > Anatomy - External: Coloration; Individual
     > Identification; Photographs - Fluke Coloration
     > Patterns; Atlantic - Western North

204. ---------- and S. Kraus.
     > 1979, August. Photographic identification of
     > individual humpback whales (*Megaptera
     > novae-angliae*): Evaluation and analysis of
     > the technique. Springfield, VA: NTIS PB-298
     > 740, Report No. MMC-77/17, 20p., 2 tables, 4
     > appendices, 10 references.
     >
     > Anatomy - External: Coloration; Individual
     > Identification; Photographs - Fluke Coloration
     > Patterns
     >
     > *Abstract entitled Photo-identification of
     > humpback whales appeared in Rep. int. Whal.
     > Commn 29, Cambridge, 1979, p. 139
     >
     > Source: WR September, 1980

205. ----------; P. Harcourt; J. Perkins; and S. Kraus.
     > In press. Humpback whales in the Western North
     > Atlantic Ocean: A fluke catalogue of
     > individuals identified by means of fluke
     > photographs, 2nd edition. Bar Harbor, ME:
     > College of the Atlantic.
     >
     > Anatomy - External: Coloration; Individual
     > Identification; Photographs - Fluke Coloration
     > Patterns; Atlantic - Western North
     >
     > Source: S. Katona, pers. comm., 1980

206. Kawakami, T.
     > 1958, February. (On the tagged whales caught off

Okinawa) Geiken Tsushin, No. 77: 413-414, 1
figure. (Japanese)

Distribution; Migration; Tagging/Marking; Pacific - Western North: Japanese Waters

207.   --------- and T. Ichihara.
1958, June. Japanese whale marking in the North
Pacific in 1956 and 1958. Norsk Hvalfangsttid., 47(6): 285-291, 3 tables, 1 figure,
2 references. (English and Norwegian)

Migration; Tagging/Marking; Pacific - North:
Bering Sea; Pacific - Western North:
Japanese Waters

208.   Kellogg, R.
1928, June 30. What is known of the migrations of
some of the whalebone whales. Annual Report
of the Smithsonian Institution, 467-494 + 2
plates; 6 figures, 46 references.

BA 1931(#22371)

Distribution; Migration; Photographs - Male
Ventral View

209.   Kenny, R.D.; D.R. Goodale; G.P. Scott; and H.E. Winn.
1980. (Abstract) Spatial and temporal distribution
of humpback whales in the CETAP study area.
Boston: Humpback whales of the Western
North Atlantic Workshop, November 17-21.

Distribution; Atlantic - Western North

210.   Klima, N.
1978, December. Comparison of early development
of sternum and clavicle in striped dolphin and
in humpback whale. Scient. Repts. Whales
Res. Inst., Tokyo, No. 30: 253-269, 9 figures,
38 references.

Anatomy - Skeletal; Evolution; Fetus;
Photographs - Embryonic Sternum

211.   Kraus, S. and S. Katona. (eds.)
1977. Humpback whales (*Megaptera novaeangliae*)
in the Western North Atlantic - a catalogue
of identified individuals. Bar Harbor, ME:
College of the Atlantic, 26p., numerous

unnumbered figures, 3 appendices, 2 references.

Anatomy - External: Coloration; Individual Identification; Photographs - Dorsal ID, Fluke and Flipper Coloration Patterns; Atlantic - Western North

212. Kükenthal, W.
1914. Die entwicklung der auberen korperform der bartenwale. (The development of the external body form of the humpback whale) *In:* Untersuchungen au walen (Zweiter Teil.) (Investigation of the whale 'second part') Jena. Z. Naturw., 51: 2-72 + 3 plates (26 figures); 10 unnumbered tables, 7 figures, 37 references. (German)

Fetus

213. ---------
1921, July 14. Die brustflosse des buckelwales (*Megaptera nodosa* Bonnat.) und ihre entwicklung. (The pectoral fin of humpbacks '*Megaptera nodosa* Bonnat.' and their development) Sbet. preuss. Akad. Wiss., 36: 568-588, 12 figures, 8 references. (German)

Anatomy - Flippers/Flukes; Anatomy - Skeletal; Evolution

214. Kulikov, A.N. and M.V. Ivashin.
1959. K voprosu o tsikle razmnozheniya finvalon (*Balaenoptera physalus* L.) i gorbatykh kitov (*Megaptera nodosa* Bonn.) Atlantocheskogo sektora Antarktiki. (On the problems of the reproduction cycle of the fin whale '*Balaenoptera physalus* L.' and the humpback whale '*Megaptera nodosa* Bonn.' of the Atlantic sector of the Antarctic) Zool. Zh., 38(1): 123-125, 2 figures, 9 references. (Russian, English summary)

Anatomy - Reproductive; Behavior - Reproductive; Antarctic (2,3)

215. Lawton, W.; D. Rice; A. Wolman; and H. Winn.
1979. (Abstract) Occurrence of Southeastern Alaskan humpback whales, *Megaptera novaeangliae*, in Mexican coastal waters. Seattle: Abstracts from presentations at the third

biennial conference of the biology of marine mammals, October 7-11, p. 35.

Anatomy - External: Coloration; Distribution; Individual Identification; Migration; Pacific - Eastern North: Alaskan, Mexican Waters

216.   Lehman, L.D.; F.E. Dwulet; B.N. Jones; R.A. Bogardt, Jr.; S.T. Krueckeberg; R.B. Visscher; and F.R.N. Gurd.
1979, September 5. Complete amino acid sequence of the major component myoglobin from the humpback whale, *Megaptera novaeangliae.* Biochemistry, 17(18):3736-3739, 1 table, 4 figures, 28 references.

BA 1979 v. 67 (#36-76)

Biochemical Studies

217.   Lesch, P. and K. Bernhard.
1966, July 11. Zur kenntnis der lipide aus dem gehirn eines buckelwals (*Megaptera novae-angliae* Borowsky). (Information on the lipids from the brain of the humpback whale '*Megaptera novaeangliae* Borowsky') Helv Chim Acta, 49(5): 1607-1611, 6 tables, 13 references. (German, English summary)

BA 1967 (#38336)

Anatomy - Nervous; Biochemical Studies

218.   Levenson, C.
1969, May. Behavioral, physical, and acoustic characteristics of humpback whales (*Meg-aptera novaeangliae*) at Argus Island. Washington, D.C.: Naval Oceanographic Office, Informal Report No 69-54, 13p. 6 figures, 3 references.

Behavior - Acoustic: Vocalization; Behavior - Diving; Behavior - Respiratory; Photographs - Dorsal Fins and Fluke; Atlantic - Western North: Bermuda Waters

219.   ----------
1970, June. (Letter) Speed of a humpback whale observed from an oceanographic aircraft. Deep Sea Res., 17(3): 675-676 + 1 figure; 1 figure, 3 references.

Behavior - Locomotion; Atlantic - Western North: Bermuda Waters

220. ---------
1972, October 25. Characteristics of sounds produced by humpback whales (*Megaptera novaeangliae*). Washington, D.C.: Naval Ocean-ographic Office, NAVOCEANO Technical Note No. 7700-6-72, 10p. + 9 figures; 13 references.

Acoustics - Ambient Noise Levels; Behavior - Acoustic: Echolocation, Vocalization; Behavior - Locomotion; Atlantic - Western North: Bermuda Waters

221. --------- and W.T. Leapley.
1978, August. Distribution of humpback whales (*Megaptera novaeangliae*) in the Caribbean determined by a rapid acoustic method. J. Fish. Res. Bd Can., 35(8): 1150-1152, 2 figures, 11 references. (French and English abstracts)

BA 1979 v. 67 (#20905)

Behavior - Acoustic: Vocalization; Distribution; Atlantic - Western North: Caribbean Waters

*Preliminary paper appeared as: Humpback whale distribution in the Eastern Caribbean determined acoustically from an oceanographic aircraft. Washington, D.C.: Naval Oceanographic Office, NAVOCEANO Technical Note No. 3700-46-76, 1976, 6p., 2 figures, 9 references.

Source: Reference #373

222. ---------
1978, Fall. (Abstract) Source level and bistatic target strength of the humpback whale (*Megaptera novaeangliae*) measured from an oceanographic aircraft. J. acoust. Soc. Am., 64(supp. 1): S97, abstract #JJ17.

Behavior - Acoustic: Vocalization

Paper delivered at the Acoustical Society of America and the Acoustical Society of Japan

Joint Meeting, Honolulu, November 27-December 1, 1978.

223. Lien, J. and B. Merdsoy.
     1979, June-July. The humpback is not over the hump. Nat. Hist., 88(6): 46-49, 2 unnumbered figures.

     Entrapments - Net; Atlantic - Western North: Canadian Waters

224. ---------; J. Perkins; B. Merdsoy; and S. Johnson.
     1979. (Abstract) Baleen whales collisions with inshore fishing gear in Newfoundland. Seattle: Abstracts from presentations at the third biennial conference of the biology of marine mammals, October 7-11, p. 37.

     Entrapments - Net; Atlantic - Western North: Canadian Waters

225. ---------; S. Johnson; B. Davis; and S. Gray.
     1980. (Abstract) Humpback whale collisions with inshore fishing gear. Boston: Humpback whales of the Western North Atlantic Workshop, November 17-21.

     Entrapments - Net; Atlantic - Western North: Canadian Waters

226. Lillie, D.G.
     1915. Cetacea. British Antarctic Terra Nova Expedition. Natural History Report on Zoology, 1(3): 85-124 + 8 plates (32 figures); 2 tables, 14 figures, 67 references.

     Anatomy - Ear; Anatomy - External: Coloration, Hair, Scars; Anatomy - Reproductive; Anatomy - Skeletal; Behavior - Locomotion; Behavior - Object Interaction: Rocks; Distribution; Entrapments - Net; Fetus; Hearing - Acoustic Physiology; Migration; Photographs - Ear Bones, Fetus, Ventral Coloration Patterns; Physiology - Lactation; Antarctic (5); Pacific - Western South: New Zealand Waters

227. Love, R.H.
     1973, November. Target strengths of humpback whales *Megaptera novaeangliae*. J. acoust.

Soc. Am., 54(5): 1312-1315, 1 table, 1 figure, 7 references.

BA 1974 v. 58 (#7130)

Behavior - Acoustic: Vocalization; Atlantic - Western North: Bermuda Waters

228. Machin, D. and B.L. Kitchenham.
1971, November 19. A multivariate study of the external measurements of the humpback whale (*Megaptera novaeangliae*). J. zool. Soc. Lond., 165(3): 415-421, 5 tables, 1 figure, 3 references.

BA 1972 v. 53 (#53735)

Measuring Techniques; Population Dynamics

229. Mackintosh, N.A.
1942, June. The southern stocks of whalebone whales. 'Discovery' Rep., 22: 197-300, 30 tables, 9 figures, 78 references.

Distribution; Fetus; Food; Migration; Tagging/ Marking; Whaling - Catch Statistics; Antarctic (1-6); Atlantic - South; Pacific - South

230.* Malm, A.W.
1870. Om några delar af skelette utaf en år 1803 uti Goteborgs skargård strandad Megaptera (med Traesnit). (On a skeleton of a humpback stranded on the outlying rocks of Goteborg in 1803) Goteborgs K. Vetensk.-o. VitterhSamh. Handl., 10: 94-99. (Swedish)

Anatomy - Skeletal; Strandings; Atlantic - Eastern North: Scandinavian Waters

Source: Reference #195

231. Marine Mammal Commission.
1980, February. Humpback whales in Glacier Bay National Monument, Alaska. Springfield, VA: NTIS, PB80-141559, Report No. MMC-79/01, 44p., 3 appendices, 8 tables, 6 figures, 2 references.

Behavior - Acoustic: Vocalization; Behavior - Feeding; Behavior - Surface Dependent:

Breach, Flippering, Lobtail; Distribution; Food; Harassment; Individual Identification; Oceanography; Population Dynamics; Stock Assessment; Tagging/Marking; Pacific - Eastern North: Alaskan Waters

232. Marr, J.

1957, April. Further comments on the occurrence of *Euphausia superba* in a humpback whale caught off Southern Queensland. Norsk Hvalfangsttid., 46(4): 181-182, 3 references. (English and Norwegian)

Behavior - Feeding; Food; Pacific - Western South: Australian Waters

233. Martin, K.R.

1980. (Abstract) Humpback whales and whaling: A look at historical habitats and abundance levels of the humpback whale in the Western North Atlantic. Boston: Humpback whales of the Western North Atlantic Workshop, November 17-21.

Stock Assessment; Whaling - Whaling Material; Atlantic - Western North

234. Mathew, A.P.

1948, August. Stranding of a whale (*Megaptera nodosa*) on the Travancore Coast in 1943. J. Bombay nat. Hist. Soc., 47(4): 732-733.

Entrapments - Net; Strandings; Indian - Western North: Arabian Sea

235. Matsuura, Y.

1935, September. (The distribution and habits of humpback whales in the adjacent waters of Japan) Bull. Jap. Soc. scient. Fish., 4(3): 161-170, 9 tables, 7 figures, 26 references. (Japanese, English summary)

Behavior - Reproductive; Distribution; Fetus; Whaling - Catch Statistics; Pacific - Western North: Japanese Waters

236. ---------

1938, July. (Statistical studies of whale fetuses. II. The humpback whale in the Antarctic) Bull. Jap. Soc. scient. Fish., 7(2): 72-74, 3 tables, 2 figures, 5 references. (Japanese, English summary)

Fetus; Antarctic

237. ---------
1940, July. (The constitution of whale populations in the Antarctic) Bull. Jap. Soc. scient. Fish., 9(2): 51-60, 8 tables, 4 figures, 12 references. (Japanese, English summary)

Age Determination; Anatomy - External: Coloration; Anatomy - Reproductive; Antarctic (4)

238. Matthews, L.H.
1937, December. The humpback whale, *Megaptera nodosa*. 'Discovery' Rep., 17: 7-92 + 1 plate (2 figures); 24 tables, 1 unnumbered appendix table, 84 figures, 25 references.

BA 1939 (#1657)

Age Determination; Anatomy - External: Coloration, Hair, Scars; Anatomy - Reproductive; Distribution; Fetus; Food; Growth Rates; Migration; Parasites - Ectoparasites; Parasites - Endoparasites; Photographs - Coloration Patterns, Female Antero-Ventral View; Whaling - Catch Statistics; Atlantic - Eastern South: South African Waters; Atlantic - Western South: South Georgia Waters; Pacific - Western South: New Zealand Waters

239. Maul, G.E. and D.E. Sergeant.
1977. New cetacean records from Madeira. Boca-giana - Museu Municipal do Funchal, 10(43): 1-8, 1 table, 5 figures, 10 references.

Distribution; Photographs - Skull; Atlantic - Eastern North: Madeira Waters

240. Mayo, C.A.
1980. (Abstract) Observations on humpback whales in Massachusetts Bay and Cape Cod Bay. Boston: Humpback whales of the Western North Atlantic Workshop, November 17-21.

Anatomy - External; Coloration; Distribution; Individual Identification; Population Dynamics; Atlantic - Western North: U.S. Waters

241. Mead, J.G.
1980. (Abstract) *Megaptera novaeangliae* strandings along the Atlantic coast of the United States from 1970-Nov. 1980. Boston: Humpback whales of the Western North Atlantic Workshop, November 17-21.

Entrapments - Net; Strandings; Atlantic - Western North: U.S. Waters

242. Miller, J.A.
1979, January 13. A whale of a song. Science News, 115(2): 26-27, 2 unnumbered figures.

Behavior - Acoustic: Vocalization; Photographs - Underwater View

243. Minasian, S.M.
1977, May-June. Wings in the water. Pacific Discovery, 30(3): 28-29, 3 unnumbered figures.

Behavior - Species Interaction: Spotted Dolphin; Photographs - Underwater View; Pacific - North: Hawaiian Waters

244. Mitchell, E.
1973. Draft report on humpback whales taken under special scientific permit by eastern Canadian land stations, 1969-1971. Rep. int. Whal. Commn 23, London, Appendix IV, Annex M, p. 138-154, 7 tables, 2 figures, 11 references.

Anatomy - Reproductive; Fetus; Growth Rates; Parasites - Ectoparasites; Parasites - Endoparasites; Pollutants; Stock Assessment; Tagging/Marking; Whaling - Shore Stations; Atlantic - Western North: Canadian Waters

245. Mitchell, J.H.
1978, July. Humpback passage. Massachusetts Audubon, 17(10): 4-7, 7 unnumbered figures.

Individual Identification; Photographs - Diving Sequence; Atlantic - Western North: U.S. Waters

246. Mobius, K.
       1893, December. Über den fang und die verwee-
       thung der walfische in Japan. (On the capture
       and utilization of whales in Japan) Math. u.
       naturwiss. Mitth., No. 10: 649-668, 9 figures,
       18 references. (German)

       Anatomy - Skeletal; Parasites - Ecto-
       parasites; Whaling - Whaling Material; Pacific
       - Western North: Japanese Waters

247. Mörch, J.A.
       1911, April 4. On the natural history of whalebone
       whales. Proc. zool. Soc. Lond., 47(paper 30):
       661-670, 4 figures, 1 reference.

       Anatomy - External: Coloration; Distribution;
       Fetus; Migration; Parasites - Ectoparasites;
       Atlantic - Western South: Falkland, South
       Shetland Waters

248. Morrison, P.
       1962, August. Body temperatures in some
       Australian mammals. III. Cetacea (*Meg-
       aptera*). Biol. Bull. Mar. biol. lab., Woods
       Hole, 123(1): 154-169, 9 tables, 6 figures, 36
       references.

       Physiology - Thermoregulation; Pacific -
       Western South: Australian Waters

249.* Myers, R.G.
       1920. A chemical study of whale blood. J. biol.
       Chem., 41: 137-143.

       CA 1920 v. 14 (p. 764)

       Anatomy - Circulatory; Biochemical Studies

250. Nasu, K.
       1974, July. (The story of the humpback whale that
       surfaced in the Satsuma Peninsula) Geiken
       Tsushin, No. 275: 54-57, 4 plates, 2 figures.
       (Japanese)

       Distribution; Pacific - Western North: Jap-
       anese Waters

251. National Marine Fisheries Service (Staff). Western
     Pacific Program Office. Honolulu
       1978, April. Regulations and enforcement relating

to the humpback whale (*Megaptera novae-angliae*) in Hawaii. *In*: Report on a workshop on problems related to humpback whales (*Megaptera novaeangliae*) in Hawaii. K.S. Norris and R.R. Reeves (eds.), Springfield, VA: NTIS, PB-280 794, Report No. MMC-77/03, appendix 12, p. 86-88.

Conservation; Harassment; Pacific - North: Hawaiian Waters

252. Nemoto, T.
1957, June. Foods of baleen whales in the Northern Pacific. Scient. Repts. Whales Res. Inst. Tokyo, No. 12: 33-89, 15 tables, 26 figures, 74 references.

Food; Oceanography; Whaling - Catch Statistics; Pacific - North; Pacific - North: Bering Sea

253. ---------
1959, September. Food of baleen whales with reference to whale movements. Scient. Repts. Whales Res. Inst., Tokyo, No. 14: 149-290 + 1 plate (14 figures); 43 tables, 2 appendix tables, 48 figures, 145 references.

Anatomy - Baleen; Food; Oceanography; Antarctic (1, 4-6); Pacific - North; Pacific - North: Bering Sea; Pacific - Western North: Japanese Waters

254. ---------
1964, March. School of baleen whales in the feeding areas. Scient. Repts. Whales Res. Inst., Tokyo, No. 18: 89-110, 22 tables, 7 figures, 26 references.

Population Dynamics; Whaling - Catch Statistics; Pacific - North

255. ---------
1978, December. Humpback whales observed within the continental shelf waters of the Eastern Bering Sea. Scient. Repts. Whales Res. Inst., Tokyo, No. 30: 245-247, 1 figure, 10 references.

Anatomy - External: Coloration; Behavior - Species Interaction: Kittiwake, Sooty Shear-

water; Distribution; Food; Pacific - North:
Bering Sea

256. Nishiwaki, M. and K. Hayashi.
        1950, February. Copulation of humpback whales.
        In: Biological survey of fin and blue whales
        taken in the Antarctic season 1947-1948 by
        the Japanese fleet October 1948. Scient.
        Repts. Whales Res. Inst., Tokyo, No. 3: 183-
        185, 4 figures.

        Behavior - Reproductive; Behavior - Surface
        Dependent: Mating Activity; Antarctic

257. ---------
        1959, September. Humpback whales in Ryukyuan
        waters. Scient. Repts. Whales Res. Inst.,
        Tokyo, No. 14: 49-87, 6 tables, 9 appendix
        tables, 19 figures, 33 references.

        Age Determination; Anatomy - External:
        Coloration; Anatomy - Circulatory; Anatomy -
        Reproductive; Fetus; Growth Rates; Migra-
        tion; Parasites - Ectoparasites; Parasites -
        Endoparasites; Physiology - Lactation; Physi-
        ology - Ovulation; Whaling - Catch Statistics;
        Pacific - Western North: Japanese Waters

        Preliminary paper appeared as: (Humpback
        whales in Ryukyuan Island waters) Geiken
        Tsushin, No. 93 (June, 1959), p. 677-690, 4
        unnumbered tables, 11 figures. (Japanese)

258. ---------
        1960, November. Ryukyuan humpback whaling in
        1960. Scient. Repts. Whales Res. Inst., Tokyo,
        No. 15: 1-15, 3 tables, 8 figures, 5
        references.

        Age Determination; Anatomy - Circulatory;
        Anatomy - External: Coloration; Anatomy -
        Reproductive; Food; Migration; Parasites -
        Ectoparasites; Tagging/Marking; Whaling -
        Catch Statistics; Pacific - Western North:
        Japanese Waters

        Preliminary paper appeared as: (On the
        results of Ryukyuan whaling in 1960) Geiken
        Tsushin, No. 107 (July, 1960), p. 133-139, 3
        tables, 5 figures. (Japanese)

259.   ----------
1962, March. Ryukyuan whaling in 1961. Scient. Repts. Whales Res. Inst., Tokyo, No. 16: 19-28, 4 tables, 3 figures, 2 references.

Anatomy - Circulatory; Anatomy - External: Coloration; Distribution; Migration; Tagging/ Marking; Whaling - Catch Statistics; Pacific - Western North: Japanese Waters

260.   Norris, K.S. and R.R. Reeves (eds.).
1978, April. Report on a workshop on problems related to humpback whales (*Megaptera novaeangliae*) in Hawaii. Springfield, VA: NTIS, PB-280 794, Report No. MMC-77/03, 90p.

Pacific - North: Hawaiian Waters

See papers and abstracts in this bibliography by: L.M. Herman; National Marine Fisheries Service (Staff); R. Payne; D.W. Rice; E.W. Shallenberger; H.E. Winn; and A.A. Wolman.

261.   Norwegian Delegation.
1949, October. The catch of humpback whales. Norsk Hvalfangsttid., 38(10): 436-437, 439-442, 4 unnumbered tables.

Whaling - Catch Statistics; Whaling - Quotas

262.   Ogawa, T.
1953, June. On the presence and disappearance of the hind limb in cetacean embryos. Scient. Repts. Whales Res. Inst., Tokyo, No. 8: 127-132, 3 figures, 6 references.

Fetus

263.   Ohsumi, S.
1979, July. (Protection of humpback whales in Hawaii) Geiken Tsushin, No. 327: 48-51, 2 figures. (Japanese)

Conservation; Distribution; Pacific - North: Hawaiian Waters

264.   Ommanney, F.D.
1933, December. Whaling in the Dominion of New Zealand. 'Discovery' Rep., 7: 239-252 + 3 plates (10 figures); 1 figure, 10 references.

Distribution; Entrapments - Net; Fetus; Migration; Photographs - Net Entrapments; Whaling - Shore Stations; Pacific - Western South: New Zealand Waters

265.   Omura, H.
1953, June. Biological study on humpback whales in the Antarctic whaling Areas IV and V. Scient. Repts. Whales Res. Inst., Tokyo, No. 8: 81-102, 8 tables, 14 figures, 12 references.

Age Determination; Anatomy - External: Coloration; Anatomy - Reproductive; Distribution; Fetus; Migration; Population Dynamics; Whaling - Catch Statistics; Antarctic (4,5)

266.   ---------
1964, March. A systematic study of the hyoid bones in baleen whales. Scient. Repts. Whales Res. Inst., Tokyo, No. 18: 149-170 + 15 plates (182 figures); 2 tables, 11 figures, 11 references.

Anatomy - Skeletal; Photographs - Ear Bones

267.   --------- and S. Ohsumi.
1964, April. A review of Japanese whale marking in the North Pacific to the end of 1962, with some information on marking in the Antarctic. Norsk Hvalfangsttid., 53(2): 90-112, 12 tables, 11 unnumbered appendix tables, 5 figures, 11 references. (English and Norwegian)

Distribution; Migration; Tagging/Marking; Antarctic; Pacific - North; Pacific - Western North

268.   Overholtz, W.J. and J.R. Nicolas.
1979, January. Apparent feeding by the fin whale, *Balaenoptera physalus*, and humpback whale, *Megaptera novaeangliae*, on the American Sand Lance, *Ammodytes americanus*, in the Northwest Atlantic. Fishery Bull. Fish Wildl. Serv. U.S., 77(1): 285-287, 10 references.

Behavior - Feeding; Behavior - Species Interaction: Fin Whale, Great Black-Backed Gull, Herring Gull; Food; Atlantic - Western North: U.S. Waters

268A. Paterson, R.
　　　　1980, April. Encouraging sightings of humpback whales off east coast. Aust. Fish., p. 8-10, 3 tables, 3 references.

　　　　Distribution; Migration; Pacific - Western South: Australian Waters

269. Payne, K.
　　　　1978. 'Forward' to special issue on humpback whales. Waters - Journal of the Vancouver Aquarium, 3(4th quarter): 2-3, 1 unnumbered figure.

　　　　Conservation

270. ---------; P. Tyack; and R. Payne.
　　　　1979. (Abstract) Progressive changes in songs of humpback whales. Seattle: Abstracts from presentations at the third biennial conference of the biology of marine mammals, October 7-11, p. 45.

　　　　Behavior - Acoustic: Vocalization; Atlantic - Western North: Bermuda Waters; Pacific - North: Hawaiian Waters

271. ---------
　　　　1980. (Abstract) Variations in the songs of humpback whales (*Megaptera novaeangliae*). *In*: Abstracts of papers of the 146th national meeting, 3-8 January 1980, San Francisco, California. A. Herschman (ed.), Washington, D.C.: American Association for the Advancement of Science, AAAS Publication 80-2, p. 40.

　　　　Behavior - Acoustic: Vocalization; Population Dynamics; Atlantic - Western North: Bermuda Waters; Pacific - North: Hawaiian Waters; Pacific - Eastern North: Mexican Waters

272. Payne, R.S.
　　　　1970. Songs of the humpback whale. Phonograph record with accompanying 36p. book. Del Mar, CA: CRM Books; New York: Capitol Records, SWR-11. (English and Japanese text)

　　　　Behavior - Acoustic: Vocalization; Conservation; International Whaling Commission;

Whaling - Catch Statistics; Atlantic - Western North: Bermuda Waters

273.   --------- and S. McVay.
1971, August 13. Songs of humpback whales. Science, 173(3997): 585-597, 11 figures, 17 references.

BA 1971 (#134339)

Behavior - Acoustic: Vocalization; Behavior - Surface Dependent: Breach, Flippering, Lobtail; Atlantic - Western North: Bermuda Waters

274.   ---------
1978, April. Behavior and vocalizations of humpback whales (*Megaptera* sp.). *In*: Report on a workshop on problems related to humpback whales (*Megaptera novaeangliae*) in Hawaii. K.S. Norris and R.R. Reeves (eds.), Springfield, VA: NTIS, PB-280 794, Report No. MMC-77/03, appendix 6, p. 56-78, 7 plates, 1 table, 15 references.

Behavior - Acoustic: Vocalization; Behavior - Agonistic; Behavior - Locomotion; Behavior - Reproductive; Behavior - Respiratory; Behavior - Surface Dependent: Breach, Flippering, Lobtail; Migration; Pacific - North: Hawaiian Waters

275.   ---------
1978, April. A note on harassment. *In*: Report on a workshop on problems related to humpback whales (*Megaptera novaeangliae*) in Hawaii. K.S. Norris and R.R. Reeves (eds.), Springfield, VA: NTIS, PB-280 794, Report No. MMC-77/03, appendix 13, p. 89-90.

Harassment; Atlantic - Western North: Bermuda Waters; Pacific - North: Hawaiian Waters

276.   ---------
1979, January. Humpbacks: Their mysterious songs. Natn. geogr. Mag., 155(1): 18-25 + sound sheet; 7 unnumbered figures.

Behavior - Acoustic: Vocalization; Behavior - Feeding; Conservation; Photographs - Feeding;

Atlantic - Western North: Bermuda Waters; Pacific - North: Hawaiian Waters; Pacific - Eastern North: Alaskan Waters

277. ---------
1979. The songs of whales. *In*: Project Interspeak. T. Wilkes (ed.), Hollywood, CA: Project Interspeak, Inc., p. 65-68.

Acoustics - Underwater Sound Transmission; Behavior - Acoustic: Vocalization; Atlantic - Western North: Bermuda Waters

278. ---------
1979. (Abstract) Humpback whale songs as an indicator of "stocks." Seattle: Abstracts from presentations of the third biennial conference on the biology of marine mammals, October 7-11, p. 46.

Behavior - Acoustic: Vocalization; Population Dynamics; Stock Assessment; Atlantic - Western North: Bermuda Waters; Pacific - North: Hawaiian Waters; Pacific - Eastern North: Mexican Waters

279. Pedersen, T.
1952, July. A note on humpback oil and on the milk and milk fat from this species (*Megaptera nodosa*). Norsk Hvalfangstiid., 41(7): 375-378, 3 tables, 10 references. (English and Norwegian)

Biochemical studies

280. Perkins, J.S. and H. Whitehead.
1977, September. Observations on three species of baleen whales off northern Newfoundland and adjacent waters. J. Fish. Res. Bd Can., 34(9): 1436-1440, 3 tables, 4 figures, 17 references. (English and French summaries)

Anatomy - External: Coloration; Behavior - Acoustic: Vocalization; Behavior - Species Interaction: Gull, Tern; Individual Identification; Photographs - Dorsal Fin ID, Fluke Coloration Patterns; Stock Assessment; Atlantic - Western North: Canadian Waters

281. --------- and P.C. Beamish.
1979, May. Net entanglements of baleen whales in

the inshore fishery of Newfoundland. J. Fish.
Res. Bd Can., 36(5): 521-528, 5 tables, 5
figures, 27 references. (English and French
summaries)

Behavior - Acoustic: Playback; Behavior -
Feeding; Entrapments - Net; Food; Tagging/
Marking; Atlantic - Western North: Canadian
Waters

282. Pike, G.C.
        1953, July. Colour pattern of humpback whales
        from the coast of British Columbia. J. Fish.
        Res. Bd Can., 10(6): 320-325, 3 tables, 1
        figure, 5 references.

        BA 1954 (#17463)

        Anatomy - External: Coloration; Photographs
        - Ventral View Coloration Patterns; Atlantic
        - Eastern South: South African Waters;
        Atlantic - Western South: South Georgia
        Waters; Indian - Western North: Madagascar
        Waters; Pacific - Eastern North: Canadian
        Waters; Pacific - Western South: New
        Zealand Waters

283. Pilleri, G.
        1966. Hirnlipom beim buckelwal, *Megaptera
        novaeangliae*. (Brain lipoma in the humpback
        whale, *Megaptera novaeangliae*) Path. Veter-
        inaria, 3(4): 341-349, 8 figures, 14
        references. (German, English summary)

        BA 1967 (#18625)

        Anatomy - Nervous; Photographs - Brain
        Anatomy; Physiology - Disease

284. ---------
        1966. Morphologie des gehirnes des buckelwals
        Megaptera novaeangliae Borowski (Cetacea,
        Mysticeti, Balaenopteridae). (Brain morph-
        ology of the humpback whale, *Megaptera
        novaeangliae* Borowski 'Cetacea, Mysticeti,
        Balaenopteridae') J. Hirnforsch., 8: 437-491,
        5 tables, 57 figures, 27 references. (German,
        English summary)

        Anatomy - Nervous; Photographs - Barnacles,
        Brain Anatomy, Breach, Cyamids

285.   ---------
     1966. Zum hirnbau und verhalten des buckelwals, *Megaptera novaeangliae* Borowski (Cetacea, Mysticeti, Balaenopteridae). (On the brain structure and behavior of humpback whales, *Megaptera novaeangliae* Borowski 'Cetacea, Mysticeti, Balaenopteridae') Acta. Anat., 64(1-3): 256-262, 1 table, 5 figures, 11 references. (German, English and French summaries)

     BA 1967 (#106408)

     Anatomy - Nervous; Photographs - Brain Anatomy, Breach

286.   ---------
     1966, April. Note on the brain anatomy of the humpback whale, *Megaptera novaeangliae*. Revue suisse Zool., 73(7): 161-165 + 3 plates (6 figures); 2 figures, 3 references.

     Anatomy - Nervous; Photographs - Brain Anatomy

287.   Pouchet, G.
     1885, December 7. Sur l'échouement d'une *Mégaptère* près de la Seyne. (On the stranding of a humpback near the Seyne) C.r. Hebd. Séance. Acad. Sci., Paris, 101(23): 1172. (French)

     Distribution; Strandings; Mediterranean Sea

288.   ---------
     1892, May 9. Sur un échouement de cétacé de la 11^{3e} olympiade. (On the stranded cetacea of the 11^{3e} olympiade) C.r. Hebd. Séance. Acad. Sci., Paris, 114(19): 1077-1079. (French)

     Distribution; Strandings; Indian - Western North: Persian Gulf; Mediterranean Sea

289.   ---------
     1892, May 14. Note sur la baleine observée par Néarque. (Note on the whale observed by Nearque) C.R. Soc. Biol., 9th series, 4: 422-423, 2 references. (French)

     Distribution; Mediterranean Sea

290. ---------
          1893, April 13. Un échouage de cétacéa sur la
          côte d'Arabie en l'an 300 de l'Hegyre (922).
          (A stranding of a whale on the coast of
          Arabia in the year 300 of the Hegira '922
          AD') *In*: Anciens echouages de cetaces du
          IXe au XVIIe siecle. (Ancient strandings of
          cetaceans of the 9th to 17th centuries) C.R.
          Soc. Biol., 9th series, 5: 97-104. (French)

          Strandings; Indian - Western North: Persian
          Gulf

291. Price, W.S. and D. Gaskin.
          1980. (Abstract) Preliminary estimates of North
          Atlantic population of the humpback whale.
          Boston: Humpback whales of the Western
          North Atlantic Workshop, November 17-21.

          Stock Assessment; Atlantic - Western North:
          Caribbean Waters

292.* Rawitz, B.
          1897. Ueber Norwegische bartenwale. (On the
          Norwegian humpack whale) Sitz. Ber. Ges.
          nat. Fr. Berlin, p. 146-150. (German)

          Anatomy - External: Coloration

          Source: Reference #349

293. ---------
          1900. Ueber *Megaptera boops*, Febr., nebstt
          bemerkungen zur biologie der Norwegischen
          mystacoceten. (Concerning *Megaptera boops*,
          Febr., and notes on the biology of Norwegian
          Mysticeti) Arch. Naturgesch., 66: 71-114 + 1
          plate (2 figures); 36 references. (German)

          Behavior - Acoustic: Vocalization; Behavior -
          Respiratory; Photographs - Antero-Ventral
          View

294. ---------
          1906. Beiträge zur mikroskopischen anatomie der
          cetaceen. V. Uber den feineren bau der haare
          von *Megaptera boops* Fabr. und *Phocaena
          communis* Cuv. (Concerning the microscopic
          anatomy of the cetacean. V. On the structure
          of the hair of *Megaptera boops* Fabr. and
          Phocaena communis Cuv.) Int. Mschr. Anat.

Physiol., 23: 19-38 + 6 figures, 10 references. (German)

Anatomy - External: Hair

295. Rayner, G.W.
      1948, May. Whale marking. II. Distribution of blue, fin, and humpback whales marked from 1932 to 1938. 'Discovery' Rep., 25: 31-38 + 18 plates; 1 unnumbered table, 1 figure.

      Tagging/Marking; Antarctic (1-4)

296. Rice, D.W.
      1974. Whales and whale research in the Eastern North Pacific. *In:* The whale problem - A status report. W.E. Schevill (ed.), Cambridge, MA: Harvard University Press, p. 170-195, 5 tables, 26 references.

      Distribution; Migration; Whaling - Catch Statistics; Pacific - Eastern North: Mexican Waters

297. --------- and A.A. Wolman.
      1978, April. Humpback whale census in Hawaiian waters - February 1977. *In:* Report on a workshop on problems related to humpback whales (*Megaptera novaeangliae*) in Hawaii. K.S. Norris and R.R. Reeves (eds.), Springfield, VA: NTIS, PB-280 794, Report No. MMC-77/03, appendix 5, p. 45-53 + 6 figures; 3 references.

      Behavior - Species Interaction: Pygmy Killer Whale; Distribution; Population Dynamics; Stock Assessment; Pacific - North: Hawaiian Waters

298. ---------
      1978, April. The humpback whale in the North Pacific: Distribution, exploitation, and numbers. *In:* Report on a workshop on problems related to humpback whales (*Megaptera novaeangliae*) in Hawaii. K.S. Norris and R.R. Reeves (eds.), Springfield, VA: NTIS, PB-280 794, Report No. MMC-77/03, appendix 4, p. 29-44, 2 tables, 4 figures, 16 references.

      Distribution; Migration; Stock Assessment;

Tagging/Marking; Whaling - Catch Statistics;
Whaling - Shore Stations; Pacific - North

299. Richard, J.
      1936. Documents sur les cétacés et pinnipèdes
      provenant des campagnes du Prince Albert 1er
      de Monaco. (Documents on the cetaceans and
      pinnipeds seen during the cruise of Prince
      Albert 1st of Monaco) Resultats des
      campagnes scientifiques accomplies sur son
      Yacht par Albert 1er (Monaco) (Results of
      the scientific cruises accomplished on his
      yacht by Prince Albert 1st of 'Monaco') 94:
      1-71 + 8 plates (numerous figures). (French)

      Distribution; Photographs - Humpbacks in the
      Mediterranean; Mediterranean Sea

300. Ridewood, W.G.
      1922. Observations on the skull in foetal
      specimens of whales of the genre *Megaptera*
      and *Balaenoptera*. Phil. Trans. R. Soc., series
      B, 211: 209-272, 1 unnumbered table, 16
      figures, 35 references.

      Anatomy - Skeletal; Fetus

301. Riese, W.
      1928, May 15. Über das vorderhirn des walfotus
      (*Megaptera boops*). (On the brain of a whale
      fetus '*Megaptera boops*') Anat. Anz. Jena,
      65(14-15): 255-260, 4 figures, 4 references.
      (German)

      Anatomy - Nervous; Fetus; Photographs -
      Fetal Brain Anatomy

302. Risting, S.
      1912. Knølhvalen. (The humpback whale) Norsk
      FiskTid., 31(11): 437-449, 5 figures.*
      (Norwegian)

      Anatomy - External: Coloration; Distribution;
      Migration

      Translation available from: U.S. Hydrographic
      Office, Washington, D.C. (Translation #290).
      Also translated in Hinton, M.A.C. Reports on
      papers left by the late Major G.E.H. Barrett-
      Hamilton relating to whales of South Georgia.
      London, 1925, p. 57-209.

*The 5 figures are not included in translation #290. I have not seen the Hinton translation.

303. Robins, J.P.
       1954, January 30. Ovulation and pregnancy corpora lutea in the ovaries of the humpback whale. Nature, 173(4396): 201-203, 4 figures, 5 references.

       Anatomy - Reproductive; Photographs - Corpus Luteum

304. ----------
       1960, March. Age studies on the female humpback whale, *Megaptera nodosa* (Bonnaterre), in East Australia waters. Aust. J. mar. Freshwat. Res., 11(1): 1-13, 6 tables, 7 figures, 14 references.

       BA 1960 (#40318)

       Age Determination; Anatomy - Reproductive; Pacific - Western South: Australian Waters

305. Romanes, G.J.
       1945, October. Some features of the spinal nervous system of the foetal whale (*Megaptera nodosa*). J. Anat., 79(4): 145-156, 1 table, 6 figures, 30 references.

       BA 1946 (#10343)

       Anatomy - Nervous; Fetus

305A. Rorvik, C.J.
       1980. Whales and whaling off Mozambique. Rep. int. Whal. Commn 30, Cambridge, SC/31/Doc 15, p. 223-225, 2 tables, 2 figures, 4 references.

       Distribution; Migration; Whaling - Catch Statistics; Whaling - Shore Stations; Indian Ocean - Western South: Mozambique Waters

306.* Rovnin, A.A
       1966. (Atavistic ear of the humpback whale) Izv. Tikhookean. nauchno-issled. Inst. ryb. Khoz. Okeanogr., 58: 243-244, 1 figure. (Russian)

       BA 1968 (#32632)

Anatomy - Ear

307.* Ruud, J.R.
1937, April. Knobhvalen, *Megaptera nodosa*, Bonn.
(The humpback whale, *Megaptera nodosa*,
Bonn.) Norsk Hvalfangsttid., 26(4): 113 +.
(Norwegian)

Source: Reference #345

308. Sars, G.O.
1880. Fortsatte bidrag til kundskaben om vore
bardehvalen. "finhvalen" og "knolhvalen."
(Contribution to the knowledge of the
whalebone whales. "finwhale" and "humpback") Forh. VidenskSelsk. Krist., No. 12: 1-
20 + 3 plates. (Norwegian)

309. Scammon, C.M.
1869, April. On the cetaceans of the western
coast of North America. Proc. natn. Acad.
Sci., p. 13-69, 3 figures, 24 references.

Anatomy - External: Coloration; Behavior -
Reproductive; Behavior - Surface Dependent:
Breach, Flippering, Lobtail; Distribution;
Parasites - Ectoparasites; Pacific - Eastern
North: Mexican, U.S. Waters; Pacific -
Eastern South: Peruvian Waters

309A. Scheffer, V.B.
1939, September-December. Organisms collected
from whales in the Aleutian Islands. Murrelet,
20(3): 67-69, 5 figures.

Parasites - Ectoparasites; Photographs -
Barnacle, Cyamid; Pacific - North: Aleutian
Waters

310. Schevill, W.E.
1957, Summer-Autumn. A breaching humpback.
Oceanus, 5(3-4): 17: 1 unnumbered figure.

Behavior - Surface Dependent: Breach,
Flippering, Lobtail; Photographs - Breach;
Atlantic - Western North: U.S. Waters

311. --------- and R.H. Backus.
1960, May 20. Daily patrol of a *Megaptera*. J.
Mammal., 41(2): 279-281, 1 figure.

Anatomy - External: Coloration; Behavior - Lone Whale; Behavior - Surface Dependent: Breach, Lobtail, Spyhop; Individual Identification; Atlantic - Western North: U.S. Waters

312.   --------- and W.A. Watkins.
1962. Whale and porpoise voices. Woods Hole: Woods Hole Oceanographic Institution, 24p. + record; 35 figures, 8 references.

Behavior - Acoustic: Vocalization; Atlantic - Western North: U.S. Waters

313.   Schreiber, O.W.
1952, January. (Abstract) Some sounds from marine life in the Hawaiian area. J., acoust. Soc. Am., 24(1): p. 116, abstract #C7.

Behavior - Acoustic: Vocalization; Pacific - North: Hawaiian Waters

314.*   Scott, G.P. and H.E. Winn.
1978. Assessment of humpback whale (*Megaptera novaeangliae*) stocks using vertical photographs. Proc. PECORA IV Sym., p. 235-243.

Census Techniques; Stock Assessment; Atlantic - Western North: Caribbean Waters

Source: Reference #316

Abstract of this paper appeared as: An aerial sampling method for humpback whales, *Megaptera novaeangliae*, using vertical photographs. San Diego: Proceedings (abstracts) of the second conference on the biology of marine mammals, December 12-15, p. 14.

315.   ---------
1980. (Abstract) Aerial and shipboard estimates of humpback whale abundance on Silver and Navidad Banks, West Indies. Boston: Humpback whales of the Western North Atlantic Workshop, November 17-21.

Stock Assessment; Atlantic - Western North: Caribbean Waters

316.   --------- and H.E. Winn.
1980. Comparative evaluation of aerial and

shipboard sampling techniques for estimating the abundance of humpback whales (*Megaptera novaeangliae*). Springfield, VA: NTIS, PB81-109 852, Report No. MMC-77/24, 96p. 19 figures, 14 tables, 106 references.

Behavior - Surface Dependent; Census Techniques; Population Dynamics; Stock Assessment; Atlantic - Western North: Caribbean Waters

Abstract of this paper appeared as: Scott, G.P. Aerial and Shipboard estimates of humpback whale abundance on Silver and Navidad Banks, West Indies. Boston: Humpback whales of the Western North Atlantic Workshop, November 17-21, 1980.

317. ---------
    1980. (Abstract) Humpback whale group sizes and distance between calling individuals relative to local density. Boston: Humpback whales of the Western North Atlantic Workshop, November 17-21.

    Behavior - Acoustic: Vocalization; Population Dynamics; Atlantic - Western North: Caribbean Waters

318. ---------
    1980. (Abstract) Localized distribution, activity, and movements of humpback whales on certain banks in the West Indies. Boston: Humpback whales of the Western North Atlantic Workshop, November 17-21.

    Population Dynamics; Atlantic - Western North: Caribbean Waters

319. ---------
    1980. (Abstract) Possible effects of the Bequia whale fishery on the stock of humpback whales in the Lower Lesser Antilles. Boston: Humpback whales of the Western North Atlantic Workshop, November 17-21.

    Stock Assessment; Whaling - Whaling Material; Atlantc - Western North: Caribbean Waters

320. Shallenberger, E.W.
      1976, November. Report to Seaflight & Sea Grant
      on the population and distribution of
      humpback whales in Hawaii. 13p. + 3 tables.

      Census   Techniques;   Distribution;   Stock
      Assessment; Pacific - North: Hawaiian Waters

321. ----------
      1977. Humpback whales in Hawaii - population &
      distribution. *In*: Oceans '77 conference
      record, vol. 2, New York: IEEE; Washington,
      D.C.: MTS, 7p., 5 tables, 1 figure, 2
      unnumbered figures, 9 references.

      Census Techniques; Distribution; Individual
      Identification; Stock Assessment; Pacific -
      North: Hawaiian Waters

322. ----------
      1977. (Abstract) The effect of human generated
      activities on the Hawaiian humpback whale.
      San Diego: Proceedings (abstracts) of the
      second conference on the biology of marine
      mammals, December 12-15, p. 63.

      Harassment; Pacific - North: Hawaiian Waters

323. ----------
      1978, April. Activities possibly affecting the
      welfare of humpback whales. *In*: Report on a
      workshop on problems related to humpback
      whales (*Megaptera novaeangliae*) in Hawaii.
      K.S. Norris and R.R. Reeves (eds.),
      Springfield, VA: NTIS, PB-280 794, Report
      No. MMC-77/03, appendix 11, p. 81-85.

      Conservation; Harassment; Pacific - North:
      Hawaiian Waters

324.* ----------
      1979. Workshop on humpback whales in Hawaii.
      Rep. int. Whal. Commn 29, Cambridge, p. 139-
      140.

      Pacific - North: Hawaiian Waters

      Source: WR September, 1980

325. Slijper, E.J.; W.L. Van Utrecht; and C. Naaktgeboren.
      1964. Remarks on the distribution and migration

of whales, based on observations from Netherlands ships. Bijdragen Tot De Dierkunde, 34: 3-93, 10 charts, 8 tables, 26 figures, 144 references.

Distribution; Migration; Strandings

326.   Soares, M.; E. Shallenberger; and R. Antinoja.
1977. (Abstract) The Hawaiian humpback whale: Population characteristics. San Diego: Proceedings (abstracts) of the second conference on the biology of marine mammals, December 12-15, p. 57.

Stock Assessment; Pacific - North: Hawaiian Waters

327.   Souverbie.
1877. Ossements de *Mégaptère* au musée de Bordeaux. (Skeletal remains of a humpback at the Bordeaux Museum) J. Zool., 6: 279-280. (French)

Anatomy - Skeletal

328.   Steinitz, W.
1920, February. Untersuchungen über die entwicklung des auges vom buckelwal (*Megaptera nodosa*). (Investigation on the development of the eyes of the humpback whale '*Megaptera nodosa*') Jena Z. Naturw., 56: 119-154, 17 tables, 10 figures, 13 references. (German)

Anatomy - Eye; Evolution

329.   Struthers, J.
1885, February. On the rudimentary hind limb of *Megaptera longimana*. Am. Nat., 19(2): 124-125.

Anatomy - Skeletal; Atlantic - Eastern North: British Isles Waters

330.   ---------
1885, February. On finger muscles in *Megaptera longimana* and in other whales. Am. Nat., 19(2): 126-127.

Anatomy - Flippers/Flukes; Anatomy - Muscles

331.   ---------

1887, October. On some points in the anatomy of a *Megaptera longimana*. Part 1. History and external characters. J. Anat. Physiol. Lond., 22(new series, vol. 2, 1st part): 109-125 + 2 plates (5 figures); 1 unnumbered table, 13 references.

Anatomy - External: Coloration, Hair, Scars; Anatomy - Reproductive

332.   ---------

1888, January. On some points in the anatomy of a *Megaptera longimana*. Part 2. The limbs. J. Anat. Physiol. Lond., 22(new series, vol. 2, 2nd part): 240-282 + 3 plates (11 figures); 5 tables, 11 references.

Anatomy - Circulatory; Anatomy - Flippers/ Flukes; Anatomy - Muscles; Anatomy - Skeletal

333.   ---------

1888, April. On some points in the anatomy of a *Megaptera longimana*. Part 3. The vertebral column. J. Anat. Physiol. Lond., 22(new series, vol. 2, 3rd part): 441-460, 3 tables, 4 references.

Anatomy - Skeletal

334.   ---------

1888, July. On some points in the anatomy of a *Megaptera longimana*. Part 3 (cont.). The vertebral column. J. Anat. Physiol. Lond., 22(new series, vol. 2, 4th part): 629-654, 1 table.

Anatomy - Skeletal

335.   ---------

1888, October. On some points in the anatomy of a *Megaptera longimana*. Part 3 (cont.). The vertebral column. J. Anat. Physiol. Lond., 23(new series, vol. 3, 1st part): 124-163 + 1 plate (2 figures); 4 tables, 14 references.

Anatomy - Skeletal

336.   ---------

1889, January. On some points in the anatomy of a

Megaptera longimana. Part 4. The skull. J. Anat. Physiol. Lond., 23(new series, vol. 3, 2nd part): 308-335, 1 table, 6 references.

Anatomy - Skeletal

337. ---------
     1889, April. On some points in the anatomy of a *Megaptera longimana.* Part 4 (cont.). Ear bone, mandible, and hyoid. J. Anat. Physiol. Lond., 23(new series, vol. 3, 3rd part): 358-373, 2 tables, 3 references.

     Anatomy - Ear; Anatomy - Skeletal

338.* ---------
     1889. Memoir on the anatomy of a humpback whale, *Megaptera longimana.* Edinburgh: Maclachlan and Stewart, 188p.

     Entries #s331-337 in book form

     Source: ZR 1887

339. Stump, C.W.; J.P. Robins; and M.L. Garde.
     1960, December. The development of the embryo and membranes of the humpback whale, *Megaptera nodosa* (Bonnaterre). Aust. J. mar. Freshwat. Res., 11(3): 365-386 + 15 plates (70 figures); 2 references.

     BA 1961 (#31167)

     Anatomy - Reproductive; Fetus; Photographs - Embryo and Membranes; Pacific - Western South: Australian Waters

340. Symons, H.W. and R.D. Weston.
     1957, May. An underfished humpback population? Norsk Hvalfangsttid., 46(5): 231-238, 5 tables, 4 references. (English and Norwegian)

     Distribution; Tagging/Marking; Whaling - Catch Statistics; Antarctic (1-4); Pacific - Eastern South: Chilean Waters

341. --------- ---------
     1958, February. Studies on the humpback whale (*Megaptera nodosa*) in the Bellingshausen Sea. Norsk Hvalfangsttid., 47(2): 53-81, 8 plates,

10 tables, 5 figures, 20 references. (English and Norwegian)

Age Determination; Anatomy - External: Coloration; Anatomy - Reproductive; Food; Parasites - Ectoparasites; Photographs - Ear Plugs, Ovaries; Population Dynamics; Whaling - Catch Statistics; Antarctic (1)

342.   Thompson, P.O.; W.C. Cummings; and S.J. Kennison.
1977, Fall. (Abstract) Sound production of humpback whales, *Megaptera novaeangliae*, in Alaskan waters. J. acoust. Soc. Am., 62(supp. 1): S89, abstract #KK22.

Behavior - Acoustic: Vocalization; Behavior - Feeding; Behavior - Surface Dependent: Flippering, Lobtail; Pacific - Eastern North: Alaskan Waters

343.   --------- ---------; A.J. Perrone; and S.J. Kennison.
1977. (Abstract) Humpback whale sounds in Alaska, Hawaii and Western North Atlantic. San Diego: Proceedings (abstracts) of the second conference on the biology of marine mammals, December 12-15, p. 39.

Behavior - Acoustic: Vocalization; Atlantic - Western North: Bermuda Waters; Pacific - North: Hawaiian Waters; Pacific - Eastern North: Alaskan Waters

344.* Thompson, T.J. and H.E. Winn.
1977. (Abstract) Temporal aspects of the humpback whale song. University Park, PA: Abstracts of papers presented at the Animal Behavior Society meeting, p. 147.

Behavior - Acoustic: Vocalization

Source: Reference #373

345.   Tomilin, A.G.
1967. Mammals of the U.S.S.R. and adjacent countries. Vol. 9. Cetacea. Jerusalem: Israel Program for Scientific Translations, 755p., 196 tables, 146 figures, numerous references. (translated by O. Ronen from the 1957 Russian edition)

Anatomy - External: Coloration; Anatomy - Reproductive; Anatomy - Skeletal; Behavior - Acoustic: Vocalization; Behavior - Feeding; Behavior - Locomotion; Behavior - Reproductive; Behavior - Respiratory; Behavior - Surface Dependent: Breach, Flippering, Lobtail, Spyhop; Classification; Distribution; Entrapments - Net; Fetus; Food; Migration; Parasites - Ectoparasites; Parasites - Endoparasites; Photographs - List of Important Photographs, Skeletal Remains; Strandings; Whaling - Catch Statistics

346. Townsend, C.H.
        1935, April 3. The distribution of certain whales as shown by logbook records of American whaleships. Zoologica, N.Y., 19(1): 1-50 + 4 plates, 2 figures, 3 references.

        Distribution; Whaling - Catch Statistics; Whaling - Logbooks/Journals

347.* Toyanna, Y.
        1924. Investigation of the fatty acids of whale oil. Chem Umsch. Geb. Fette, 31: 221-227, 238, 249.

        CA 1925 v. 19 (p. 410)

        Biochemical Studies

348.* ---------- and T. Ishikawa.
        1934. (Identification of gadoleic acid in sei-whale and humpback whale oils) J. Soc. chem. Ind. Japan, 37(supp.): 534-536. (Japanese)

        CA 1935 v. 29 (p. 367)

        Biochemical Studies

349. True, F.W.
        1904. The whalebone whales of the Western North Atlantic compared with those occurring in European waters with some observations on the species of the North Pacific. Smithson. Contr. Knowl., 33: 332p. + 50 plates (188 figures); numerous unnumbered tables, 97 figures, 92 references.

        Anatomy - External: Coloration, Scars; Anatomy - Flippers/Flukes; Anatomy -

Reproductive; Anatomy - Skeletal; Classi-
fication; Photographs - Antero-Ventral View,
Dermal Tubercles, Ventral View of Male and
Female; Strandings; Atlantic - Western North;
Atlantic - Eastern North: European Waters;
Pacific - North

350. Tyack, P.
1979. (Abstract) Interactions between singing
humpback whales and nearby whales. Seattle:
Abstracts from presentations at the third
biennial conference of the biology of marine
mammals, October 7-11, p. 61.

Behavior - Acoustic: Vocalization; Individual
Identification; Population Dynamics; Pacific -
North: Hawaiian Waters

351. ---------
1980. (Abstract) The function of song in
humpback whales, *Megaptera novaeangliae. In*:
Abstracts of papers of the 146th national
meeting, 3-8 January 1980, San Francisco,
California. A. Herschman (ed.), Washington,
D.C.: American Association for the
Advancement of Science, AAAS Publication
80-2, p. 41.

Behavior - Acoustic: Vocalization; Behavior -
Agonistic; Behavior - Reproductive; Pacific -
North: Hawaiian Waters

352. Vanhoffen, E.
1900, March 5. Berichtigung zu dem aufsatz des
Herrn B. Rawitz "Über *Megaptera boops*
Fabr." (Correction to the article by Mr. B.
Rawitz 'on *Megaptera boops* Fabr.') Zool.
Anz., 23(609, paper 2): 114-116, 1 reference.
(German)

353. Vladykov, V.D.
1958. (Abstract) Baleine à bosse (*Megaptera
novaeangliae*) échouée à Lauzon en Septem-
bre, 1957. (Humpback whale '*Megaptera
novaeangliae*' stranded at Lauzon in September, 1957) Ann. ACFAS, 24: 80, abstract #21,
section III, biologie B.

Strandings; Atlantic - Western North:
Canadian Waters

Paper presented at: 25th Congress of the
Association Canadienne-Francaise pour l'Advancement des Sciences Montreal.

353A. Votrogov, L.M. and M.V. Ivashin.
　　　1980. Sightings of fin and humpback whales in the
　　　　Bering and Chukchi Seas. Rep. int. Whal.
　　　　Commn 30, Cambridge, SC/31/Doc 20, p. 247-
　　　　248, 4 references.

　　　　Distribution; Arctic Ocean - Chukchi Sea;
　　　　Pacific - North: Bering Sea

354.* Wang, P.
　　　1978. (Studies on the baleen whales in the Yellow
　　　　Sea) Acta zool. sin., 24(3): 269-277.
　　　　(Chinese, English summary)

　　　　BA 1979 v. 68 (#20573)

　　　　Distribution; Pacific - Western North: Yellow
　　　　Sea

355.　Warren, R.
　　　1893, May. Hump-backed whale on the coast of
　　　　Sligo. Zoologist, 3rd series, 17(197): 188-189.

　　　　Anatomy - External: Coloration; Parasites -
　　　　Ectoparasites; Strandings; Atlantic - Eastern
　　　　North: British Isles Waters

356.　Watkins, W.A.
　　　1967, November 20. Air-borne sounds of the
　　　　humpback whale, *Megaptera novaeangliae*. J.
　　　　Mammal., 48(4): 573-578, 3 figures, 20
　　　　references.

　　　　BA 1968 (#49196)

　　　　Behavior - Acoustic: Vocalization; Behavior -
　　　　Respiratory; Atlantic - Western North: U.S.
　　　　Waters

357.　--------- and J.H. Johnson.
　　　1977. (Abstract) Tests of radio whale tag on
　　　　carcasses and on free swimming whales. San
　　　　Diego: Proceedings (abstracts) of the second
　　　　conference on the biology of marine mammals,
　　　　December 12-15, p. 16.

Tagging/Marking; Telemetry; Pacific - Eastern North: Alaskan Waters

358.   --------- ---------; and D. Wartzok.
1978, August. Radio tagging report of finback and humpback whales. Woods Hole: Woods Hole Oceanographic Institution, WHOI-78-51, 13p., 6 references.

Tagging/Marking; Telemetry

358A.   --------- and W.E. Schevill.
1979, February. Aerial observation of feeding behavior in four baleen whales: *Eubalaena glacialis, Balaenoptera borealis, Megaptera novaeangliae,* and *Balaenoptera physalus.* J. Mammal., 60(1): 155-163, 4 figures, 15 references.

Behavior - Feeding; Behavior - Species Interaction: Fin Whale; Atlantic - Western North: U.S. Waters

359.   Wheeler, J.F.G.
1941, October. On a humpback whale taken at Bermuda. Proc. zool. Soc. Lond., series B, 111(1-2): 37-38, 3 references.

BA 1944 (#15272)

Anatomy - External: Coloration, Scars; Parasites - Ectoparasites; Atlantic - Western North: Bermuda Waters

360.   ---------
1944, February. On a humpback whale taken at Bermuda in 1942. Proc. zool. Soc. Lond., series A, 113(3-4): 121-125, 2 tables, 1 figure, 8 references.

BA 1944 (#20931)

Age Determination; Anatomy - External: Coloration, Hair, Scars; Behavior - Respiratory; Fetus; Growth Rates; Parasites - Ectoparasites; Atlantic - Western North: Bermuda Waters

361.   Whitehead, H.; P. Harcourt; K. Ingham; and H. Clark.
1980, May. The migration of humpback whales past the Bay de Verde Peninsula, Newfoundland,

during June and July, 1978. Can. J. Zool., 58(5): 687-692, 7 figures, 7 references. (English and French summary)

BA 1980 v. 70 (#42363)

Anatomy - External: Coloration; Behavior - Locomotion; Distribution; Food; Individual Identification; Migration; Stock Assessment; Atlantic - Western North: Canadian Waters

362.   ---------; B. Baxter; and G. Nichols, Jr.
        1980. Preliminary report on the populations of humpback whales on Silver, Navidad and Mouchoir Banks during the winter 1977-1978. Boston: Ocean Research and Education Society, Expedition #s5-6, 8p. + 6 tables, 3 figures, 7 references.

        Distribution; Population Dynamics; Stock Assessment; Atlantic - Western North: Caribbean Waters

363.*   ---------
        1980. (Summary) Humpback whale songs. Salford, England: Program of the annual meeting of the British Association for the Advancement of Science, September 1-5.

        Behavior - Acoustic: Vocalization

        Source: CPI v. 8, September 1980

364.   Wilks, L.
        1978. The winter whales of Hawaii. Part three: Getting to know some whales. Waters - Journal of the Vancouver Aquarium, 3(4th quarter): 26-30, 7 unnumbered figures.

        Anatomy - External: Scars; Individual Identification; Photographs - Scars, Underwater View; Pacific - North: Hawaiian Waters

365.   Williamson, G.R.
        1961, August. Winter sightings of a humpback suckling its calf on the Grand Bank of Newfoundland. Norsk Hvalfangsttid., 50(8): 335-336, 339-341, 1 table, 2 unnumbered figures, 11 references. (English and Norwegian)

Behavior - Nursing; Behavior - Surface Dependent: Flippering; Food; Atlantic - Western North: Canadian Waters

366. Winn, H.E.; P.J. Perkins; and T.C. Poulter.
1971. Sounds of the humpback whale. Menlo Park, CA: Proceedings of the 7th annual conference on biological sonar and diving mammals, Stanford Research Institute, p. 39-52 + 5 tracings; 3 tables, 2 figures, 10 references.

Behavior - Acoustic: Vocalization; Atlantic - Western North: Bermuda, Caribbean Waters

367. ---------; W.L. Bischoff; and A.G. Taruski.
1973. Cytological sexing of cetacea. Marine Biology, 23(4): 343-346, 1 table, 2 figures, 12 references.

BA 1974 v. 57 (#59828)

Anatomy - External: Skin; Anatomy - Reproductive; Sexing Techniques

368. ----------
1974. (Abstract) Geographic variation and behavioral correlates of the call of the humpback whale. Champaign-Urbana, IL: Abstracts of papers presented at the Animal Behavior Society meetings, May 24-27, paper #98, session K.

Behavior - Acoustic: Vocalization; Atlantic - Western North: Caribbean Waters

369. ---------; R.K. Edel; and A.G. Taruski.
1975, April. Population estimates of the humpback whale (*Megaptera novaeangliae*) in the West Indies by visual and acoustic techniques. J. Fish. Res. Bd Can., 32(4): 499-506, 1 table, 4 figures, 14 references. (English and French summaries)

BA 1975 v. 60 (#18982)

Behavior - Acoustic: Vocalization; Census Techniques; Stock Assessment; Atlantic - Western North: Caribbean Waters

370. ----------
1975. (Abstract) Dialects and social organization

of humpback whale. Santa Cruz: Conference
on the biology and conservation of marine
mammals - abstracts, December 4-7, p. 48.

Behavior - Acoustic: Vocalization; Atlantic -
Western North: Caribbean Waters

371.   --------- and G.P. Scott.
1977. (Abstract) Evidence for three substocks of
humpback whales (*Megaptera novaeangliae*) in
the Western North Atlantic. San Diego:
Proceedings (abstracts) of the second
conference on the biology of marine mammals,
December 12-15, p. 15.

Stock Assessment; Atlantic - Western North

372.   --------- and D.W. Rice.
1978. Humpback whales in the Sea of Cortez.
Currents, 2(4): 5,7.

Anatomy - External: Coloration; Behavior -
Acoustic: Vocalization; Individual Identi-
fication; Pacific - Eastern North: Mexican
Waters

373.   --------- and L.K. Winn.
1978. The song of the humpback whale *Megaptera
novaeangliae* in the West Indies. Marine
Biology, 47(2): 97-114, 4 tables, 5 figures, 19
references.

Behavior - Acoustic: Vocalization; Behavior -
Respiratory; Atlantic - Western North:
Caribbean Waters

374.   ---------
1978, April. Environmental correlates of the
humpback whale in the tropical winter calving
grounds. *In*: Report on a workshop on
problems related to humpback whales
(*Megaptera novaeangliae*) in Hawaii. K.S.
Norris and R.R. Reeves (eds.), Springfield,
VA: NTIS, PB-280 794, Report No. MMC-
77/03, appendix 10, p. 80.

Distribution; Oceanography

375.*   ---------
1979. (Abstract) Comparative aerial-shipboard
surveys of humpback whales in Silver and

Navidad Banks, West Indies. Rep. int. Whal. Commn 29, Cambridge, p. 140.

Census Techniques; Atlantic - Western North: Caribbean Waters

Source: WR September, 1980

376.   ---------; P. Beamish; and P.J. Perkins.
1979. Sounds of two entrapped humpback whales (*Megaptera novaeangliae*) in Newfoundland. Marine Biology, 55(2): 151-155, 3 tables, 2 figures, 9 references.

Behavior - Acoustic: Vocalization; Entrapments - Net; Atlantic - Western North: Canadian Waters

377.   --------- and T.J. Thompson.
1979. (Abstract) Comparison of humpback whale sounds across the northern hemisphere. Seattle: Abstracts from presentations at the third biennial conference of the biology of marine mammals, October 7-11, p. 62.

Behavior - Acoustic: Vocalization; Atlantic - Eastern North: Cape Verde Waters; Atlantic - Western North: Caribbean Waters; Pacific - North: Hawaiian Waters; Pacific - Eastern North: Mexican Waters

378.   ---------
1980. (Abstract) Distribution and population estimates on the Virgin Island Banks. Boston: Humpback whales of the Western North Atlantic Workshop, November 17-21.

Distribution; Population Dynamics; Stock Assessment; Atlantic - Western North: Caribbean Waters

379.   --------- and G.P. Scott.
1980. (Abstract) The humpback whale (*Megaptera novaeangliae*): Present knowledge and future trends in research with special reference to the Western North Atlantic. Boston: Humpback whales of the Western North Atlantic Workshop, November 17-21.

380.   --------- ---------
1980. (Abstract) A model for stock structuring of

humpback whales in the Western North Atlantic. Boston: Humpback whales of the Western North Atlantic Workshop, November 17-21.

Population Dynamics; Stock Assessment; Atlantic - Western North

381.   --------- and L.K. Winn.
        1980. (Abstract) Sound production during feeding of humpback whales. Boston: Humpback whales of the Western North Atlantic Workshop, November 17-21.

        Behavior - Acoustic: Vocalization; Behavior - Feeding; Atlantic - Western North: Caribbean Waters

382.   ---------
        1980. (Abstract) Sounds and visual sightings: An independent measure of shipboard visual sightability. Boston: Humpback whales of the Western North Atlantic Workshop, November 17-21.

        Behavior - Acoustic: Vocalization; Census Techniques; Stock Assessment; Atlantic - Western North: Caribbean Waters

383.* Wolman, A.A.
        1972. Humpback whale. *In*: Baleen whales in Eastern North Pacific and Arctic waters. A. Seed (ed.), Seattle: Pacific Search Press, p. 38-42.

        Source: Pacific Search Press Catalog

384.   --------- and C.M. Jurasz.
        1977, July. Humpback whales Hawaii: Vessel census, 1976. Mar. Fish. Rev., 39(7): 1-5, 3 tables, 5 figures, 8 references.

        ASFA 1974 (#39361Q8)

        Anatomy - External: Coloration; Behavior - Species Interaction: Dolphin, Pygmy Killer Whale; Distribution; Harassment; Population Dynamics; Pacific - North: Hawaiian Waters

385.   ---------
        1978. Humpback whales. *In*: Marine mammals of

Eastern North Pacific and Arctic Waters. D. Haley (ed.), Seattle: Pacific Search Press, p. 46-53, 8 unnumbered figures.

Anatomy - External: Coloration, Scars; Behavior - Feeding; Behavior - Surface Dependent: Breach; Distribution; Individual Identification; Migration; Parasites - Ecto-parasites; Photographs - Breach, Feeding, Fluke Coloration Patterns; Tagging/Marking; Pacific - North: Bering Sea; Pacific - Eastern North

386.   Yoshikawa, Y. and T. Suzuki.
1963, February. The lamination of the masseter of the humpback whale. Sci. Repts. Whales Res. Inst., Tokyo, No. 17: 49-52, 2 figures, 9 references.

Anatomy - Muscles; Fetus

387.   Zeitlin, H.
1971. Whale watch - a month in the life of a humpback. Nature Canada, p. 6-11, 7 unnumbered figures.

Behavior - Acoustic: Echolocation; Entrap-ments - Net; Atlantic - Western North: Canadian Waters

388.   Zemsky, V.A. and G.A. Budylenko.
1970. Siamskie bliznetsky u gorbatatogo kita. (Siamese twins in the humpback whale) Trudȳ Atlant. nauchno-issled. Inst. rȳb. Khoz. Okeanogr., 29: 225-230, 1 unnumbered table, 5 figures, 1 reference. (Russian)

Fetus; Photographs - Twin Fetus; Antarctic (5)

389.   --------- ---------
1973, March. "Siamskie Bliznetsy" u kita-gor-bacha. (Humpback whale's "siamese twins") Priroda, 3: 124-125, 1 unnumbered table, 1 unnumbered figure. (Russian)

Fetus; Photographs - Twin Fetus; Antarctic (5)

390.   Zenkovich, B.A.
1937. Gorbatyi ili dlinnorukii kit (*Megaptera*

nodosa Bonnaterre, 1789). (The humpback whale '*Megaptera nodosa* Bonnaterre, 1789') Akademiia Nauk. SSSR. Dal'nevostochnyi Filial, 37(27): 37-62, 16 tables, 1 figure, 16 references. (Russian)

ArB 1954 (#27572)

Classification; Distribution; Food; Growth Rates; Photographs - Antero-Ventral View; Arctic Ocean - Chukchi Sea; Pacific - North: Bering Sea; Pacific - Western North: Okhotsk Sea

Author Index

Subject Index

Geographic Index

Journal Index

Right Whale Bibliography

1. Anonymous.
    1907, April. The museum's new whales. Am. Mus.
    J., 7(4): 53-56, 4 unnumbered figures.

    Anatomy - Skeletal; Atlantic - Western
    North: U.S. Waters

2. ---------
    1970, December 19. Conservation can succeed.
    Nature 228(5277): 1140-1141.

    Conservation; Stock Assessment; Atlantic -
    Eastern South: South African Waters

3. ---------
    1972, November-December. Southern right whales
    discovered off Argentina. Mar. Fish. Rev.,
    34(11-12): 13, 1 unnumbered figure.

    Behavior - Acoustic: Vocalization; Behavior -
    Reproductive; Photographs - Callosities,
    Antero-Dorsal View; Atlantic - Western
    South: Argentine Waters

4. ---------
    1976, Fall. Whalewatchers rescue mother and calf.
    Mainstream - Animal Protection Institute of
    America, 7(4): 31, 3 unnumbered figures.

    Anatomy - External: Callosities; Entrap-
    ments - Net; Photographs - Callosities, Net
    Entrapment; Atlantic - Western North:
    Canadian Waters

5. ---------
    1978, August. Two Atlantic right whales...South
    Carolina.... Mar. Fish. Rev., 38(8): 40.

    Distribution; Atlantic - Western North: U.S.
    Waters

6. Abel, P.
    1907. Die morphologie der huftbeinrudimente der
    cetaceen. (The morphology of the rudimentary
    hind limb of cetaceans) Denkschr. Akad.
    Wiss., Wien, 81: 139-195, 56 figures, 130
    references. (German)

Anatomy - Skeletal

7.   Allen, G.M.
        1916, September. The whalebone whales of New
        England. Mem. Boston Soc. nat. Hist., 8(2):
        107-322 + 9 plates (29 figures); 99
        references.

        Anatomy - External: Callosities, Coloration,
        Hair; Anatomy - Skeletal; Behavior - Care
        Giving; Classification; Food; Parasites -
        Ectoparasites; Whaling - Catch Statistics;
        Whaling - Shore Stations; Atlantic - Western
        North: U.S. Waters

8.   Allen, J.A.
        1883, June 29. The right whale of the North
        Atlantic. Science, 1(21): 598-599.

        Classification; Atlantic - North

        A review of Holder's paper (reference #102)

9.   ---------
        1908. The North Atlantic right whale and its near
        allies. Bull. Am. Mus. nat. Hist., 24(paper
        18): 277-329 + 6 plates; 1 figure, 130
        references.

        Anatomy - External: Callosities; Anatomy -
        Skeletal; Classification; Distribution; Para-
        sites - Ectoparasites; Atlantic - North

10. Andrews, R.C.
        1908. Notes upon the external and internal
        anatomy of *Balaena glacialis* Bonn. Bull Am.
        Mus. nat. Hist., 24(paper 10): 171-182, 6
        figures.

        Anatomy - External: Callosities, Coloration,
        Scars; Anatomy - Eye; Anatomy - Skeletal;
        Photographs - Skeletal Remains; Atlantic -
        Western North: U.S. Waters

11. ---------
        1909. Further notes on *Eubalaena glacialis*. Bull.
        Am. Mus. nat. Hist., 26(paper 21): 273-275 +
        4 plates (8 figures).

        Anatomy - External: Callosities, Coloration;
        Parasites - Ectoparasites; Photographs -

Anterior View, Callosities; Atlantic - Western
North: U.S. Waters

12.   Arnold, P.W. and D.E. Gaskin.
           1972, October. Sight records of right whales
           (*Eubalaena glacialis*) and finback whales
           (*Balaenoptera physalus*) from the lower Bay
           of Fundy. J. Fish. Res. Bd Can., 29(10):
           1477-1478, 1 figure, 4 references. (English
           and French summaries)

           Anatomy - External: Callosities, Scars;
           Behavior - Lone Whale; Behavior -
           Respiratory; Behavior - Surface Dependent:
           Breach, Flippering, Surface Active; Distri-
           bution; Individual Identification; Atlantic -
           Western North: Canadian Waters

13.   Augustin, W.
           1913. Die from variabilitat der beckenknocken bei
           Nord Atlantischen bartenwalen. (The variable
           form of the pelvic bone of the North Atlantic
           right whale) Zool. Jb., 35: 533-580. (German)

           Anatomy - Skeletal

14.*   Ayers, W.D.
           1886, August 21. The Atlantic right whales. Sci.
           Am., 55(8): 117.

           Source: Nesheim, K. Whales and whaling: A
           checklist of articles appearing in American
           and British periodicals 1667-1930. SLS 200
           Bibliography, June 1961.

15.   Bailey, A.M. and J.H. Sorenson.
           1961. Subantarctic Campbell Island. Proc. Denver
           Mus. nat. Hist., No. 10, 305p.

           Anatomy - External: Callosities, Coloration;
           Behavior - Locomotion; Behavior - Shallow
           Water; Behavior - Surface Dependent:
           Breach, Flippering, Mating Activity, Surface
           Active; Pacific - Western South: New Zealand
           Waters

16.   Beneden, P.J. van.
           1865, December 2. Note sur les cétacés. (Note
           on the cetaceans) Bull. Acad. r. Belg. Cl.
           Sci., 2nd series, 20(12): 851-854, 1
           unnumbered figure, 2 references. (French)

Classification; Distribution

17.   ---------

1867, January 5. Notice sur la découverte d'un os de Baleine, à Furnes. (Notice on the discovery of a whale bone, at Furnes) Bull. Acad. r. Belg. Cl. Sci., 2nd series, 23(1): 13-21, 6 references. (French)

Anatomy - Skeletal; Classification; Atlantic - Eastern North: European Waters

18.   ---------

1868, January 4. Les baleines et leur distribution geographique. (The whales and their geographic distribution) Bull. Acad. r. Belg. Cl. Sci., 2nd series, 25(1): 9-21 + foldout map; 5 references. (French)

Distribution

Also published (with minor variations) in: Annls. Soc. Nat. Charente-Marit., 5th series, 9 (1868), p. 43-52. (foldout map not present)

19.   ---------

1868, August 1. Sur le bonnet et quelques organes d'un foetus de baleine de Gröenland. (On the bonnet and some organs of a fetus of a Greenland whale) Bull. Acad. r. Belg. Cl. Sci., 2nd series, 26(8): 186-195, 3 unnumbered figures, 15 references. (French)

Anatomy - External: Callosities; Fetus

20.   ---------

1870, December 13. Observations sur l'ostéographie des cétacés. (Observations on the osteography of cetaceans) Bull. Acad. r. Belg. Cl. Sci., 2nd series, 30(12): 380-388, 3 references. (French)

Anatomy - Skeletal; Classification

21.   ---------

1872. Les Balenides fossiles d'Anvers. (The fossil whales of Anvers) J. Zool., 1: 407-419, 7 references. (French)

Anatomy - Skeletal; Classification

22.   ---------

1877, June 2. Un mot sur une baleine capturée
dans la Méditerranée. (A word on a whale
captured in the Mediterranean) Bull. Acad. r.
Belg. Cl. Sci., 2nd series, 43(6): 741-745, 2
references. (French)

Classification;   Distribution;   Mediterranean
Waters

23.   ---------

1880, April 11. Une baleine de 50 pieds (mesure
Americaine) est venue échouer sur le côtes
de Charleston, État de la Caroline du
Sud....(A whale of 50 feet 'American measure'
is stranded on the coast of South Carolina....)
Bull. Acad. r. Belg. Cl. Sci., 2nd series,
49(5): 313-315. (French)

Classification; Distribution; Strandings; Atlantic - Western North: U.S. Waters

24.   ---------

1882, November 4. Note sur des ossements de la
baleine de biscaye au musée de la Rochelle.
(Note on the remains of the biscay whale at
the Rochelle Museum) Bull. Acad. r. Belg. Cl.
Sci., 3rd series, 4(11): 407-414 + 1 plate; 2
references. (French)

Anatomy - Skeletal

25.   ---------

1884, April 5. Sur la présence aux temps anciens
et modernes de la baleine de biscaye (ou
nordcaper), aux côtes de Norwege. (On the
presence in ancient and modern times of the
biscay whale 'or nordcaper', on the coasts of
Norway) Bull. Acad. r. Belg. Cl. Sci., 3rd
series, 7(4): 288-290. (French)

Distribution;   Atlantic   -   Eastern   North:
Scandinavian Waters

26.   ---------

1885, April 4. Sur l'apparition d'une petite gamme
de vraires baleines sur les côtes est des
États-Unis d'Amérique. (On the appearance of
a small group of true whales on the coasts of
the United States of America) Bull. Acad. r.

Belg. Cl. Sci., 3rd series, 9(4): 212-214, 1 reference. (French)

Classification;   Distribution;   Atlantic   - Western North: U.S. Waters

27. _____

1886. Histoire naturelle de la baleine des basques (*Balaena biscayensis*). (Natural history of the Basque whale '*Balaena biscayensis*') Mem. cour. Acad. r. Sci. Belg., 38(5): 1-44, 65 references. (French)

Anatomy - Skeletal; Classification; Distribution; Parasites - Ectoparasites; Strandings; Whaling - Shore Stations; Atlantic - North

28. _____

1890, July 5. Une coronule de la Baie de Saint-Laurent. (A coronule of the St. Lawrence Bay) Bull. Acad. r. Belg. Cl. Sci., 3rd series, 20(7): 49-54 + 1 plate; 6 references. (French)

Parasites - Ectoparasites

A discussion of J.W. Dawson's paper (Humpback reference #122)

29. Berzin, A.A. and A.A. Rovnin.

1966. The distribution and migration of whales in the northeastern part of the Pacific, Chuckchee and Bering Seas. Isv. tikhookean. naucho-issled. Inst. rȳb. Khoz. Okeanogr., 58: 179-207, 2 tables, 8 figures, 47 references.

Distribution; Antarctic (5); Indian - Eastern South: Australian Waters; Pacific - North: Okhotsk Sea

Translation available from - Office of International Fisheries, Washington, D.C. 20235

30. _____

1978, July-August. Distribution and number of whales in the Pacific whose capture is prohibited. Soviet J. mar. Biol., 4(4): 738-743, 4 figures, 15 references.

Distribution; Antarctic (5); Indian - Eastern
South: Australian Waters; Pacific - North:
Okhotsk Sea

31. Best, P.B.
     1970. Exploitation and recovery of right whales
     Eubalaena australis off the Cape Province.
     Cape Town: Department of Industries,
     Division of Sea Fisheries Investigational
     Report No. 80, 20p., 3 tables, 8 figures, 30
     references.

     BA 1971 (#19026)

     Anatomy - External: Callosities, Coloration;
     Behavior - Reproductive; Behavior - Respir-
     atory; Behavior - Species Interaction: Sperm
     Whales; Behavior - Surface Dependent:
     Flippering, Lobtail; Distribution; Fetus;
     Migration; Parasites - Ectoparasites; Photo-
     graphs - Adult and Calf, Breach; Population
     Dynamics; Stock Assessment; Strandings;
     Whaling - Catch Statistics; Whaling - Shore
     Stations; Atlantic - Eastern South: South
     African Waters

32. ---------
     1973. Status of whale stocks off South Africa,
     1971. Rep. int. Whal. Commn 23, London,
     Appendix 4, Annex J, p. 115-126, 6 tables, 5
     unnumbered tables, 2 figures, 8 references.

     Distribution; Stock Assessment; Atlantic -
     Eastern South: South African Waters

33. --------- and M.J. Roscoe.
     1974. Survey of right whales off South Africa,
     1972, with observations from Tristan da
     Cunha, 1971/72. Rep. int. Whal. Commn 24,
     London, Appendix 4, Annex P, p. 136-141, 2
     tables, 2 figures, 4 references.

     Distribution; Population Dynamics; Stock
     Assessment; Atlantic - Eastern South: South
     African Waters; Atlantic - South: Tristan da
     Cunha Waters

34. ---------
     1975. Status of whale stocks off South Africa,
     1973. Rep. int. Whal. Commn 25, London,

Appendix 5, Annex Q, p. 198-207, 8 tables, 3 unnumbered tables, 8 references.

Distribution; Stock Assessment; Atlantic - South: Tristan da Cunha Waters

35.   ----------
1976. Status of whale stocks off South Africa, 1974. Rep. int. Whal. Commn 26, London, SC/27/Doc 4, p. 264-286, 7 tables, 1 figure, 14 references.

Distribution; Atlantic - South: Gough, Tristan da Cunha Waters; Indian - Western South: Crozet, Marion Waters

36.   ----------
1977. Status of whale stocks off South Africa, 1975. Rep. int. Whal. Commn 27, Cambridge, SC/28/Doc 4, p. 116-121, 7 tables, 5 references.

Distribution; Stock Assessment; Atlantic - Eastern South: South African Waters

37.   ---------- and R.M. McCully.
1979, July. Zygomycosis (phycomycosis) in a right whale (*Eubalaena australis*). J. comp. Pathol. Ther., 89(3): 341-348, 10 figures, 13 references.

Anatomy - Muscles; Photographs - Muscle Lesion; Physiology - Disease; Strandings; Atlantic - Eastern South: South African Waters

38.   Bruce, W.S.
1915. Some observations on Antarctic cetacea. Scottish National Antarctic Expedition, 4(part 20): 491-505 + 2 plates; 1 unnumbered figure.

Photographs - Ventral Coloration Pattern

39.   Buchet, G.
1895. De la baleine des Basques dans les eaux islandaises et de l'aspect des grands cetaces a la mer. (Of the Basque whale in Icelandic waters and aspects of the larger cetaceans of the sea) Mem. Soc. zool. Fr., 8: 229-231 + 3 plates. (French)

Distribution; Photographs - Anterior View, Dorsal View, Ventral View

40. Caldwell, D.K. and M.C. Caldwell.
1971, May 14. Sounds produced by two rare cetaceans stranded in Florida. Cetology, No. 4: 6p., 10 figures, 7 references.

Behavior - Acoustic: Vocalizations; Strandings; Atlantic - Western North: U.S. Waters

41. Capellini, M.
1877, January-June. Sur une baleine proprement dite pechee dans le Golfe de Tarente. (On a true right whale fished in the Gulf of Tarente) C. r. hebd. Seance. Acad. Sci., Paris, 84(10): 1043. (French)

Anatomy - Skeletal; Classification; Mediterranean Sea

42. ---------
1877. Della balene di Tarento confrontata con quelle della Nuova Zelanda e con talune fossili del Belgio e della Toscana. (The Tarento whale compared with the New Zealand whale and with some fossils from Belgium and from Tuscany) Atti Accad. Sci. Ist. Bologna Memorie, 3rd series, 7: 1-34 + 3 plates. (Italian)

Anatomy - Skeletal; Classification

43. Castello, H.P. and M.C. Pinedo.
1979, May. Southern right whales (*Eubalaena australis*) along the southern Brazilian coast. J. Mammal., 60(2): 429-430, 13 references.

Distribution; Strandings; Whaling - Shore Stations; Atlantic - Western South: Brazilian Waters

44.* Chapman, F.P.
1977. Some notes on early whaling in False Bay. Bull. Simon's Town Historical Soc., 9(4): 132-160, 30 references + newspaper files and archival records.

Atlantic - Eastern South: South African Waters

Source: G.J.B. Ross, pers. comm. 1980.

45. Chittleborough, R.G.
1956, August. Southern right whale in Australian waters. J. Mammal., 37(3): 456-457.

Anatomy - External: Callosities; Behavior - Object Interaction: Boat; Behavior - Respiratory; Behavior - Shallow Water; Distribution; Whaling - Shore Stations; Indian - Eastern South: Australian Waters; Pacific - Western South: Australian Waters

46. Clark, C.W. and J.M. Clark.
1980, February 8. Sound playback experiments with southern right whales (*Eubalaena australis*). Science, 207(4431): 663-665, 1 table, 1 figure, 12 references.

Anatomy - External: Callosities; Behavior - Acoustic: Playback, Vocalization; Individual Identification; Atlantic - Western South: Argentine Waters

Abstract of this paper appeared in: Abstracts from presentations at the 3rd biennial conference of the biology of marine mammals, Seattle, October 7-11, 1979, p. 9.

47. ---------
1980, August. A real-time direction finding device for determining the bearing to the underwater sounds of southern right whales, *Eubalaena australis*. J. acoust. Soc. Am., 68(2): 508-511, 5 figures, 16 references.

Behavior - Acoustic: Vocalization; Oceanography; Atlantic - Western South: Argentine Waters

Abstract of this paper appeared in: Abstracts from presentations at the 3rd biennial conference of the biology of marine mammals, Seattle, October 7-11, 1979, p. 10.

48. Clark, E.S., Jr.
1958, March. Right whale (*Eubalaena glacialis*) enters Cape Cod Canal, Massachusetts, U.S.A. Norsk Hvalfangsttid., 47(3): 138-143, 2 unnumbered figures. (English and Norwegian)

Anatomy - External: Callosities; Behavior - Feeding; Behavior - Locomotion; Behavior - Lone Whale; Behavior - Shallow Water; Disribution; Atlantic - Western North: U.S. Waters

49.   Clarke, R.
1965, June. Southern right whales on the coast of Chile. Norsk Hvalfangsttid., 54(6): 121-128, 1 table, 2 figures, 30 references.

Anatomy - External: Callosities, Coloration; Behavior - Reproductive; Behavior - Respiratory; Behavior - Surface Dependent: Flippering, Fluke Up, Spyhop; Distribution; Whaling - Catch Statistics; Whaling - Shore Stations; Pacific - Eastern South: Chilean Waters

50.   Collett, R.
1909, January 12. A few notes on the whale *Balaena glacialis* and its capture in recent years in the North Atlantic by Norwegian whalers. Proc. zool. Soc. Lond., No. 7: 91-97 + 3 plates; 2 unnumbered tables, 1 figure.

Anatomy - External: Callosities, Coloration; Behavior - Reproductive; Behavior - Respiratory; Behavior - Surface Dependent: Breach; Fetus; Food; Parasites - Ectoparasites; Photographs - Antero-Lateral View, Ventral Views of Male and Female; Whaling - Whaling Material; Atlantic - North

51.   Condy, P.R. and A. Burger.
1975, November. A southern right whale at Marion Island. S. Afr. J. Sci., 71(11): 349, 3 references.

Behavior - Lone Whale; Behavior - Shallow Water; Behavior - Surface Dependent: Flippering; Distribution; Indian - Western South: Marion Waters

52.   Cope, E.D.
1865, August. Note on a species of whale occurring on the coasts of the United States. Proc. Acad. nat. Sci., Philad., p. 168-169, 5 references.

Anatomy - Skeletal; Classification; Atlantic - Western North: U.S. Waters

53. ---------
1874, June 16. On the *Balaena cisarctica*. Proc. Acad. nat. Sci., Philad., p. 89.

Anatomy - External: Coloration; Atlantic - Western North: U.S. Waters

54. Cummings, W.C. and L.A. Philippi
1970, May. Whale phonations in repetitive stanzas. San Diego: Naval Undersea Research and Development Center, NUC TP 196, 4p., 2 references.

Behavior - Acoustic: Vocalization; Atlantic - Western North: Canadian Waters

55. ----------; J.F. Fish; P.O. Thompson; and J.R. Jehl, Jr.
1971, November-December. Bioacoustics of marine mammals off Argentina: R/V Hero Cruise 71-3. Antarctic J. U.S., 6(6): 266-268, 2 unnumbered figures, 2 references.

Behavior - Acoustic: Vocalization; Behavior - Agonistic; Behavior - Feeding; Behavior - Shallow Water; Behavior - Species Interaction: Brown-Hooded Gull, Kelp Gull, Killer Whale; Behavior - Surface Dependent: Fluke Up, Mating Activity; Atlantic - Western South: Argentine Waters

56. ---------- ---------- ----------
1972, March 15. Sound production and other behavior of southern right whales, *Eubalaena glacialis*. Trans. S. Diego Soc. nat. Hist., 17(1): 1-14, 1 table, 6 figures, 21 references.

BA 1973 v. 55 (#46859)

Behavior - Acoustic: Playback, Vocalization; Behavior - Agonistic; Behavior - Feeding; Behavior - Locomotion; Behavior - Object Interaction: Boat; Behavior - Respiratory; Behavior - Shallow Water; Behavior - Species Interaction: Brown-Hooded Gull, Kelp Gull, Killer Whale; Behavior - Surface Dependent: Breach, Flippering, Fluke Up, Mating Activity, Spyhop; Photographs - Antero-

Dorsal View, Diving Sequence, Fluke; Atlantic - Western South: Argentine Waters

57. ---------; P.O. Thompson; and J.F. Fish.
1974, March-April. Behavior of southern right whales: R/V Hero Cruise 72-3. Antarctic J. U.S., 9(2): 33-38, 4 figures, 9 references.

Anatomy - External: Callosities; Behavior - Acoustic: Playback, Vocalization; Behavior - Reproductive; Behavior - Species Interaction: Brown-Hooded Gull; Behavior - Surface Dependent: Breach, Mating Activity; Photographs - Antero-Dorsal View, Callosities; Atlantic - Western South: Argentine Waters

58. Daugherty, A.E.
1968, July. Pacific right whale. Anchor 2(7): 35.

Distribution; Pacific - Eastern North: U.S. Waters

Reprinted from: Daugherty, A.E. 1965. Marine mammals of California. Sacramento: Department of Fish and Game, p. 17.

59. Donnelly, B.G.
1967, May. Observations on the mating behaviour of the southern right whale *Eubalaena australis*. S. Afr. J. Sci., 63(5): 176-181, 6 plates, 3 figures, 4 references.

BA 1967 (#99471)

Anatomy - External: Scars; Behavior - Object Interaction: Boat; Behavior - Reproductive; Behavior - Species Interaction: Bottlenose Dolphin, Killer Whale; Behavior - Surface Dependent: Breach, Mating Activity; Oceanography; Photographs - Breach, Callosities, Fluke, Mating Behavior; Atlantic - Eastern South: South African Waters

60. ---------
1969, February. Further observations on the southern right whale, *Eubalaena australis*, in South African waters. J. Reprod. Fert. supp. 6: 347-352 + 2 plates (5 figures); 1 figure, 8 references.

Anatomy - External: Scars; Behavior - Reproductive; Behavior - Respiratory; Behavior - Surface Dependent: Breach, Mating Activity, Spyhop; Photographs - Antero-Dorsal View, Callosities, Mating Behavior, Scars, Spyhop; Atlantic - Eastern South: South African Waters

61. Doran, A.
1877. Capture of a right whale in the Mediterranean. Ann. Mag. nat. Hist., 20(118, paper 41): 328-331.

Review of data presented in Capellini (Reference #42)

62. Duguy, R. and R. Gautier.
1967, March. De la baleine des Basques échouée a l'ile de Ré en 1680. (Of the Basque whale stranded at the Isle of Re in 1680) Annls. Soc. Sci. nat. Charente-Marit., 4(7): 1-2. (French)

Distribution; Strandings; Atlantic - Eastern North: European Waters

The date "1680" given in the title of the published paper is in error and should read "1860". (R. Duguy, pers. comm., 1980)

63. Elliott, H.F.I.
1953, March. The fauna of Tristan da Cunha. Oryx, 2(1): 41-53.

Anatomy - External: Callosities; Behavior - Species Interaction: Man; Atlantic - South: Tristan da Cunha Waters

64. Ellis, R.
1980, June-July. Stalking the right whale in Patagonia. National Wildlife, 18(6): 33-35.

65. Eschricht, D.F. and J. Reinhardt.
1866. On the Greenland right whale (*Balaena mysticetus*, Linn.), with especial reference to its geographical distribution and migrations in times past and present, and to its external and internal characteristics. *In*: Recent memoirs on the cetacea. W.H. Flower (ed.) London: The Ray Society, p. 1-150 + 6 plates

(27 figures); 181 references. (Translated from
the 1861 Danish edition)

Anatomy - Baleen; Anatomy - External:
Callosities; Behavior - Reproductive; Para-
sites - Ectoparasites; Pacific - North:
Hawaiian Waters

66. Fischer, P.
1870. Melanges cétologiques. (Miscellaneous cetological works) Act. Soc. linn. Bordeaux, 3rd
series, 27(7, part 1): 5-22 + 2 plates (5
figures); 50 references. (section 1: Sur un
foetus de baleine australe. 'On a fetus of the
southern right whale' p. 5-9) (French)

Fetus; Migration; Atlantic - Eastern South:
South African Waters; Atlantic - South:
Tristan da Cunha Waters

67. ---------
1871, January-June. Sur une baleine des basques
(*Baleana biscayensis*). (On the Basque whale
'*Balaena biscayensis*') C.r. hebd. Séance.
Acad. Sci., Paris, 72(11): 298-300. (French)

Anatomy - Skeletal; Classification; Atlantic -
Eastern North: European Waters

68. ---------
1871, May-August. Documents pour servir a
l'histoire de la baleine des basques (*Balaena
biscayensis*). (Documents to present the
history of the Basque whale '*Balaena
biscayensis*') Annls. Sci. nat., 5th series,
15(paper 3): 1-20, 2 unnumbered tables, 64
references. (French)

Anatomy - External: Callosities; Anatomy -
Skeletal; Classification; Parasites - Ectoparasites; Strandings

69. ---------
1881. Cétacés du sud-ouest de la France. (Whales
of the south-west of France) Act. Soc. linn.
Bordeaux, 4th series, 35(5): 219p. + 8 plates
(40 figures); 2 tables, 5 unnumbered tables, 7
figures, 318 references. (French)

Anatomy - Skeletal; Classification; Distribution; Migration; Parasites - Ectoparasites;

Atlantic - Eastern North: European Waters; Mediterranean Sea

70. Fraser, F.C.
1937. Early Japanese whaling. Proc. Linn. Soc. Lond., 31(12, part 1): 19-21 + 1 plate (2 figures).

Entrapments - Net; Whaling - Whaling Material; Pacific - Western South: New Zealand Waters

71. Gasco, F.
1878. Intorno alla balena press in Tarento nel Febbrajo 1877. (About the whale caught at Tarento in February 1877) Atti. Acad. Sci. fis. mat., Napoli, 7(16): 1-47 + 9 plates (61 figures); 58 references. (Italian)

Anatomy - Baleen; Anatomy - Circulation; Anatomy - Skeletal; Classification; Mediterranean Sea

72. ---------
1878, September 9. La *Balaena (Macleayius) australiensis* du musee de Paris, comparee a la *Balaena biscayensis* de l'Universite de Naples. (The *Balaena 'Macleayius' australiensis* from the Paris Museum compared to the *Balaena biscayensis* of the University of Naples) C.r. hebd. Seance. Acad., Paris, 87(11): 410-412. (French)

Anatomy - Skeletal; Classification; Mediterranean Sea

Translated and republished in: Ann. Mag. nat. Hist., 5th series, 2:12 (December, 1878), p. 495-497.

73. ---------
1879, July 11. La *Balaena macleayius* del museo di Parigi (*Macleayius australiensis* Gray = *B. australis* Desmoulins) descritta. (The *Balaena macleayius* in the Paris Museum '*Macleayius australiensis* Gray = *B. australis* Desmoulins' described) Annali Mus. civ. Stor. nat. Giacomo Doria, 14: 509-551, 15 unnumbered tables, 2 unnumbered figures, 37 references. (Italian)

Anatomy - Skeletal; Classification

74. ----------
    1879, August 7. Il balenotto catturato nel 1854 a
    San Sebastiano (Spagna)(*Balaena biscayensis*,
    Eschricht) per la prima volta descritto. (The
    whale captured in 1854 at San Sebastiano
    'Spain' '*Balaena biscayensis*, Eschricht' des-
    cribed for the first time) Annali Mus. civ.
    Stor. nat. Giacomo Doria, 14: 573-608, 3
    unnumbered tables, 36 references. (Italian)

    Anatomy - Skeletal; Classification; Parasites
    - Ectoparasites; Atlantic - Eastern North:
    European Waters

75. Gaskin, D.E.
    1964, July. Return of the southern right whale
    (*Eubalaena australis* Desm.) to New Zealand
    waters, 1963. Tuatara, 12(2): 115-118, 1
    unnumbered figure, 5 references.

    BA 1965 (#36666)

    Anatomy - External: Callosities; Distri-
    bution; Whaling - Shore Stations; Pacific -
    Western South: New Zealand Waters

76. ----------
    1968. The New Zealand cetacea. Wellington, N.Z.:
    New Zealand Marine Department, Fisheries
    Research Division, Fisheries Research
    Bulletin No. 1 (New Series), 92p., 30 tables,
    80 figures, 311 references.

    Behavior - Surface Dependent: Mating
    Activity; Distribution; Migration; Stock
    Assessment; Whaling - Catch Statistics;
    Pacific - Western South: New Zealand Waters

77. ---------- and G.J.D. Smith.
    1979, December. Observation of marine mammals,
    birds and environmental conditions in the
    Head Harbour region of the Bay of Fundy. *In*:
    Evaluation of recent data relative to
    potential oil spills in the Passamaquoddy area.
    D.J. Scarratt (ed.), Ottawa: Fisheries and
    Environment Canada, Fisheries and Marine
    Service, Technical Report No. 901: 69-86.

Behavior - Nursing; Distribution; Entrapments - Net; Atlantic - Western North:
Canadian Waters

78. Gervais, P.
1871, January-June. Remarques sur l'anatomie des
cétacés de la division des balenides, tirées de
l'examen des pieces relatives à ces animaux
qui sont conservées au museum d'histoire
naturelle. (Remarks on the anatomy of the
cetaceans of the division Balaenidae, based
on the examination of fragments relative to
these animals which are conserved at the
museum of natural history) C.r. hebd. Séance.
Acad. Sci., Paris, 72(22): 663-672, 9
references. (French)

Anatomy - Skeletal; Classification

79. ---------
1877. Documents nouveaux relatifs aux balenides.
(New documents relative to the whales) J.
Zool., 6: 285-288 + 4 plates (12 figures); 18
references. (French)

Anatomy - Ear; Anatomy - Skeletal

80. Giglioli, H.H.
1882, March 30. The Basque whale in the
Mediterranean. Nature, 25(648): 505.

Distribution; Mediterranean Sea

Letter in response to Markham's (Reference
#113) February 16, 1882 paper in Nature,
informing Markham and the readership of the
right whale in the Mediterranean described by
Gasco (Reference #71) in 1878.

81. Gilmore, R.M.
1956, July-August. Rare right whale visits
California. Pacific Discovery, 9(4): 20-25, 9
unnumbered figures.

Behavior - Surface Dependent: Breach,
Flippering, Lobtail; Distribution; Photographs
- Breach; Pacific - Eastern North: U.S.
Waters

82. ---------
1969, November-December. Populations, distri-

bution, and behavior of whales in the Western South Atlantic: Cruise 69-3 of R/V Hero. Antarctic J. U.S., 4(6): 307-308, 3 unnumbered figures.

Behavior - Reproductive; Distribution; Photographs - Mating Behavior; Atlantic - Western South: Argentine Waters

83. ---------
1978. Right whales. *In:* Marine mammals of Eastern North Pacific and Arctic waters. D. Haley (ed.), Seattle: Pacific Search Press, p. 62-69, 3 unnumbered figures.

Anatomy - Baleen; Anatomy - External: Callosities; Behavior - Acoustic: Vocalization; Behavior - Species Interaction: Killer Whale; Behavior - Surface Dependent: Breach; Distribution; Migration; Photographs - Breach; Pacific - Eastern North; Pacific - North: Bering Sea

84.* Gomez de Llarena, J.
1952. Restos de una ballena en la playa de Deva. (Remains of a whale on the beach of Deva) Munibe, 4: 220-223. (Spanish)

Anatomy - Skeletal; Distribution; Atlantic - Eastern North: European Waters

Source: Casinos, A. and J.R. Vericad. The cetaceans of the Spanish coasts: A survey. Mammalia, 40:2 (1976), p. 267-289.

85. Graells, M.P.
1889. Las ballenas en las costas oceanicas de España - noticias recogidas é investigaciones hechas. (The whales on the oceanic coasts of Spain - records and investigations) Mems. R. Acad. Cienc. exact. fis. nat. Madr., 5(13, part 3), 115p. + 9 plates (36 figures); 46 references. (Spanish)

Distribution; Atlantic - Eastern North: European Waters

86. Gray, J.E.
1864, April 26. Note on the bonnet of the right whale. Proc. zool. Soc. Lond., No. 11(paper 1): 170-171.

Anatomy - External: Callosities; Parasites - Ectoparasites; Pacific - North: Hawaiian Waters

87. ---------
1864, May 24. On the cetacea which have been observed in the seas surrounding the British Islands. Proc. zool. Soc. Lond., No. 14: 195-248, 2 unnumbered tables, 24 figures.

Anatomy - Skeletal; Classification; Atlantic - Eastern North: British Isles Waters

88. ---------
1864, November 22. Notice of the atlas and other cervical vertebrae of a right whale in the Museum of Sydney, New South Wales. Proc. zool. Soc. Lond., Nos. 37-38(paper 1): 587-594, 5 figures.

Anatomy - Skeletal; Classification

89. ---------
1868. On the geographical distribution of the Balaenidae or right whales. Ann. Mag. nat. Hist., 4th series, 1(4, paper 31): 242-247.

Distribution

90. ---------
1870, September. Observations on the whales described in the 'Osteographie'. Ann. Mag. nat. Hist., 4th series, 6(33, paper 17): 193-204.

Anatomy - Skeletal; Classification

91. ---------
1873. Notice of the skeleton of the New Zealand right whale (*Macleayius australiensis*). Trans. Proc. N.Z. Inst., 6(2, paper 17): 90-92 + 2 plates (4 figures).

Anatomy - Skeletal; Classification

92. ---------
1873. On the *Macleayius australiensis* from New Zealand. Ann. Mag. nat. Hist., 4th series, 11(61, paper 10): 75-76.

Anatomy - Skeletal; Classification

93. ---------

1873, January 21. Notice of the skeleton of the New Zealand right whale (*Macleayius australiensis*) and other whales, and other New Zealand mammalia. Proc. zool. Soc. Lond., No. 9(paper 6): 129-145, 5 figures, 2 references.

Anatomy - Skeletal; Classification

94. Guldberg, G.A.

1884. Sur la présence aux temps anciens et modernes de la baleine de biscaye (ou nordcaper) sur les côtes de Norwége. (On the presence in ancient and modern times of the biscaye whale 'or nordcaper', on the coasts of Norway) Bull. Acad. r. Belg. Cl. Sci., 3rd series, 7(4): 374-402, 2 references. (French)

Anatomy - Skeletal; Distribution; Atlantic - Eastern North: Scandinavian Waters

95. ---------

1884, June 12. The north cape whale. Nature, 30(763): 148-149.

Distribution; Atlantic - North; Atlantic - Eastern North: European Waters; Atlantic - Western North: U.S. Waters

96. ---------

1889. Nordkapern eller biskayerhvalen. (Nordcaper or biscaye whale) Naturen, 13: 1-11 + 1 plate; 5 references. (Norwegian)

97. ---------

1891. Bidrag til naiere kundskab on Atlanterhavets rethval, Eubalaena biscayensis, Eschricht. (Contribution to the knowledge of the Atlantic Ocean nordcaper 'right whale', *Eubalaena biscayensis*, Eschricht) Fohr. VidenskSelsk. Krist., No. 8: 1-14, 10 references. (Norwegian)

98. ---------

1893, May 20. Zur kenntniss des nordkapers (*Eubalaena biscayensis* Eschr). (Toward the knowledge of the nordcapers '*Eubalaena biscayensis* Eschr.') Zool. Jb., 7(1): 1-22 + 2 plates (7 figures); 23 references. (German)

ArB 1953 (#6353)

Anatomy - Skeletal; Photographs - Antero-Dorsal, and Antero-Ventral Views of Male

99. ---------
1903, December 15. Ueber die wanderungen verschiedener bartenwale. (On the various migration routes of the baleen whale) Biol. Zbl., 23(24): 803-816, 7 references. (continued in 24(11), June 1, 1904, p. 371-374, 2 references.) (German)

Distribution; Migration

Part 1 contains right whale information

100. Harrison, J.P.
1954, April. An 1849 statement on the habits of right whales by Captain Daniel McKenzie of New Bedford. American Neptune, 14(2): 139-141.

Behavior - Feeding; Behavior - Reproductive; Behavior - Shallow Water; Distribution; Whaling - Logbooks/Journals; Atlantic - South: Gough and Tristan da Cunha Waters; Atlantic - Western South: Brazilian and Falkland Waters; Pacific - Western South: New Zealand Waters

101. Herman, L.M.; C.S. Baker; P.H. Forestell; and R.C. Antinoja.
1980, May 31. Right whale *Balaena glacialis* sightings near Hawaii: A clue to the wintering grounds? Mar. Ecol. Prog. Ser., 2: 271-275, 2 figures, 24 references.

Anatomy - External: Callosities, Coloration; Behavior - Acoustic: Vocalization; Behavior - Species Interaction: Bottlenose Dolphin, Humpback Whale; Behavior - Surface Dependent: Flippering, Surface Active; Distribution; Individual Identification; Photographs - Callosities, Fluke; Pacific - North: Hawaiian Waters

Preliminary report appeared as Baker, C.S.; P.H. Forestell; R.C. Antinoja; and L.M. Herman. (Abstract) Interactions of the Hawaiian humpback whale, *Megaptera novae-angliae*, with the right whale, *Balaena glacialis*, and odontocete cetaceans. Seattle: Abstracts from presentations at the third biennial conference of the biology of marine mammals, October 7-11, 1979, p. 2.

102. Holder, J.B.
     1883, May 1. The Atlantic right whale: A contribution embracing an examination of I. The exterior characters and osteology of a cisarctic right whale - male. II. The exterior characters of a cisarctic right whale - female. III. The osteology of a cisarctic right whale - sex not known. To which is added a consise resume of historical mention relating to the present and allied species. Bull. Am. Mus. nat. Hist., 1(4, paper 6): 97-137 + 4 plates; 63 references.

     Anatomy - External: Callosities; Anatomy - Skeletal; Behavior - Reproductive; Classification; Atlantic - Western North: U.S. Waters

103. Imaizumi, Y.
     1958. (On a mounted skeleton of *Eubalaena glacialis japonica* in the National Science Museum) Journal of the Tokyo National Science Museum - Natural Science and Museum, 25: 13-15, 1 unnumbered figure. (Japanese)

     Anatomy - Skeletal; Photographs - Skeletal Remains

104. Ivanova, E.I.
     1961. K morfologii yaponskogo kita (*Eubalaena sieboldi* Gray). (Contribution to the morphology of '*Eubalaena sieboldi* Gray') Trudȳ Inst. Morf. Zhivot., 34: 216-225, 1 table, 3 references. (Russian)

     BA 1964 (#26737)

     Anatomy - Skeletal; Classification

105. Iwasa, M.
      1934, July. Two species of whale-lice (Amphipoda,
      Cyamidae) parasitic on a right whale. J. Fac.
      Sci. Hokkaido Univ., 6th series, Zoology,
      3(1): 33-39 + 4 plates (54 figures); 7
      references.

      Parasites - Ectoparasites

106. Kamiya, T.
      1958, September. How to count the renculi of
      cetacean kidneys with special regard to the
      kidney of the right whale. Scient. Repts.
      Whales Res. Inst., Tokyo, No. 13: 253-267, 4
      tables, 2 appendix tables, 2 figures, 11
      references.

      Anatomy - Endocrine

107. Klumov, S.K.
      1962. Gladkie kity tikhogo okeana. (The right
      whales in the Pacific Ocean) *In*: Biolog-
      icheskie Issledovaniya marya (Plankton)
      (Biological marine studies 'plankton') P.I.
      Usachev (ed.), Trudȳ Inst. Okeanogr., 58:
      202-297, 16 tables, 51 figures, 70 references.
      (Russian, English summary)

      BA 1964 (#8698), 1965 (#4568)

      Anatomy - External: Callosities, Coloration;
      Anatomy - Reproductive; Distribution; Food;
      Growth Rates; Migration; Parasites - Ecto-
      parasites; Photographs - Male Ventral
      Coloration Pattern; Pacific - North; Pacific -
      Western North

108. Koeford, E.
      1905, February. Notiser om nordkaper og kaskelot.
      (Notice on the right whale and the sperm
      whale) Naturen, 29(2): 54-56, 2 figures.
      (Norwegian)

      Photographs - Antero-Lateral View

109. Liouville, J.
      1913. La balaena franche (*Balaena australis*
      Desm.) dans l'Antarctique.--L'y a-t-on jamais
      indis-cutablement observée?--ses migrations.
      (The right whale '*Balaena australis* Desm.' in
      the Antarctic -- Has anyone ever observed

one there? -- migrations) *In*: Cétaces de l'Antarctique (Baleinopteres, Ziphiided, Delphinides). Deuxième Expédition Antarctique Française (1908-1910) Commandée par Le Dr. Jean Charcot. Sciences Naturelles: Documents Scientifiques. Paris: Masson et c^{le}, p. 5-27. (French)

Distribution; Fetus; Migration; Antarctic

110. Lönnberg, E.
1906. Contributions to the fauna of South Georgia. 1. Taxonomic and biological notes on vertebrates. K. Svenska. Vetensk-Akad. Handl., 40(5): 41-49.

Anatomy - Baleen; Anatomy - External: Callosities, Coloration, Hair; Behavior - Feeding; Behavior - Respiratory; Fetus; Parasites - Ectoparasites; Atlantic - Western South: South Georgia Waters

Balaena australis p. 41-49. Entire paper not available to me.

111. ----------
1923. Cetological notes. Ark. Zool., 15(4, paper 24): 1-18, 6 figures, 3 references.

Anatomy - Flippers/Flukes; Anatomy - Skeletal; Photographs - Flipper, Skeletal Remains

112. Manigault, G.E.
1885, September. The black right whale captured in Charleston Harbor January, 1880. Proceedings of the Elliot Society, p. 98-104, 3 figures, 1 reference.

Anatomy - External: Coloration; Anatomy - Skeletal; Classification; Atlantic - Western North: U.S. Waters

113. Markham, C.R.
1881, December 13. On the whale-fishery of the Basques Provinces of Spain. Proc. zool. Soc. Lond., No. 62(paper 1): p. 969-976.

Whaling - Catch Statistics; Whaling - Shore Stations; Whaling - Whaling Material; Atlantic - Eastern North: European Waters

Revised edition of this paper published in
Nature, 25: 642 (February 16, 1882), p. 365-
368.

114. Matsuura, Y.
1936. (Studies on the right whale, *Balaena
glacialis* Bonnaterre, in the adjacent waters
of Japan) Botany Zool., Tokyo, 4(4): 696-702,
3 tables, 2 unnumbered tables, 3 figures, 16
references. (Japanese)

Distribution; Whaling - Catch Statistics;
Pacific - Western North: Japanese Waters

115. Matthews, L.H.
1938, April. Notes on the southern right whale,
Eubalaena australis. 'Discovery' Rep., 17:
169-182 + 6 plates (19 figures); 2 tables, 15
references.

Anatomy - Baleen; Anatomy - External:
Callosities, Coloration, Hair; Anatomy -
Reproductive; Behvior - Object Interaction:
Rocks; Behavior - Shallow Water; Behavior -
Surface Dependent: Breach, Flippering; Para-
sites - Ectoparasites; Parasites - Endo-
parasites; Photographs - Baleen Plates,
Female Dorsal View and Ventral Coloration
Pattern, Male Ventral Coloration Pattern;
Physiology - Lactation; Atlantic - Eastern
South: South African Waters; Atlantic -
Western South: South Georgia Waters

116. Mermoz, J.F.
1980. Preliminary report on the southern right
whale in the Southwestern Atlantic. Rep. int.
Whal. Commn 30, Cambridge, SC/31/Doc 4, p.
183-186, 1 figure, 31 references.

Behavior - Surface Dependent: Flippering,
Fluke Up; Distribution; Food; Migration;
Strandings; Atlantic - Western South:
Argentine and Uruguayan Waters

117. Miranda-Ribeiro, A. de.
1931, December. Notes cétologiques. (Cetological
notes) Bull. Soc. port. Sci. nat., 11(11): 145-
153, 1 unnumbered table, 4 figures, 6
references. (Portuguese)

Anatomy - Skeletal; Classification; Photographs - Skeletal Remains

118. ---------
1932. As pretensas especies de baleias lisas do Atlantico. (The apparent species of the smooth whales of the Atlantic) Boln. Mus. nac. Rio de Janeiro, 8: 1-11 + 7 plates; 26 references. (Portuguese)

Anatomy - External: Callosities; Anatomy - Skeletal; Classification; Photographs - Callosities, Dorsal and Lateral Views, Skeletal Remains

119. Mitchell, E.
1974. Trophic relationships and competition for food in Northwest Atlantic whales. *In*: Proceedings of the Canadian Society of Zoologists Annual Meeting, June 2-5, M.D.B. Burt (ed.), p. 123-133, 4 figures, 21 references.

Behavior - Feeding; Food; Stock Assessment; Atlantic - Western North

Contains a translation of a portion of Omura's 1974 book - (Ecology of the whale) - discussing the competition for food between the right whale and other baleen whales in the southern hemisphere (Reference #135)

120. Moore, J.C. and E. Clark.
1963, July 19. Discovery of right whales in the Gulf of Mexico. Science, 141(3577): 269, 6 references.

BA 1964 (#26744)

Anatomy - External: Callosities, Coloration; Distribution; Photographs - Adult and Calf; Atlantic - Western North: Gulf of Mexico

121. Moses, S.T.
1947. Whales in Baroda (Orca stranded at Aramda in 1943 and Balaena at Gajana in 1944). Bull. Dept. Fish. Baroda, 10: 1-5, 8 references.

Anatomy - Skeletal; Distribution; Strandings; Indian - Western North: Arabian Sea

Paper on this subject delivered and published as an abstract in the Proceedings of the 34th Indian Science Congress, 3 (1948), p. 188, Abstract #62.

122. Muller, J.
1954, February 22. Observations on the orbital region of the skull of the Mystacoceti Zool. Meded., Leiden, 32(33): 279-290, 2 tables, 3 figures, 29 references.

Anatomy - Skeletal; Classification

123. Neave, D.J. and B.S. Wright.
1968, May. Seasonal migrations of the harbor porpoise (*Phocoena phocoena*) and other cetacea in the Bay of Fundy. J. Mammal., 49(2): 259-264, 1 table, 4 figures, 10 references.

Distribution; Migration; Atlantic - Western North: Canadian Waters

W.E. Schevill comments on this paper in J. Mammal., 49:4 (November 26, 1968), p. 794-796.

Neave and Wright respond in J. Mammal., 50:3 (August, 1969), p. 653-654.

124. Nemoto, T.
1959, September. Food of baleen whales with reference to whale movement. Scient. Repts. Whales Res. Inst., Tokyo, No. 14: 149-290 + 1 plate (14 figures); 43 tables, 2 appendix tables, 48 figures, 145 references.

Behavior - Feeding; Food

125. ---------
1964, March. School of baleen whales in the feeding areas. Scient. Repts. Whales Res. Inst., Tokyo, No. 18: 89-110, 22 tables, 7 figures, 26 references.

Population Dynamics; Pacific - North

126. --------- and T. Kasuya.
1965, April. Foods of baleen whales taken in the Gulf of Alaska of the North Pacific. Scient.

Repts. Whales Res. Inst., Tokyo, No. 19: 45-
51, 7 figures, 9 references.

Behavior - Feeding; Food; Pacific - Eastern
North: Alaskan Waters

127. Newton, E.T.
    1886. A contribution to the history of the cetacea
    of the Norfolk Forest Bed. Q. Jl. geol. Soc.
    Lond., 42(paper 26): 316-324 + 1 plate (8
    figures); 2 unnumbered tables, 1 reference.

    Anatomy - Skeletal; Classification

128. Nishiwaki, M. and Y. Hasagawa.
    1969, June. The discovery of the right whale skull
    in the Kisagata Shell Bed. Scient. Repts.
    Whales Res. Inst., Tokyo, No. 21: 79-84 + 3
    plates (10 figures); 1 table, 2 figures, 10
    references.

    BA 1971 (#100368)

    Anatomy - Skeletal; Photographs - Skeletal
    Remains

129. Ommanney, F.D.
    1933, December. Whaling in the Dominion of New
    Zealand. 'Discovery' Rep., No. 7: 239-252 + 3
    plates (10 figures); 1 figure, 10 references.

    Behavior - Object Interaction: Rocks;
    Behavior - Reproductive; Behavior - Shallow
    Water; Migration; Parasites - Ectoparasites;
    Whaling - Shore Stations; Whaling - Whaling
    Material; Pacific - Western South: New
    Zealand Waters

130. Omura, H.
    1957, July. Report on two right whales caught off
    Japan for scientific purposes under Article
    VIII of the International Convention for the
    Regulation of Whaling. Norsk Hvalfangsttid.,
    46(7): 374-386, 389, 390, 2 tables, 23
    figures. (English and Norwegian)

    Anatomy - External: Coloration, Scars;
    Anatomy - Skeletal; Food; Parasites -
    Ectoparasites; Photographs - Cyamids, Female
    Ventral Coloration Pattern, Male Lateral

View, Scratch Marks, Skeletal Remains; Pacific - Western North: Japanese Waters

Preliminary paper by M. Nishiwaki published in Geiken Tsushin, No. 64 (January, 1957), p. 160-179, unnumbered tables, 25 unnumbered plates, 3 figures. (Japanese)

131. ---------

1958, September. North Pacific right whale. Scient. Repts. Whales Res. Inst., Tokyo, No. 13: 1-52 + 8 plates (20 figures); 18 tables, 27 figures, 35 references.

Anatomy - Baleen; Anatomy - External: Callosities, Coloration, Hair, Scars; Anatomy - Reproductive; Anatomy - Skeletal; Classification; Distribution; Food; Migration; Parasites - Ectoparasites; Photographs - Baleen Plates, Callosities, Coloration Pattern, Cyamids, Hair, Scars, Skeletal Remains, Ventral Views of Male and Female; Whaling - Catch Statistics; Pacific - North

Paper based on this reference published in: Geiken Tsushin, No. 88 (January 1959), p. 589-616, 1 table, 8 figures. (Japanese)

132. ---------

1964, March. A systematic study of the hyoid bones in baleen whales. Scient. Repts. Whales Res. Inst., Tokyo, No. 18: 149-170 + 15 plates (182 figures); 2 tables, 11 figures.

Anatomy - Skeletal

133. ---------; S. Ohsumi; T. Nemoto; K. Naso; and T. Kasuya.

1969, June. Black right whales in the North Pacific. Scient. Repts. Whales Res. Inst., Tokyo, No. 21: 1-78 + 18 plates (45 figures); 38 tables, 4 appendix tables, 27 figures, 64 references.

BA 1971 (#100790)

Age Determination; Anatomy - Baleen; Anatomy - External: Callosities, Coloration, Hair, Scars; Anatomy - Reproductive; Anatomy - Skeletal; Behavior - Feeding; Behavior - Respiratory; Behavior - Species

Interaction: Dolphin; Behavior - Surface
Dependent: Breach; Classification; Distri-
bution; Fetus; Food; Growth Rates; Parasites
- Ectoparasites; Photographs - Callosities,
Ear Bones, Fetus, Hair, Skeletal Remains;
Weight; Whaling - Catch Statistics; Pacific -
North

Preliminary papers published by T. Kasuya in
Geiken Tsushin, No. 134 (October, 1962), p.
193-210, 16 tables, 24 plates, 4 figures; T.
Nemoto in Geiken Tsushin, No. 146 (October,
1963), p. 173-192, 14 tables, 24 plates, 4
figures; S. Ohsumi in Geiken Tsushin, No. 122
(October, 1961), p. 177-203, 15 tables, 36
plates, 4 figures; H. Omura in Geiken Tsushin,
No. 88 (January, 1959), p. 598-616, 1 table, 8
figures; and H. Omura and S. Kawamura in
Geiken Tsushin, No. 205 (September-October,
1968), p. 199-220, 7 tables, 13 plates, 5
figures. (All papers published in Japanese)

Later papers published by H. Omura in Geiken
Tsushin, No. 223 (March, 1970), p. 1-5, 2
tables, 7 figures; and Geiken Tsushin, No. 224
(April, 1970), p. 106, 3 tables, 3 figures.
(Both papers published in Japanese)

134.   ---------; M. Nishiwaki; and T. Kasuya.
        1971, September. Further studies on two skeletons
        of the black right whale in the North Pacific.
        Scient. Repts. Whales Res. Inst., Tokyo, No.
        23: 71-81, 4 tables, 2 appendix tables, 3
        figures, 4 references.

        BA 1973 v. 53 (#65360)

        Anatomy - Skeletal; Photographs - Skeletal
        Remains; Sexual Dimorphism

135.*   ---------
        1974. (Ecology of the Whale) Kyoritsu Shippan K.
        K., 186p. (Japanese)

        Behavior - Feeding; Distribution; Stock
        Assessment

        Source: Reference #119.

136.   Payne, R. and K. Payne.
        1971, Winter. Underwater sounds of southern right

whales. Zoologica, N.Y., 56(4): 159-165, 1
plate, 2 figures, 13 references.

BA 1973 v. 55 (#28752)

Behavior - Acoustic: Vocalization; Behavior -
Greeting; Behavior - Lone Whale; Behavior -
Nocturnal; Photographs - Adult, Calf, and
Third Whale; Atlantic - Western South:
Argentine Waters

137. ----------
1972, Spring. Report from Patagonia: The right
whales. New York Zoological Society Newsletter, foldout.

Anatomy - External: Callosities; Behavior -
Object Interaction: Seaweed; Behavior - Play;
Behavior - Reproductive; Behavior - Surface
Dependent: Flippering, Mating Activity; Con-
servation; Individual Identification; Photo-
graphs - Adult and Calf, Callosities, Flipper,
Skeletal Remains; Whaling - Shore Stations;
Atlantic - Western South: Argentine Waters

138. ----------
1972, October. Swimming with Patagonia's right
whales. Natn. geogr. Mag., 142(4): 576-587,
10 unnumbered figures.

Anatomy - External: Callosities; Behavior -
Acoustic: Vocalization; Behavior - Agonistic;
Behavior - Object Interaction: Boat,
Seaweed; Behavior - Play; Behavior -
Reproductive; Behavior - Respiratory;
Behavior - Surface Dependent: Fluke Up,
Mating Activity; Conservation; Individual
Identification; Parasites - Ectoparasites;
Photographs - Adult and Calf, Callosities,
Cyamids, Mating Behavior, Underwater View;
Atlantic - Western South: Argentine Waters

139. ----------
1974, April. A playground for whales. Animal
Kingdom, 77(2): 7-12, 6 unnumbered figures.

Anatomy - External: Callosities, Coloration;
Behavior - Acoustic: Vocalization; Behavior -
Feeding; Behavior - Object Interaction:
Seaweed, Sonobuoys, Tide Gauges; Behavior -
Species Interaction: Porpoise, Sea Lion;

Behavior - Surface Dependent: Breach, Flippering, Lobtail; Conservation; Individual Identification; Photographs - Adult and Calf, Callosities, Feeding; Atlantic - Western South: Argentine Waters

140. ----------
1976, March. At home with right whales. Natn. geogr. Mag., 149(3): 322-339, 21 unnumbered figures.

Anatomy - External: Callosities, Coloration; Behavior - Agonistic; Behavior - Reproductive; Behavior - Shallow Water; Behavior - Surface Dependent: Breach, Flippering, Lobtail, Mating Activity; Conservation; Individual Identification; Measuring Techniques; Parasites - Ectoparasites; Photographs - Breach, Callosities, Cyamids, Mating Behavior, Underwater View; Atlantic - Western South: Argentine Waters

141. ---------- and K. Payne.
1977. Wettlauf mit den Jagern verhaltensforschung an glattwalen. (Race with the hunters - behavioral investigations of right whales) Bild der Wissenschaft, p. 42-49, 5 unnumbered figures. (German)

Anatomy - External: Callosities, Coloration; Behavior - Greeting; Behavior - Object Interaction: Seaweed; Behavior - Play; Behavior - Surface Dependent: Breach, Flippering, Lobtail; Individual Identification; Measuring Techniques; Photographs - Adult and Calf, Baleen, Callosities, Coloration Pattern, Measuring Technique; Atlantic - South: Tristan da Cunha Waters; Atlantic - Western South: Argentine Waters

142. ---------- and E.M. Dorsey.
1979. (Abstract) Sexual dimorphism in the callosities of southern right whales. Seattle: Abstracts from presentations at the third biennial conference of the biology of marine mammals, October 7-11, p. 46.

Anatomy - External: Callosities, Scars; Behavior - Agonistic; Individual Identification; Sexual Dimorphism; Atlantic - Western South: Argentine Waters

143.   ---------

1980. (Abstract) Behavior of southern right whales, *Balaena glacialis australis*. *In*: Abstracts of papers of the 146th national meeting 3-8 January, San Francisco, California. A. Herschman (ed.), Washington D.C.: American Association for the Advancement of Science, AAAS Publication 80-2, p. 41.

Anatomy - External: Callosities; Behavior - Play; Behavior - Reproductive; Behavior - Surface Dependent: Breach, Flippering, Lobtail; Individual Identification; Oceanography; Population Dynamics; Stock Assessment; Atlantic - Western South: Argentine Waters

144.   Pilleri, G.

1964. Morphologie des gehirned des "southern right whale", *Eubalaena australis* Desmoulins 1822 (cetacea, mysticeti, balaenidae). (Brain morphology of the 'southern right whale', *Eubalaena australis* Desmoulins 1822 'cetacea, mysticeti, balaenidae') Acta Zool. Stock., 45(3): 245-272, 22 figures, 21 references. (German)

Anatomy - Nervous; Photographs - Brain Anatomy

145.   Pouchet, G. and H. Beauregard.

1888, March 19. Sur la presence de deux baleines franches dans les eaux d'Alger. (On the presence of two right whales in Algerian waters) C.r. Hebd. Seance. Acad. Sci., Paris, 106(12): 875-876. (French)

Distribution; Mediterranean Sea

146.   ---------

1890, December 13. A propos de deux photographies de baleine franche. (An observation on two photographs of the right whale) C.R. Soc. Biol., 9th series, 2(37): 705-708, 5 references. (French)

Anatomy - External: Callosities; Classification; Parasites - Ectoparasites; Atlantic - Western North: U.S. Waters; Mediterranean Sea

147.   ---------- and H. Beauregard.
         1891, December 7. Nouvelle liste d'echouements
         de grands cetaces sur la cote francaise. (New
         list of strandings of large cetaceans on the
         French coast) C.r. Hebd. Seance. Acad. Sci.,
         Paris, 113(23): 810-813. (French)

         Distribution; Strandings; Mediterranean Sea

148.   Reeves, R.R.
         1975. A whale in Newark Bay. New Jersey Out-
         doors, 2(5): 18-20.

         Distribution; Atlantic - Western North: U.S.
         Waters

149.   ----------
         1975, August-September. The right whale. The
         Conservationist, 30(1): 32-33, 45.

         Behavior   -   Object   Interaction: Buoys;
         Behavior - Species Interaction: Leatherback
         Turtle;   Distribution;   Atlantic   -   Western
         North: U.S. Waters

150.   ----------; J.G. Mead; and S. Katona.
         1978. The right whale, *Eubalaena glacialis*, in the
         Western   North   Atlantic.   Rep.   int.   Whal.
         Commn 28, Cambridge, SC/29/Doc 44, p. 303-
         312,   1   table,   3   unnumbered   figures,   60
         references.

         Behavior - Acoustic: Vocalization; Behavior -
         Species Interaction: Killer Whale; Conser-
         vation;   Distribution;   Entrapments   -   Net;
         Individual   Identification;   Migration;   Pollu-
         tants; Strandings; Whaling - Shore Stations;
         Atlantic - Western North: U.S. Waters

151.   Rice, D.W. and C.H. Fiscus.
         1968, September-October. Right   whales   in   the
         Southeastern   North   Pacific.   Norsk   Hval-
         fangsttid.,   57(5):   105-107,   1   figure,   13
         references.

         Anatomy - External: Callosities; Distribution;
         Photographs - Anterior Dorsal View, Callo-
         sities; Pacific - Eastern North: Mexican, U.S.
         Waters

152. Ridewood, W.G.
          1901, February 5. On the structure of the horny
          excrescence, known as the 'bonnot' of the
          southern right whale '*Balaena australis*'. Proc.
          zool. Soc. Lond., 1(3, paper 2): 44-47 + 1
          plate (10 figures); 10 references.

          Anatomy - External: Callosities; Parasites -
          Ectoparasites

153. Robinson, N.H.
          1979, September-October. Recent records of
          southern right whales in New South Wales.
          Victorian Nat., 96(5): 168-169, cover
          photograph, 5 references.

          Anatomy - External: Callosities, Coloration;
          Behavior - Locomotion; Distribution; Photo-
          graphs - Callosities; Pacific - Western South:
          Australian Waters

154.* Ross, G.J.B.
          1969. An unusual stranding of a southern right
          whale, *Eubalaena glacialis* at Plettenberg Bay.
          East. Cape Nat., No. 37: 28-29, 1 reference.

          Behavior - Care Giving; Strandings; Atlantic
          - Eastern South: South African Waters

          Source: G.J.B. Ross, Pers. comm., 1980.

155.* ---------
          1971. A note on early whaling at the Cape of
          Good Hope. Africana Notes and News, 19(7):
          300-302.

          Whaling - Shore Stations; Atlantic - Eastern
          South: South African Waters

          Source: G.J.B. Ross, pers. comm., 1980

156. ---------
          1973. Are the whales coming right? Afr. Wildlife,
          27: 7-9, 3 unnumbered figures, 5 references.

          Anatomy - External: Callosities; Behavior -
          Reproductive; Photographs - Callosities,
          Stranded Calf; Stock Assessment; Strandings;
          Whaling - Shore Stations; Atlantic - Eastern
          South: South African Waters; Pacific -
          Western South: New Zealand Waters

157. Rowntree, V.; J. Darling; G. Silber; and M. Ferrari.
1980, February. Rare sighting of a right whale (*Eubalaena glacialis*) in Hawaii. Can. J. Zool., 58(2): 309-312, 1 figure, 18 references. (English and French summaries)

Anatomy - External: Callosities, Coloration; Behavior - Species Interaction: Humpback Whale, Milk Fish; Distribution; Individual Identification; Migration; Photographs - Callo-sities, Coloration Pattern, Underwater View; Pacific - North: Hawaiian Waters; Pacific - Western North: Japanese Waters

157A* Ruud, J.T.
1937. Nordkaperen *Balaena glacialis* (Bonnaterre). (The northcaper *Balaena glacialis* 'Bonna-terre') Norsk Hvalfangsttid., 26(3): 270-278.

Source: Reference #133

158. Saayman, G.S. and C.K. Tayler.
1973. Some behavior patterns of the southern right whale *Eubalaena australis*. Z. Saugetierkunde, 38(3): 172-183, 11 figures, 15 references. (English and German summary)

BA 1974 v. 57 (#223)

Anatomy - External: Callosities, Coloration, Scars; Behavior - Object Interaction: Boat; Behavior - Reproductive; Behavior - Surface Dependent: Breach, Flippering, Lobtail, Mating Activity; Individual Identification; Photographs - Breach, Callosities, Flipper, Mating Behavior, Stranded Male; Strandings; Atlantic - Eastern South: South African Waters

159. Scammon, C.M.
1869, April. On the cetaceans of the western coast of North America. Proc. Acad. nat. Sci., Philad., p. 13-63, 3 figures, 24 references.

Distribution; Pacific - Eastern North: U.S. Waters

160. Schevill, W.E. and W.A. Watkins.
1962. Whale and porpoise voices. Woods Hole:

Woods Hole Oceanographic Institution, 24p. + record; 35 figures, 8 references.

Behavior - Acoustic: Vocalization; Atlantic - Western North: U.S. Waters

161. ---------
1962, December. Whale music. Oceanus, 9(2): 2-13, 23 unnumbered figures.

Behavior - Acoustic: Vocalization; Photographs - Callosities, Fluke

162. --------- and W.A. Watkins.
1966, May. Radio-tagging of whales. Woods Hole: Woods Hole Oceanographic Institution, Reference No. 66-17, 15p. + 10 figures.

Tagging/Marking; Telemetry; Atlantic - Western North: U.S. Waters

163. Schmidley, D.J.; C.O. Martin; and G.F. Collins.
1972. First occurrence of a black right whale 'Balaena glacialis' along the Texas coast. SWest. Nat., 17(2): 214-215, 1 figure.

ASFA 1973 (#3Q 3884M)

Anatomy - External: Callosities; Distribution; Photographs - Callosities, Stranded Whale; Strandings; Atlantic - Western North: Gulf of Mexico

164. Schubert, K.
1951, September. Die weideplätze der bartenwale im Sudlichen Eismeer. (The feeding grounds of the right whale in the Antarctic Seas) Fischwirtschaft, 3(9): 146-148, 5 figures, 1 reference. (German)

Food; Antarctic

165. Sears, R.
1979, September-December. An occurrence of right whale (*Eubalaena glacialis*) on the north shore of the Gulf of St. Lawrence. Nature Canada (Québec), 106(5-6): 567-568, 5 references. (English and French summary)

Distribution; Atlantic - Western North: Canadian Waters

166. Seki, Y.
1958, September. Observations on the spinal cord of the right whale. Sci. Repts. Whales Res. Inst., Tokyo, No. 13: 231-251, 15 figures, 20 references.

ArB 1961 (#61492)

Anatomy - Nervous; Photographs - Spinal Cord

167. Sergeant, D.E.; A.W. Mansfield; and B. Beck.
1970, November. Inshore records of cetacea for Eastern Canada, 1949-68. J. Fish. Res. Bd Can., 27(11): 1903-1915 + 1 plate (2 figures); 2 tables, 3 figures, 27 references.

Strandings; Atlantic - Western North: Canadian Waters

168. Shaler, N.S.
1873, January. Notes on the right and sperm whale. Am. Nat., 7(1): 1-4.

Anatomy - External: Hair; Behavior - Reproductive; Behavior - Species Interaction: Killer Whale

169. Sheldrick, M.C.
1976. A very early description of the southern right whale *Balaena australis* (Desmoulins, 1822). Mammalia, 40(1): 162-164, 4 references.

Anatomy - External: Callosities, Coloration; Parasites - Ectoparasites; Atlantic - South: Tristan da Cunha Waters

170. Skriabin, A.S.
1969. Novaya nematoda *Crassicauda costata* Sp. n.-- parazit yuzhnogo kita. (A new nematode *Crassicauda costata* Sp. n., a parasite of the southern whale) Parazitologiya, 3(3): 258-265, 1 unnumbered table, 5 figures, 11 references. (Russian, English summary)

BA 1969 (#128793)

Parasites - Ectoparasites

171. Solokov, V.E.
      1961. Stroenie i prichiny vozniknoveniya kozhnykh
      narostov u yaponskikh kitov (*Eubalaena
      glacialis sieboldii* Gray). (Structure and
      causes of the rise of dermal excrescences in
      whales '*Eubalaena glacialis sieboldii* Gray')
      Zool. Zh., 40(9): 1427-1429 + 1 plate (2
      figures); 1 figure, 8 references. (Russian,
      English summary)

      ArB 1965 (#75808); BA 1963 (#13741)

      Anatomy - External: Callosities; Photographs
      - Microscopic View of Callosities

172. Southwell, T.
      1881, February 22. On the occurrence of the
      Atlantic right whale, *Balaena biscayensis*
      (Eschricht), on the east coast of Scotland.
      Proc. nat. Hist. Soc. Glascow, 5(part 1): 66-
      69, 2 references.

      Distribution; Atlantic - Eastern North:
      British Isles Waters

      Also published (with minor variations) in:
      Trans. Norfolk Norwich Nat. Soc., 3: paper 7
      (1884), p. 228-230.

173. ---------
      1904. On the whale fishery from Scotland, with
      some account of the changes in that industry
      and of the species hunted. Ann. Scot. nat.
      Hist., 13(50): 77-90 + 1 plate (6 figures); 1
      unnumbered table, 4 references.

      Distribution; Whaling - Whaling Material;
      Atlantic - North

174. Taija, G.
      1935. (The story of whaling at Kumano-Taiji-
      Ura.), 52p., 4 figures. (Japanese)

      Entrapments - Net; Whaling - Whaling Mat-
      erial; Pacific - Western North: Japanese
      Waters

175.* Terán, M.
      1949. La *Balaena biscayensis* y los balleneros
      españoles del mar Cantábrico. (The *Balaena
      biscayensis* and the Spanish whalers of the

Bay of Biscay) Estudios geogr., 37: 639-668.
(Spanish)

Whaling - Whaling Material; Atlantic -
Eastern North: European Waters

Source: Casinos, A. and J.-R. Vericad. The
cetaceans of the Spanish coasts: A survey.
Mammalia, 40:2 (1976), p. 267-289.

176. Thompson, B.W.
1918, September. On whales landed at the Scottish
whaling stations, especially during the years
1908-1914 -- Part 1. The nordcaper. Scott.
Nat., No. 81: 197-208, 5 unnumbered tables,
1 figure, 21 references.

Anatomy - External: Callosities; Whaling -
Catch Statistics; Whaling - Shore Stations;
Atlantic - Eastern North: British Isles Waters

177. Tomilin, A.G.
1967. Mammals of the U.S.S.R. and adjacent
countries. vol. 9: Cetacea. Jerusalem: Israel
Program for Scientific Translations, 755p.,
196 tables, 146 figures, numerous references.
(translated by O. Ronen from the 1957
Russian edition)

Anatomy - Baleen; Anatomy - External:
Callosities, Coloration, Hair; Anatomy -
Flippers/Flukes; Anatomy - Skeletal; Behavior
- Locomotion; Behavior - Reproductive;
Behavior - Respiratory; Behavior - Surface
Dependent: Breach; Classification; Distri-
bution; Food; Migration; Parasites - Ecto-
parasites; Photographs - List of Most
Important; Strandings

178. Townsend, C.H.
1935, April 3. The distribution of certain whales
as shown by logbook records of American
whaleships. Zoologica, N.Y., 19(1): 1-50 + 4
plates; 2 figures, 3 references.

Distribution; Whaling - Catch Statistics;
Whaling - Logbooks/Journals

179. True, F.W.
1885. Report of a trip to Long Island in search of

skeletons of the right whale, *Balaena cisarctica*. Bull. U.S. Fish Commn, 5: 131-132.

Anatomy - Skeletal; Distribution; Atlantic - Western North: U.S. Waters

180.   ----------

1904. The whalebone whales of the Western North Atlantic compared with those occurring in European Waters; with some observations on the species of the North Pacific. Smithson. Contr. Knowl., 33, 332p. + 50 plates (188 figures); numerous unnumbered tables, 97 figures, 92 references.

Anatomy - External: Callosities, Coloration; Anatomy - Skeletal; Classification; Distribution; Parasites - Ectoparasites; Photographs - Anterior and Lateral Views, Skeletal Remains; Strandings; Atlantic - Eastern North: European Waters; Atlantic - Western North; Pacific - North

181.   Tsuyuki, H. and U. Naruse.

1963, February. Studies on the oil of black right whale in the Northern Pacific Ocean. Scient. Repts. Whales Res. Inst., Tokyo, No. 18: 171-190, 24 tables, 4 figures, 14 references.

BA 1966 (#47223)

Biochemical Studies

Preliminary papers published in: Geiken Tsushin, No. 129 (May, 1962), p. 87-95, 1 table, 2 figures; Geiken Tsushin, No. 130 (June, 1962), p. 120-123, 6 tables, 1 figure, 8 references; and Geiken Tsushin, No. 131 (July, 1962), p. 131-134, 6 tables, 1 figure, 6 references. (All papers published in Japanese)

182.   ----------; S. Itoh; U. Naruse; and A. Mochizuki.

1964. (The liver oil of the black right whale '*Eubalaena glacialis*') Nippon Daigaku Nojuigakubu Gakujutsu Kenkyu Hokoku, No. 18: 17-24. (Japanese)

CA 1964 v. 60 (p. 16285)

Biochemical Studies

183.* ---------; U. Naruse; A. Mochizuki; and S. Itoh.
1964. (Studies on liver oil of black right whale)
Bull. Coll. Agr. Vet. Med. Nihon Univ., 18:
17-24. (Japanese)

Biochemical Studies

Source: Reference #185

184. --------- ---------
1964, March. Studies on the lipids in brain of
black right whales in the Northern Pacific
Ocean. Scient. Repts. Whales Res. Inst.,
Tokyo, No. 18: 173-180, 11 tables, 2 figures,
15 references.

Anatomy - Nervous; Biochemical Studies

Preliminary paper published in: Geiken
Tsushin, No. 136 (December, 1962), p. 233-
239, 10 tables, 1 figure, 16 references.

185. --------- and. S. Itoh.
1970, June. Fatty acid components of black right
whale oil by gas chromatography. Scient.
Repts. Whales Res. Inst., Tokyo, No. 22: 165-
170, 4 tables, 6 references.

BA 1971 (#58755)

Biochemical Studies

186. Turner, W.
1913. The right whale of the North Atlantic,
Balaena biscayensis: Its skeleton described
and compared with that of the Greenland
right whale, *Balaena mysticetus*. Trans. R.
Soc. Edinb., 48(part 4, paper 33): 889-922 +
3 plates (15 figures); 2 tables, 10 figures, 56
references.

Anatomy - Baleen; Anatomy - External:
Callosities, Coloration; Anatomy - Flippers/
Flukes; Anatomy - Skeletal; Classification;
Parasites - Ectoparasites; Photographs -
Skeletal Remains, Ventral Coloration Pattern;
Atlantic - North

187. ---------
1914, December 4. The baleen whales of the South
Atlantic. Proc. R. Soc. Edinb., 35(part 1,

paper 2): 11-22, 3 unnumbered tables, 4 figures, 10 references.

Anatomy - Ear; Photographs - Ear Bone; Atlantic - Western South: South Shetland Waters

188. Wang, P.
1978. (Studies on the baleen whales in the Yellow Sea) Acta. zool. sin., 24(3): 269-277. (Chinese, English summary)

BA 1979 v. 68 (#20573)

Distribution; Pacific - Western North: Yellow Sea

189. Watkins, W.A. and W.E. Schevill.
1976, February 27. Right whale feeding and baleen rattle. J. Mammal., 57(1): 58-66, 3 figures, 31 references.

BA 1976 v. 61 (#1002)

Behavior - Feeding; Food; Photographs - Callosities, Feeding; Atlantic - Western North: U.S. Waters

Abstract of this paper appeared as: Watkins, W.A.; W.E. Schevill; and T.A. Bray. (Abstract) Right whale feeding and baleen rattle. Santa Cruz: Conference on the biology and conservation of marine mammals, December 4-7, 1975, p. 40.

190. ---------- ----------
1977, September. The development and testing of a radio whale tag. Woods Hole: Woods Hole Oceanographic Institution, 77-58, 38p., 12 figures, 13 references.

Tagging/Marking; Telemetry

191. ---------- ----------
1979, February. Aerial observation of feeding behavior in four baleen whales: *Eubalaena glacialis*, *Balaenoptera borealis*, *Megaptera novaeangliae*, and *Balaenoptera physalus*. J. Mammal., 60(1): 155-163, 4 figures, 15 references.

Behavior - Feeding; Atlantic - Western North: U.S. Waters

192. Whitehead, H. and R. Payne.
1976. New techniques for assessing populations of right whales without killing them. *In:* Proceedings of the Scientific Consultation of Marine Mammals. Rome: UNFAO, ACMRR/ MM/SC/79, 35p. + 10 tables and 14 figures; 12 references.

Anatomy - External: Callosities; Fetus; Growth Rates; Individual Identification; Atlantic - Western South: Argentine Waters

193. Yamada, M. and F. Yoshizaki.
1959, September. Osseous labyrinth of cetacea. Scient. Repts. Whales Res. Inst., Tokyo, No. 14: 291-304 + 3 plates (10 figures); 6 figures, 4 references.

Anatomy - Ear

194. Yamamoto, Y. and H. Hiruta.
1978, December. Stranding of a black right whale at Kumomi, southwestern coast of Izu Peninsula. Scient. Repts. Whales Res. Inst., Tokyo, No. 30: 249-251, 1 table, 3 figures.

Anatomy - External: Callosities, Coloration; Harrassment; Photographs - Stranded Whale; Strandings; Pacific - Western North: Japanese Waters

Author Index

Anon., 1-5
Abel, O., 6
Allen, G.M., 7
Allen, J.A., 8, 9
Andrews, R.C., 10, 11
Antinoja, R.C., 101
Arnold, P.W., 12
Augustin, W., 13
Ayers, W.O., 14

Bailey, A.M., 15
Baker, C.S., 101
Beauregard, H., 145, 147
Beck, B., 167
Beneden, P.-j. van, 16-28
Berzin, A.A., 29, 30
Best, P.B., 31-37
Bruce, W.S., 38
Buchet, G., 39
Burger, A., 51

Caldwell, D.K., 40
Caldwell, M.C., 40
Capellini, M., 41, 42
Castello, H.P., 43
Chapman, F.P., 44
Chittleborough, R.G., 45
Clark, C.W., 46, 47
Clark, E., 120
Clark, E.S., Jr., 48
Clark, J.M., 46
Clarke, R., 49
Collett, R.A., 50
Collins, G.F., 163
Condy, P.R., 51
Cope, E.D., 52, 53
Cummings, W.C., 54-57

Darling, J., 157
Daugherty, A.E., 58
Donnelly, B.G., 59, 60
Doran, A., 61
Dorsey, E.M., 142
Duguy, R., 62

Elliott, H.F.I., 63
Ellis, R., 64
Eschricht, D.F., 65

Ferrari, M., 157
Fischer, P., 66-69
Fiscus, C.H., 151
Fish, J.F., 55-57
Forestell, P.H., 101
Fraser, F.C., 70

Gasco, F., 71-74
Gaskin, D.E., 12, 75-77
Gautier, R., 62
Gervais, P., 78, 79
Giglioli, H.H., 80
Gilmore, R.M., 81-83
Gómez de Llarena, J., 84
Graells, M.P., 85
Gray, J.E., 86-93
Guldberg, G.A., 94-99

Harrison, J.P., 100
Hasagawa, Y., 128
Herman, L.M., 101
Hiruta, H., 194
Holder, J.B., 102

Imaizumi, Y., 103
Itoh, S., 182, 183, 185
Ivanova, E.I., 104
Iwasa, M., 105

Jehl, J.R., Jr., 55

Kamiya, T., 106
Kasuya, T., 133, 134
Katona, S., 150
Kawamura, S., 133
Klumov, S.K., 107
Koeford, E., 108

Liouville, J., 109
Lönnberg, E., 110-111

McCully, R.M., 37
Manigault, G.F., 112
Mansfield, A.W., 167
Markham, C.R., 113
Martin, C.O., 163
Matsuura, Y., 114
Matthews, L.H., 115

Subject Index

Age Determination, 133
Anatomy
    Baleen, 65, 71, 83, 110s*, 115s, 131, 133, 177, 186
    Circulatory, 71
    Ear, 79, 187s, 193
    Endocrine, 106
    External
        Callosities, 4, 7, 9-12, 15s, 19, 31s, 45s, 46s, 48, 49s,
            50, 57s, 63s, 65, 68, 75s, 83, 86, 101, 102, 107,
            110s, 115s, 118, 120, 131, 133, 137s-141s, 142s,
            143s, 146, 151, 152s, 153s, 156s, 157, 158s, 163,
            169s, 171, 176, 177, 180, 186, 192s, 194
        Coloration, 7, 10, 11, 15s, 31s, 49s, 50, 53, 101, 107,
            110s, 112, 115s, 120, 130-133, 139s-141s, 153s, 157,
            158s, 169s, 177, 180, 186, 194
        Hair, 7, 110s, 115s, 131, 133, 168, 177
        Scars, 10, 12, 59s, 60s, 130, 131, 133, 142s, 158s
    Eye, 10
    Flippers/Flukes, 11, 177, 186
    Muscles, 37s
    Nervous, 144s, 166, 184
    Reproductive, 107, 115s, 131, 133
    Skeletal, 1, 6, 7, 9, 10, 13, 17, 20, 21, 24, 27, 41, 42, 52,
        67-69, 71, 72, 72s, 73s, 74, 78, 79, 84, 87, 88s, 90,
        91s-93s, 94, 98, 102-104, 111, 112, 117, 118, 121s,
        122, 127, 128, 130-134, 177, 179, 180, 186

Behavior
    Acoustic
        Playback, 46s, 56s, 57s
        Vocalization, 3s, 40, 46s, 47s, 54, 55s-57s, 83, 101,
            136s, 138s, 139s, 150, 160, 161
    Agonistic, 55s, 56s, 138s, 140s, 142s
    Care Giving, 7, 154s
    Feeding, 48, 55s, 56s, 100s, 110s, 119s, 124, 126, 133,
        135s, 139s, 189, 191
    Greeting, 136s, 141s
    Locomotion, 15s, 48, 56s, 153s, 177
    Lone Whale, 12, 48, 51s, 136s
    Nocturnal, 136s
    Nursing, 77
    Object Interaction:
        Boat, 45s, 56s, 59s, 138s, 158s
        Buoys, 149
        Rocks, 115s, 129s
        Seaweed, 137s-139s, 141s

---

* The s denotes a southern right whale.

Geographic Index

Journal Index

Izv. tikhookean. nauchno-issled. Inst. rȳb Khoz. Okeanogr., 29

J. acoust. Soc. Am., 47
J. comp. Pathol. Ther., 37
J. Fac. Sci. Hokkaido Univ. - Zoology, 105
J. Fish. Res. Bd Can., 12, 167
J. Mammal., 43, 45, 123, 189, 191
J. Reprod. Fert. supp., 60
J. Zool., 21, 79
Journal of the Tokyo National Science Museum - Natural Science and Museum, 103

K. Svenska. Vetensk-Akad. Handl., 110

Mainstream, 4
Mammalia, 169
Mar. Ecol. Prog. Ser., 101
Mar. Fish. Rev., 3, 5
Mem. Boston Soc. nat. Hist., 7
Mem. cour. Acad. r. Sci. Belg., 27
Mem. Soc. zool. Fr., 39
Mems. R. Acad. Cieno. exact. fis. nat. Madr., 85
Munibe, 84

National Wildlife, 64
Natn. geogr. Mag., 138, 140
Nature, 2, 80, 95
Nature Canada (Quebec), 165
Naturen, 96, 108
New Jersey Outdoors, 148
Nippon Daigaku Nojuigakubu Gakujutsu Kenkyu Hokoku, 182
Norsk Hvalfangsttid., 48, 49, 130, 151, 157A

Oceanus, 161
Oryx, 63

Pacific Discovery, 81
Parazitologiya, 170
Proc. Acad. nat. Sci., Philad., 52, 53, 159
Proc. Denver Mus. nat. Hist., 15
Proc. Linn. Soc. Lond., 70
Proc. nat. Hist. Soc. Glascow, 172
Proc. R. Soc. Edinb., 187
Proc. zool. Soc. Lond., 50, 86-88, 93, 113, 152
Proceedings of the Elliot Society, 112

Q. Jl. geol. Soc. Lond., 127

Rep. int. Whal. Commn, 32-36, 116, 150

Author Index

Bold numbers indicate complete references.

Subject Index